AUTOPHAGY DYSFUNCTION IN ALZHEIMER'S DISEASE AND DEMENTIA

AUTOPHAGY DYSFUNCTION IN ALZHEIMER'S DISEASE AND DEMENTIA

Edited by

TADANORI HAMANO

Associate Professor, Second Department of Internal Medicine, Faculty of Medical Sciences, University of Fukui, Fukui, Japan

Clinical Professor, Department of Neurology, University of Fukui Hospital, Fukui, Japan

TATSURO MUTOH

Professor, Department of Neurology and Neuroscience, Fujita Health University Hospital, Toyoake-City, Aichi, Japan

Department of Internal Medicine Fujita Health University Chubu International Airport Clinic, Tokoname-City, Aichi, Japan

ELSEVIER

ACADEMIC PRESS

An imprint of Elsevier

Academic Press is an imprint of Elsevier
125 London Wall, London EC2Y 5AS, United Kingdom
525 B Street, Suite 1650, San Diego, CA 92101, United States
50 Hampshire Street, 5th Floor, Cambridge, MA 02139, United States
The Boulevard, Langford Lane, Kidlington, Oxford OX5 1GB, United Kingdom

ISBN 978-0-323-89906-2

For information on all Academic Press publications
visit our website at https://www.elsevier.com/books-and-journals

Back cover image: Figure S2 (part B) reprinted from Bradlee L. Heckmann, Brett J.W. Teubner, Bart Tummers, Emilio Boada-Romero, Lacie Harris, Mao Yang, Clifford S. Guy, Stanislav S. Zakharenko, and Douglas R. Green. "LC3-Associated Endocytosis Facilitates β-Amyloid Clearance and Mitigates Neurodegeneration in Murine Alzheimer's Disease" in *Cell*, Vol 178, Issue 3, 25 July 2019, pp 536-551. e14, with permission from Elsevier.

Publisher: Nikki Levy
Acquisitions Editor: Joslyn Chaiprasert-Paguio
Editorial Project Manager: Kristi Anderson
Production Project Manager: Maria Bernard
Cover Designer: Christian Bilbow

Typeset by STRAIVE, India

Dedication

This book is dedicated to Late Professor Masaru Kuriyama, University of Fukui, Japan.

Contents

Contributors *xiii*
Foreword *xvii*
Preface *xix*
Acknowledgments *xxi*

SECTION I Degradation mechanisms of cells

1. Degradation mechanisms of cells **3**
Asako Otomo and Shinji Hadano

1. Neurons are highly polarized cells with the sophisticated trafficking system 3
2. Ubiquitin-proteasome system: UPS 4
3. Autophagy-endolysosomal system: APELS 8
4. Integration of cellular degradation systems 18

References 19

SECTION II Lysosomes

2. Lysosomes-neuronal degeneration in lysosomal storage disorders **25**
Giulia Lunghi, Emma Veronica Carsana, Nicoletta Loberto, Sandro Sonnino, and Massimo Aureli

1. Lysosomes 25
2. Lysosomal storage diseases 26
3. Impairment of lysosomal activity and alteration in the sphingolipid composition of cell membranes: A possible link with the onset of neuronal damage in LSD 31
4. Involvement of mitochondrial impairment in the onset of neurodegeneration in lysosomal storage diseases 35
5. Involvement of lysosomal impairment in the neuroinflammation 38

References 40

SECTION III The autophagic pathways

3. The autophagy pathway and its key regulators **47**
Tetsushi Kataura, Nobutaka Hattori, and Shinji Saiki

1. The autophagy machinery 47
2. Key regulators and signaling pathways of autophagy 49

3. Selective autophagy regulated by autophagic receptors 58
4. Concluding remarks 61
Acknowledgment 62
References 62

SECTION IV Amyloid beta protein and autophagy

4. Basics of amyloid β-protein in Alzheimer's disease **73**
Azusa Sugimoto and Kenjiro Ono

1. What is amyloid β? 73
2. Production and formation 74
3. Aggregation 75
4. Regulation of Aβ concentration in the brain 78
5. Decomposition and excretion 78
6. Relationship between Aβ and disease (including gene mutation) 79
7. Intervention strategy 82
References 85

5. Molecular linkages among Aβ, tau, impaired mitophagy, and mitochondrial dysfunction in Alzheimer's disease **91**
Tomas Schmauck-Medina, Thale D.J.H. Patrick-Brown, Shi-qi Zhang, Alexandra Gilbert, and Evandro F. Fang

1. Introduction 91
2. Mitochondrial dysfunction 94
3. Defective mitophagy in AD 98
4. Mitochondrial dysfunction and defective mitophagy at the circuit and behavioral level 101
5. Future research 103
Acknowledgments 103
Competing interests 104
References 104

6. Endocytosis in β-amyloid biology and Alzheimer's disease **111**
Bradlee L. Heckmann and Douglas R. Green

1. Introduction 111
2. The flavors of endocytosis in the brain 112
3. The MO(F) of Aβ 115
4. Endocytosis in a phagocytic world 117

5. Adding a modifier, a new world for endocytosis in the AD brain 120
6. Endocytosis in a starry world 124
7. Exploiting endocytosis for therapeutic gain 125
8. Recycling full circle, a summary 126
Acknowledgments 127
Disclosures 127
References 127

SECTION V Autophagy and tau protein

7. Autophagy and tau protein **135**
Tadanori Hamano and Yoshinori Endo

1. Introduction 135
2. Tau protein 137
3. Autophagy 141
4. Disturbance of the autophagy-lysosome system in AD and related disorders 145
5. Tau degradation pathway 148
6. Mitophagy and tau 150
7. Diabetes, tau, and autophagy 151
8. Propagation of tau by the disruption of autophagy 151
9. Potential of autophagy modulators as a treatment for AD 153
10. Conclusion 153
Acknowledgments 154
References 154

8. BAG3 promotes tau clearance by regulating autophagy and other vacuolar-dependent degradative processes **159**
Changyi Ji, Heng Lin, and Gail V.W. Johnson

1. Introduction 159
2. BAG3 protein 161
3. Summary 167
Acknowledgments 169
References 169

9. Tau propagation and autophagy **173**
Mohammad Nasir Uddin, Shotaro Shimonaka, and Yumiko Motoi

1. Introduction 173
2. Tau propagation mechanisms 174

3. Autophagic impairment promotes Tau aggregation 181
4. Pharmacological agents modulating Tau propagation 185
5. Conclusion 189
Author contributions 189
References 190

SECTION VI Autophagy and pathology in Alzheimer's disease

10. Granulovacuolar degeneration in neurodegeneration 197

Kouki Makioka, Tsuneo Yamazaki, Masayuki Sato, Masamitsu Takatama, Koichi Okamoto, and Yoshio Ikeda

1. Introduction 197
2. Neuropathological features of granulovacuolar degeneration (GVD) in neurons 197
3. What kind of organelle are GVBs? 199
4. GVB-like structures in oligodendroglia in multiple system atrophy 200
Acknowledgment 203
References 203

11. Autophagy dysfunction in skeletal myopathies: Inclusion body myositis and Danon disease 207

Kazuma Sugie

1. Autophagy and autophagic vacuoles 207
2. Autophagic vacuolar myopathy 208
3. Inclusion body myositis 210
4. Danon disease 215
5. Conclusion 221
References 221

SECTION VII Autophagy and other disorders causing dementia

12. Autophagy in Lewy body diseases and multiple system atrophy 227

Yasuo Miki, Kunikazu Tanji, and Koichi Wakabayashi

1. Introduction 227
2. Autophagy 227

3. Lewy body diseases 229
4. MSA 236
5. Conclusions 240
Acknowledgments 240
References 241

13. Autophagy and Huntington's disease 245
Tatsuo Mano and Atsushi Iwata

1. Introduction 245
2. Macroautophagy in HD 246
3. Physiological function of Htt in autophagy 249
4. Autophagy: A therapeutic target 251
5. Conclusion 252
References 252

SECTION VIII Drug discovery in Alzheimer's disease by modulating autophagy

14. Drug discovery in Alzheimer's disease by regulating autophagy 263
Tadanori Hamano and Yoshinori Endo

1. Introduction 263
2. Drugs that induce autophagy 263
3. Caspase activation and autophagy 282
4. Conclusion 283
References 284

15. Drug discovery in Alzheimer's disease using metal chelators: Warning toward their usage 291
Tatsuro Mutoh

1. Introduction 291
2. Abnormal metals in AD patients 292
3. Clioquinol and SMON (subacute myelo-optico-neuropathy) 294
4. Conclusion 299
Acknowledgments 300
References 300

16. Development of autophagy enhancers for Parkinson's disease therapy **305**

Yukiko Sasazawa, Nobutaka Hattori, and Shinji Saiki

1. Features of Parkinson's disease 305
2. Current PD therapy 305
3. α-Synuclein accumulation in PD 306
4. Autophagy impairment in PD 308
5. Targeting macroautophagy for potential PD therapy 310
6. Targeting lysosome for potential PD therapy 313
7. Conclusions and future perspectives 316

Acknowledgment 317

References 317

Index *321*

Contributors

Massimo Aureli
Department of Medical Biotechnology and Translational Medicine, University of Milan, Milan, Lombardy, Italy

Emma Veronica Carsana
Department of Medical Biotechnology and Translational Medicine, University of Milan, Milan, Lombardy, Italy

Yoshinori Endo
Second Department of Internal Medicine, Faculty of Medical Sciences, University of Fukui, Fukui; Department of Neurology, University of Fukui Hospital, Fukui, Japan

Evandro F. Fang
Department of Clinical Molecular Biology, University of Oslo and Akershus University Hospital, Lørenskog; The Norwegian Centre on Healthy Ageing (NO-Age), Oslo, Norway

Alexandra Gilbert
Department of Cell and Developmental Biology, UCL, London, United Kingdom

Douglas R. Green
Department of Immunology, St. Jude Children's Research Hospital, Memphis, TN, United States

Shinji Hadano
Molecular Neuropathobiology Laboratory, Department of Molecular Life Sciences, Tokai University School of Medicine, Isehara; Micro/Nano Technology Center, Tokai University, Hiratsuka; The Institute of Medical Sciences, Tokai University, Isehara; Research Center for Brain and Nervous Diseases, Tokai University Graduate School of Medicine, Isehara, Kanagawa, Japan

Tadanori Hamano
Second Department of Internal Medicine, Faculty of Medical Sciences, University of Fukui; Department of Neurology, University of Fukui Hospital, Fukui; Department of Aging and Dementia (DAD), University of Fukui; Life Science Innovation Center, University of Fukui, Fukui, Japan

Nobutaka Hattori
Department of Neurology, Juntendo University Graduate School of Medicine, Bunkyo-ku, Tokyo, Japan

Bradlee L. Heckmann
USF Health Neuroscience Institute; Department of Molecular Medicine, University of South Florida Morsani College of Medicine; Byrd Alzheimer's Center, Tampa, FL, United States

Yoshio Ikeda
Department of Neurology, Gunma University Graduate School of Medicine, Maebashi, Gunma, Japan

Atsushi Iwata
Department of Neurology, Tokyo Metropolitan Geriatric Medical Center Hospital, Tokyo, Japan

Changyi Ji
Department of Anesthesiology and Perioperative Medicine, University of Rochester Medical Center, Rochester, NY, United States

Gail V.W. Johnson
Department of Anesthesiology and Perioperative Medicine, University of Rochester Medical Center, Rochester, NY, United States

Tetsushi Kataura
Department of Neurology, Juntendo University Graduate School of Medicine, Bunkyo-ku, Tokyo, Japan

Heng Lin
Department of Anesthesiology and Perioperative Medicine, University of Rochester Medical Center, Rochester, NY, United States

Nicoletta Loberto
Department of Medical Biotechnology and Translational Medicine, University of Milan, Milan, Lombardy, Italy

Giulia Lunghi
Department of Medical Biotechnology and Translational Medicine, University of Milan, Milan, Lombardy, Italy

Kouki Makioka
Department of Neurology, Gunma University Graduate School of Medicine, Maebashi, Gunma, Japan

Tatsuo Mano
Department of Neurology, The University of Tokyo Hospital, Tokyo, Japan

Yasuo Miki
Department of Neuropathology, Institute of Brain Science, Hirosaki University Graduate School of Medicine, Hirosaki, Aomori, Japan

Yumiko Motoi
Department of Diagnosis, Prevention, and Treatment of Dementia; Department of Neurology, Graduate School of Medicine, Juntendo University, Bunkyo-ku, Tokyo, Japan

Tatsuro Mutoh
Department of Neurology and Neuroscience, Fujita Health University Hospital, Toyoake-city; Department of Internal Medicine Fujita Health University Chubu International Airport Clinic, Tokoname-City, Aichi, Japan

Koichi Okamoto
Geriatrics Research Institute and Hospital, Maebashi, Gunma, Japan

Kenjiro Ono
Division of Neurology, Department of Internal Medicine, Showa University School of Medicine, Tokyo; Department of Neurology, Kanazawa University Graduate School of Medical Sciences, Kanazawa, Japan

Asako Otomo
Molecular Neuropathobiology Laboratory, Department of Molecular Life Sciences, Tokai University School of Medicine, Isehara; Micro/Nano Technology Center, Tokai University, Hiratsuka, The Institute of Medical Sciences, Tokai University, Isehara, Kanagawa, Japan

Thale D.J.H. Patrick-Brown
Department of Clinical Molecular Biology, University of Oslo and Akershus University Hospital, Lørenskog; Division of Surgery, Inflammatory Diseases and Transplantation, Oslo University Hospital, Oslo, Norway

Shinji Saiki
Department of Neurology, Juntendo University Graduate School of Medicine, Bunkyo-ku, Tokyo, Japan

Yukiko Sasazawa
Research Institute for Diseases of Old Age, Juntendo University Graduate School of Medicine, Bunkyo-ku, Tokyo, Japan

Masayuki Sato
Department of Neurology, Gunma University Graduate School of Medicine, Maebashi, Gunma, Japan

Tomas Schmauck-Medina
Department of Clinical Molecular Biology, University of Oslo and Akershus University Hospital, Lørenskog, Norway

Shotaro Shimonaka
Department of Diagnosis, Prevention, and Treatment of Dementia; Research Institute for Diseases of Old Age, Graduate School of Medicine, Juntendo University, Bunkyo-ku, Tokyo, Japan

Sandro Sonnino
Department of Medical Biotechnology and Translational Medicine, University of Milan, Milan, Lombardy, Italy

Kazuma Sugie
Department of Neurology, Nara Medical University School of Medicine, Kashihara, Nara, Japan

Azusa Sugimoto
Division of Neurology, Department of Internal Medicine, Showa University School of Medicine, Tokyo, Japan

Masamitsu Takatama
Geriatrics Research Institute and Hospital, Maebashi, Gunma, Japan

Kunikazu Tanji
Department of Neuropathology, Institute of Brain Science, Hirosaki University Graduate School of Medicine, Hirosaki, Aomori, Japan

Mohammad Nasir Uddin
Department of Diagnosis, Prevention, and Treatment of Dementia; Genome and Regenerative Medicine Center, Graduate School of Medicine, Juntendo University, Bunkyo-ku, Tokyo, Japan

Koichi Wakabayashi
Department of Neuropathology, Institute of Brain Science, Hirosaki University Graduate School of Medicine, Hirosaki, Aomori, Japan

Tsuneo Yamazaki
Gunma University School of Health Sciences, Maebashi, Gunma, Japan

Shi-qi Zhang
Department of Clinical Molecular Biology, University of Oslo and Akershus University Hospital, Lørenskog, Norway

Foreword

It is my immense pleasure to write a foreword for the book titled *Autophagy Dysfunction in Alzheimer's Disease and Dementia* edited by Prof. Tadanori Hamano and Prof. Tatsuro Mutoh. Both are leading scientists in this innovative field and have been my sincere colleagues since 1990.

Autophagy dysfunction has been linked to the accumulation of misfolded proteins that cause the death of neurons in Alzheimer's and other neurodegenerative diseases. I have been excited to find that this book very nicely provides an overview for both basic scientists and neurologists of the mechanisms involved in the degradation of proteins as well as their dysfunction causing the misfolding/aggregation of tau, amyloid-β, and other disease-related proteins. The book also explores the possibilities of potential drug development through autophagy modulation.

With all my heart, I believe that this book, contributed by leading scientists and neurologists in the field, will be the primary source for integrating distributed basic and translational research on autophagy dysfunction as related to Alzheimer's and other neurodegenerative diseases. Finally, I congratulate the book editors and all the authors for compiling this exciting book.

Hironobu Naiki
Department of Molecular Pathology, Faculty of Medical Sciences, University of Fukui, Fukui, Japan

Preface

The major neuropathological hallmarks of Alzheimer's disease (AD) are senile plaques composed of amyloid β protein (Aβ) and neurofibrillary tangles that are composed of the highly phosphorylated tau. Usually, Aβ accumulation precedes the tau deposition. In the past, the research focus was on the overproduction or enhanced aggregation of Aβ by the genetic mutation of amyloid precursor protein (APP) or presenilin1 (PS1). Front-temporal dementia linked to chromosome17 (FTDP-17) is a neurodegenerative disorder caused by mutations in tau gene. However, most AD and dementia cases are sporadic and is not accompanied by mutations, including APP, PS1, PS2, or tau genes. Usually, Aβ accumulates when the balance between synthesis and degradation of Aβ is disrupted. It was clarified that the degradation of Aβ peptide in cerebrospinal fluid was impaired in 98% of the sporadic AD. The impaired degradation of Aβ and Tau is largely related to the pathogenesis of AD.

Basal macroautophagy, usually referred to as autophagy, is the major intracellular degradation and regenerating mechanisms. Disturbances of autophagy in AD were first identified by Nixon et al. in 2005. Since 2008, several groups, including us, have reported that tau can be degraded by autophagy. In 2008, Pickford et al. reported increased intracellular and extracellular Aβ depositions due to heterogeneous deletion of the Beclin1, autophagy-related gene. In 2011, mTORC1-dependent autophagy induction by rapamycin showed the improved cognitive function with the reduced Aβ at the early stage of AD.

This book summarizes the latest research on the implication of autophagy in pathogenesis and its prevention of AD. It also describes autophagy and other diseases causing dementia, including Huntington's disease, Parkinson's disease, or dementia with Levy bodies. The book also discusses the possibility of treatment with autophagy modulation. All the authors are leading researchers in the respective field, and we are confident that the readers will be satisfied after reading this book.

Tadanori Hamano, MD, PhD
University of Fukui, Fukui, Japan
Tatsuro Mutoh, MD, PhD
Fujita Health University Hospital, Aichi, Japan

Acknowledgments

We express our appreciation to Dr. Yoshinori Endo and Dr. Yuka Hama who served as editorial committee members. We also express our sincere appreciation to Dr. Maho Morishima of the Brain Bank for Aging Research at the Tokyo Metropolitan Geriatric Hospital and Institute of Gerontology, Japan, and Dr. Tomohisa Yamaguchi, Dr. Hirohito Sasaki, and Dr. Yuki Kitazaki at the University of Fukui for their review. We express our deepest gratitude to Professor Yasunari Nakamoto, Second Department of Internal Medicine, University of Fukui, and all the members of the Department of Neurology, University of Fukui, Japan.

We express our heartfelt gratitude to Melanie Tucker, Senior Acquisitions Editor at Neuroscience Academic Press/Elsevier, for giving us the opportunity to edit this book. We also express our deepest gratitude to Kristi Anderson, Senior Editorial Project Manager at Academic Press/Elsevier for her constant and appropriate advice.

We thank Emiko Kitagawa, Junko Nakane, and Natsumi Kitajima, Second Department of Internal Medicine, University of Fukui, Japan, for their excellent help with this project.

Finally, we express our deepest gratitude to Professor Emeritus Yasuo Ihara at the University of Tokyo, Japan, and Professor Shu-Hui Yen at Mayo Clinic Jacksonville, United States, for teaching one of us (Tadanori Hamano) about the methodology, interpretation, depth, and pleasure of Alzheimer's disease research.

Tadanori Hamano, MD, PhD
University of Fukui, Fukui, Japan

Tatsuro Mutoh, MD, PhD
Fujita Health University Hospital, Aichi, Japan

SECTION I

Degradation mechanisms of cells

CHAPTER 1

Degradation mechanisms of cells

Asako Otomo[a,b,c] and Shinji Hadano[a,b,c,d]
[a]Molecular Neuropathobiology Laboratory, Department of Molecular Life Sciences, Tokai University School of Medicine, Isehara, Kanagawa, Japan
[b]Micro/Nano Technology Center, Tokai University, Hiratsuka, Kanagawa, Japan
[c]The Institute of Medical Sciences, Tokai University, Isehara, Kanagawa, Japan
[d]Research Center for Brain and Nervous Diseases, Tokai University Graduate School of Medicine, Isehara, Kanagawa, Japan

1. Neurons are highly polarized cells with the sophisticated trafficking system

Before getting into the main topic, we briefly describe the structural and functional features of neurons and their relationship with the proteostatic and trafficking systems. Neuron is a highly polarized cell with a single long axon and several dendrites around the cell body. Intracellular transport pathways in neuron are highly organized to efficiently transport proteins and other macromolecules not only from the cell body to the tips of extended neurites but also from the distal areas to the cell body.[1] Virtually all intracellular organelles are selectively transported with the help of motor proteins moving along the polarized microtubule networks throughout neuron.[2] Indeed, protein degradation by the ubiquitin-proteasome system (UPS), which proceeds locally in neurons, depends on the correct delivery of the proteasome complex to the intended intracellular sites through such selective transport system.[3,4] On the other hand, degradation by the autophagy-endolysosomal system (APELS) is entirely dependent on the retrograde transport of autophagosomes and late endosomes, including multivesicular bodies. Their gradual degradation processes via the APELS are ultimately completed by the fusion with lysosomes that are primarily located in the cell body.[5,6] Therefore, the regulation of proteostasis and intracellular trafficking in neurons are intimately linked to each other. Most neurons in the brain do not divide during adulthood and thus generally last an entire lifetime of organisms. However, as it should be, both proteolytic and intracellular transport systems are gradually deteriorated due to aging. In fact, excessive decline of protein degradation and intracellular transport has been documented in many cellular as well as animal models of neurodegenerative diseases, including Alzheimer's disease (AD), frontotemporal

Autophagy Dysfunction in Alzheimer's Disease and Dementia
https://doi.org/10.1016/B978-0-323-89906-2.00004-6

dementia (FTD), Parkinson's disease (PD), and amyotrophic lateral sclerosis (ALS).[7–12] Here, we review general aspects of the UPS and APELS in a relationship with the intracellular trafficking system in neurons.

2. Ubiquitin-proteasome system: UPS

2.1 Molecular mechanism for the ubiquitin-dependent degradation of proteins

The UPS is a systematically regulated protein degradation system, consisting of multiple enzymatic reaction steps[13,14] (Fig. 1A). The first step involves the ubiquitination of target proteins to be degraded. A single ubiquitin (ub) molecule is covalently conjugated to a lysine (Lys) residue of the substrate protein by a series of enzymatic reactions. First, ub is attached to and activated by an E1 ub activating enzyme in an ATP-dependent manner. Then, E1 transfers the activated ub to an E2 ub conjugating enzyme via a transthiolation reaction. The ub-E2 conjugate is recognized by E3 ub ligases, and ub is transferred either directly to the substrate (for RING family E3 ub ligases) or first attached to the E3 ub ligase followed by transferring to the substrate (for HECT and RBR family E3 ub ligases). Since ubiquitination involves the attachment of ub to Lys residues on the substrates and/or ub itself, it results in the formation of the substrates that are modified with either single ub molecule (monoubiqutination) or polyubiquitin chains (polyubiquitination). Such diverse substrate-ubiquitin structures can create a multitude of distinct signals with distinct cellular responses, referred to as "ub code."[15,16] Among seven Lys residues in ub, Lys 48-linked chains are the predominant linkage type in cells (often >50% of all linkages), whose role is to target proteins to the proteasome for degradation.[15] Of note, monoubiquitinated small proteins with less than 150 amino acid residues are degraded by the UPS[17]; however, the protein degradation by the UPS generally requires tandem ubiquitin conjugation with at least two to four ubiquitin molecules.[18]

In the second step, polyubiquitinated substrates are targeted directly to or via ub-binding shuttle protein to the proteasome for proteolysis.[19] The intrinsic proteasomal (proteasome-associated) ub receptors, Rpn10 and Rpn13, can only trap polyubiquitinated substrates that are present in a proximity to the proteasome. In contrast, ub-binding shuttle proteins, which are not integral subunits of the proteasome, can recognize ubiquitinated substrates that are geometrically distant from the proteasome and recruit them to the proteosomes for degradation (Fig. 2A). The nonproteasomal UbL-UBA family of proteins, such as ubiquillins, are known as proteasomal shuttles.[20]

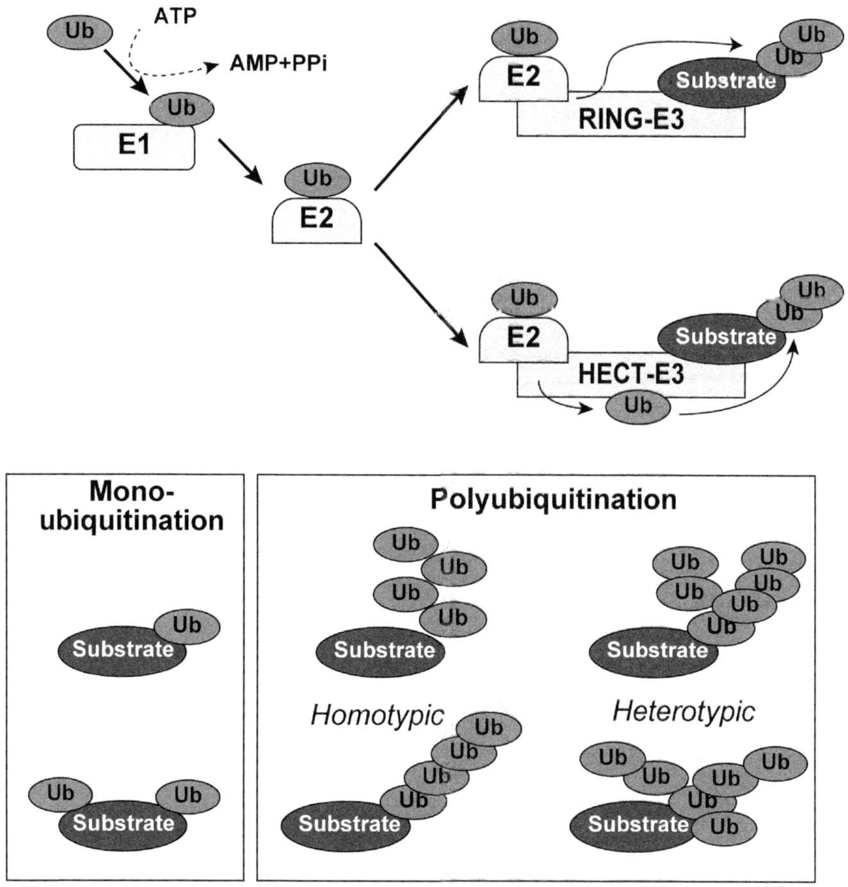

Fig. 1 The ubiquitin system. Ubiquitin (ub) is a polypeptide consisting of 76 amino acid residues, within which seven lysine residues that can be covalently conjugated by other ub molecules are included. In addition to these lysine residues, the first methionine can also be conjugated by other ub molecules. Ubiquitination of substrates by a series of enzymatic reactions. First, ub is attached to the substrate by an E1 ub activating enzyme in an ATP-dependent manner, and then transferred to an E2 ub conjugating enzyme. The ub-E2 conjugate is recognized by an E3 ub ligase. Finally, ub is transferred either directly to the substrate (for RING family E3 ub ligases) or first attached to the E3 ub ligase before being transferred to the substrate (for HECT and RBR family E3 ub ligases). Monoubiquitination and polyubiquitination of substrates both occur by this conserved system. Monoubiquitination provides signaling scaffolds for UBA-containing proteins, mediating downstream signal transduction or substrate degradation; however, monoubiquitinated small proteins with less than 150 amino acid residues are likely to be degraded by the UPS. Polyubiquitinated proteins are specifically recognized and degraded by proteasome. It is noted that structurally distinct linkage-specific homotypic and heterotypic polyubiquitin chains can covey functionally distinct intracellular signals with distinct cellular responses.

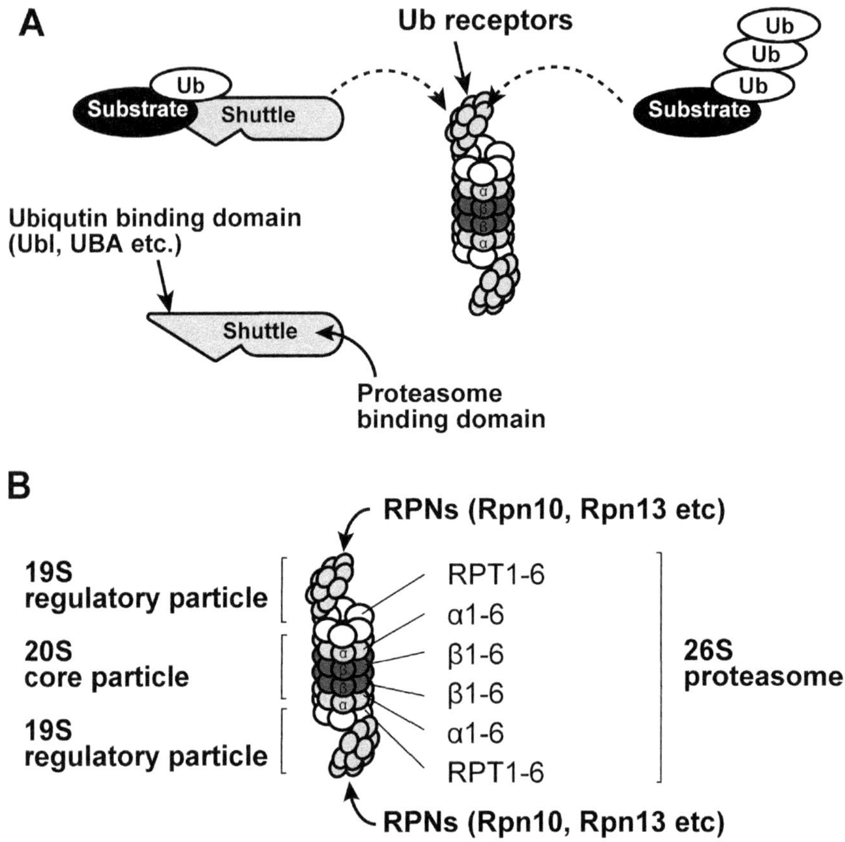

Fig. 2 Structure and function of proteasome. (A) Recognition of ubiquitinated substrates. The intrinsic proteasomal ub receptors, Rpn10 and Rpn13, directly bind to polyubiquitinated substrates and mediate their degradation by proteasomes. Ub-binding shuttle proteins, which contain both proteasome binding domain and ubiquitin-binding domain such as Ubl and UBA, recognize ubiquitinated substrates and recruit them for proteasomal degradation. (B) Schematic diagram of 26S proteasome. 26S proteasome is composed of two functionally distinct subcomplexes: the 20S core catalytic particle and the 19S regulatory particle. The 20S core particle consists of a barrel of four axially stacked rings: two outer α-rings and two inner β-rings. β-rings form a proteolytic chamber, and α-rings serve as a gate for entry into the chamber. The 19S regulatory particle contains the intrinsic proteasomal ub receptors, Rpn10 and Rpn13.

In the last step, ubiquitinated substrates are degraded by the 26S proteosome.[13,14] The 26S proteasome consists of two functionally distinct subcomplexes: the 20S core catalytic particle and the 19S regulatory particle (Fig. 2B). The 20S core particle catalyzes the protein degradation via its peptidase activities. The 19S regulatory particle caps the 20S core particle at one or both ends and captures the ubiquitinated substrates via its intrinsic proteasomal ub receptors, Rpn10 and Rpn13, facilitating the 20S-mediated protein degradation.

2.2 Functional roles of the UPS in neurons

The proteasome complex is assembled in the cytosol and shuttles between the cytosol and nucleus[21,22] (Fig. 3). Functional UPSs in the cytosol and nucleus are both essential for cell survival. Especially in neurons, the UPS

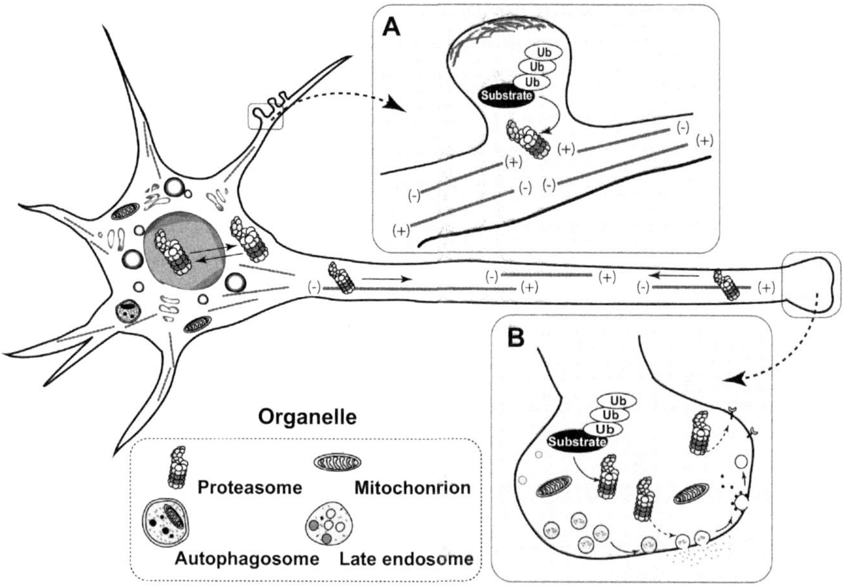

Fig. 3 The UPS in neurons. Neuron is a highly polarized cell with a single long axon and several dendrites around the cell body. Like other somatic cells, majority of lysosomes and proteasomes are present in the cell body. Proteasomes shuttle between the nucleus and cytosol in the cell body. In response to the intrinsic activity and/or extracellular stimuli, proteasomes can enter the axon and are transported either anterogradely or retrogradely in an environmental context dependent manner. (A) Proteasome function in dendrites. Proteasome plays an essential role in dendritic spine morphogenesis. Proteasomes in dendritic spines can rapidly digest postsynaptic molecules in response to neural activity to enhance the local remodeling of the protein composition of synapses, which contributes to the synaptic plasticity in neurons. (B) Proteasome function in the axon terminal. Proteasome-mediated clearance of damaged proteins plays an important role in maintaining proteostasis in the distal axon. Furthermore, since many presynaptic molecules contributing to exocytosis of synaptic vesicles are ubiquitinated, proteasomal degradation in the axonal terminal might be involved in presynaptic exocytosis.

plays a crucial role in the formation and maintenance of their unique morphology and functions. During development, the UPS activity controls the growth and pruning of both axon and dendrites.[3,4,23] In mature neurons, UPS-mediated protein degradation at synapses is critical for activity-dependent plasticity, learning and memory.[24] Interestingly, the direction of proteasome trafficking varies depending on the developmental stages of neurons. At an immature stage, a majority of proteasomes are retrogradely transported to the cell body in the axon and are eliminated from the distal axon, which is required for the proper specification of axon.[3] On the other hand, at a mature stage, the proteasomes are preferentially transported in the anterograde direction from the cell body to the axon terminal, which is regulated by proteasome inhibitor of 31 kD (PI31) that serves as an adaptor to couple proteasomes with dynein light chain proteins (DYNLL1/2).[25] Taken together, the transport of proteasomes to the appropriate locations is prerequisite not only for the development and maintenance of neuronal morphology but also for the multiple functional processes relying on the proteasome-mediated proteolytic systems in neurons. Interestingly, expression of Rnp11, a component of the 19S regulatory particle, decreases with aging in Drosophila, destabilizing the 26S proteasome complex and sequentially causes age-dependent accumulation of ubiquitinated proteins.[26] Thus, it is reasonable to assume that reduction of proteasome activity with aging is associated with decline in neuronal function, triggering neurodegeneration.

3. Autophagy-endolysosomal system: APELS

3.1 Lysosome-mediated degradation of cellular macromolecules

The lysosome is a membrane-bound organelle that contains many digestive enzymes and plays crucial roles in the intracellular degradation of various macromolecules.[27] Extracellular and membrane proteins are primarily delivered to lysosomes for degradation through the endocytic pathway, while cytoplasmic macromolecules, e.g., proteins, lipids, nucleotides, and organelles, are delivered to lysosomes through various types of the autophagic pathways (Fig. 4). Therefore, although the endocytic and autophagic pathways are distinct cellular processes, both systems ultimately merge at the lysosomes and share many protein machineries and cellular processes for degradation. After the lysosomal digestion, resulting molecules can be recycled to use either as energy sources or raw materials for anabolic reactions in cells.

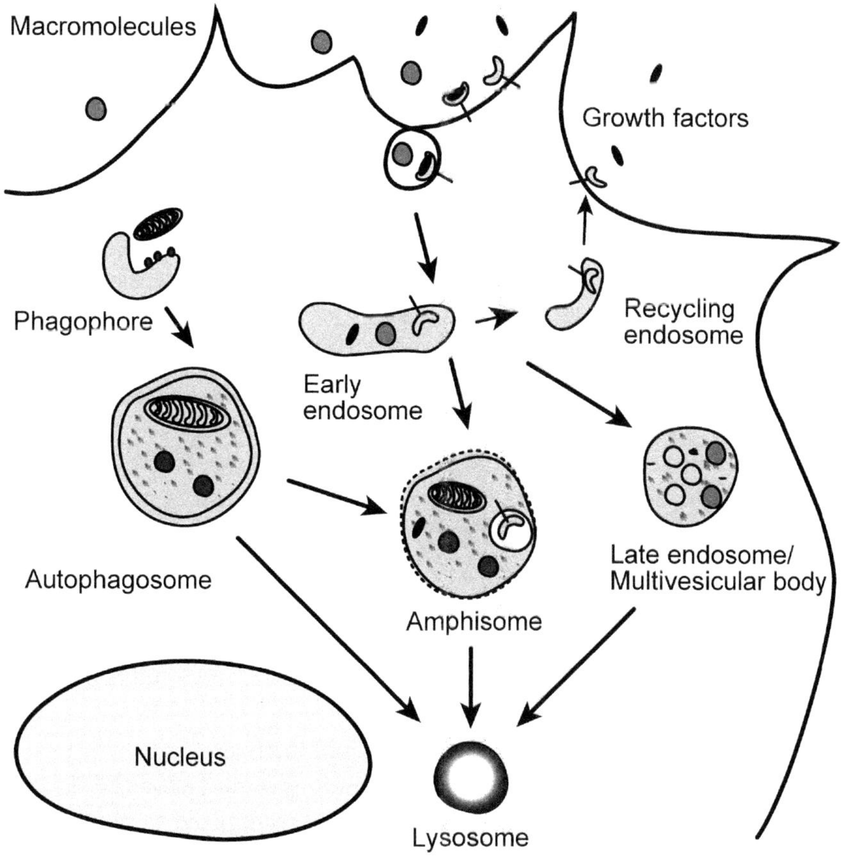

Fig. 4 The lysosome-mediated degradation system. Both extracellular and membrane-bound proteins in the plasma membrane are delivered to lysosomes via a variety of endocytic systems. Internalized molecules are transported to early endosomes, where cargo molecules are pooled and selected. The molecules that are destined for degradation are transported to late endosomes and/or multivesicular bodies. Some molecules such as receptors are transported from early endosomes to recycling endosomes and returns to the plasma membrane. On the other hand, cytoplasmic macromolecules are delivered to lysosomes through various types of autophagy. Macroautophagy is initiated by recruiting the autophagy-mediated molecules on endoplasmic reticulum membrane, from which phagophore arises and engulf the substrates for autophagy, resulting in the formation of autophagosomes. Autophagosome fuses with lysosome or first fuses with late endosome to form amphisome followed by fusion with lysosome. Ultimately, their enwrapped molecules are digested by lysosomal enzymes. The digested materials are released from lysosomes and recycled to use either as energy sources or raw materials for anabolic reactions in cells.

3.2 Endocytic pathways to the lysosome

Endocytosis is the fundamental cellular mechanism that is highly conserved among eukaryotes, through which materials outside of the cell as well as proteins embedded or anchored in the plasma membrane can be engulfed. Internalized molecules, as cargos, are first transported to early endosomes, and then are either recycled back to the cell surface or targeted to lysosomes for degradation. Early endosomes themselves, by fusing with the zymogen-containing vesicles from Golgi apparatus, can mature into late endosomes and ultimately lysosomes with concomitant intralumenal acidification by vacuolar-type ATPase (V-ATPase), through which the inactive form of acid enzymes is activated. Lysosome contains an approximately 60 active acidic enzymes such as proteases, lipases, and nuclease, which mediates degradation of various cargos and macromolecules.[28] Cargo-selective internalization and sorting are both strictly controlled, playing crucial roles not only in the maintenance of proteostatic balance but also in the regulation of signal transduction in cells. Thus far, various endocytic pathways (internalization pathways) have been identified (Table 1).[29] Among these, we here focus on three major endocytic pathways that are linked to the lysosomal degradation, i.e., clathrin-dependent endocytosis (CDE), phagocytosis, and macropinocytosis.

CDE is among the most studied endocytic pathway.[30] Growth factors such as epidermal growth factor (EGF), nerve growth factor (NGF), and brain-derived neurotrophic factor (BDNF) are recognized by the specific receptors on the cell surface and engulfed by CDE. Initially, clathrin oligomerizes into curved lattices on the inner face of the plasma membrane where ligand-bound receptors are located and induces the formation of invaginated structure called clathrin-coated pit. Once the clathrin-coated pit develops to a spherical in shape, clathrin-coated vesicles (CCVs) are pinched off from the plasma membrane by Dynamin, a membrane remodeling GTPase. After the internalization, CCVs are quickly uncoated by heat shock cognate 71 kDa protein (HSC70), a member of the heat shock and chaperone protein.[30] Resulting vesicles are transported to and fused with early endosomes, thereby delivering the cargos, e.g., the ligand-receptor complex, to early endosomes.[31] Early endosome is a specialized organelle that acts as a sorting center, where the engulfed cargos are precisely selected and enriched within a particular subcompartment of early endosome and delivered to appropriate destinations by vesicle trafficking.[31]

Unlike CDE, many endocytic pathways do not use clathrin. These types of endocytosis are referred to as clathrin-independent endocytosis (CIE).[32]

Table 1 The endocytic systems and their cargo specificity.

Name of endocytosis	Clathrin-dependent endocytosis	Caveolin-dependent endocytosis	Flotillin-associated endocytosis	Phagocytosis	Macropinocytosis	CLIC/GEEC endocytosis
Category	CDE	CIE	CIE	CIE	CIE	CIE
Coat protein	Clathrin	Caveolin	Flotillin	Actin	Actin	Actin
Name of vesicle	Clathrin-coated vesicle	Caveosome/ caveolosome	N.D.	Phagosome	Macropinosome	GEEC
Size of vesicles	100 nm	50–60 nm	N.D.	0.5–10 μm	0.5–5 μm	N.D.
Cargos	EGF receptor, transferrin receptor, LDL receptor	Aminopeptidase P, albumin, folic acid	CD59, cholera toxin	Infectious agents, apoptotic cells	Macromolecules, extracellular fluid, EGF receptor	GPI-anchored protein, CD44

CDE, clathrin-dependent endocytosis; *CIE*, clathrin-independent endocytosis; *CLIC*, clathrin-independent carriers; *GEEC*, glycosylphosphotidylinositol-anchored protein (GPI-AP) enriched compartments; *EGF*, epidermal growth factor; *LDL*, low-density lipoprotein.

Among CIE, phagocytosis and macropinocytosis are two representative macroscale endocytic systems that are tightly linked to the lysosome-mediated degradation. Both processes, which involves in the formation of large endocytosed vesicles, are triggered by and/or dependent on the actin-mediated remodeling of the plasma membrane.

Phagocytosis is a fundamental cell biological process exhibited by a wide variety of cell types. It plays an important role in engulfing large solid materials into cells. However, except for professional phagocytes, the activity of this system is generally low. Professional phagocytes, including macrophage, monocyte and neutrophil, are cells whose main tasks are the efficient elimination of infected microbes as well as the stimulation of adaptive immune responses by presenting antigens to lymphocytes.[33] Phagocytosis is activated by the interaction of cargo molecules, including not only non-self-particles and microbes but also self-entities such as apoptotic cells and cellular debris, with various specific cell surface receptors such as mannose, IgG, complement, and Fcγ receptors.[34] Once internalized, cargos are encased in the large vesicular structure called phagosomes. Phagosomes are then fused with lysosomes, and cargo molecules that are delivered from phagosomes are ultimately degraded by lysosomal enzymes.

Macropinocytosis is also a fundamental cellular system involving the engulfment of a relatively large amount of fluid as well as solid macromolecules such as albumin and bacteria.[35] Unlike phagocytosis, macropinocytosis can be induced in many cell types upon the stimulation with phorbol esters, cytokines, growth factors, and oxidative stress.[36] Macropinocytosis involves the formation of large vesicles called macropinosomes, and their contents are also degraded following the fusion with lysosomes. Interestingly, several recent studies on cancer cells have demonstrated that macropinocytosis-mediated uptake of extracellular materials coupled with the lysosome-mediated degradation of the internalized materials are essential for cancer cell survival.[37] These findings indicate that macropinocytosis has a crucial role to maintain a stable supply of energy sources in cells.

3.3 Endocytic pathways and endosome trafficking in neurons

CDE and macropinocytosis are two main endocytic pathways, both of which are linked to the lysosome-mediated degradation in neurons, merge at the early endosome. A particular population of growth factor receptors and engulfed molecules are selectively accumulated in a specific domain or compartment of early endosomes, converting early endosomes into vesicles having characteristics of late endosomes (Fig. 5). Such conversion occurred

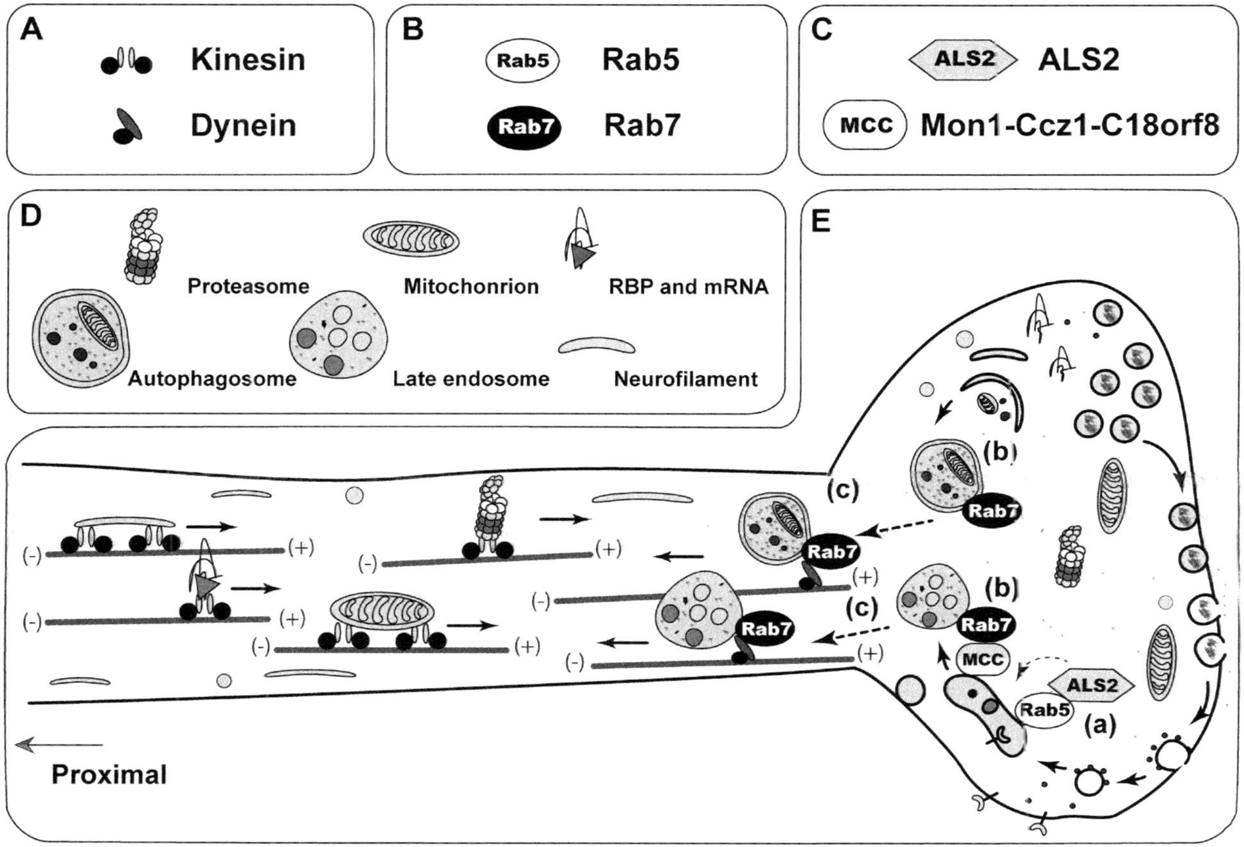

Fig. 5 See figure legend on next page.

(Continued)

Fig. 5 The APELS in neurons. (A) Motor proteins. Plus-end directed motor proteins, Kinesin, anterogradely transport cargos, while minus-end directed motor protein, Dynein, mediates the retrograde cargo transportation. (B) Rab proteins. Rab family of small GTPases are key regulators of intracellular membrane trafficking, including the formation of transport vesicles and their fusion with membranes. Rab5 is required for the fusion between endocytic vesicles and early endosomes. Rab7 regulates early endosome maturation and retrograde transport of late endosome and autophagosome. Rab proteins serve as binary switches cycling between an inactive GDP-bound form and an active GTP-bound form. When activated, Rab can recruit different sets of downstream effectors that are directly responsible for vesicle formation, fusion and transport. (C) Activators for the Rab protein. Each Rab protein has a specific guanine nucleotide exchange factor (GEF). ALS2 acts as an activator for Rab5. The trimeric Mon1-Ccz1-C18orf8 (MCC) complex activates Rab7 in mammalian cells. (D) Cargos traveling in the axon. Cargos, such as proteasome, mitochondria, mRNA associating with RBP (RNA-binding protein), autophagosome, late endosome and neurofilaments, are actively transported in the axon. (E) Late endosome and autophagosome transport in the distal axon. Organelles including mitochondria, endoplasmic reticulum, proteasome, early endosome, late endosome, and autophagosomes are present in the distal axon, while mature active lysosomes are rarely observed in the axon. Rab5 is recruited and activated by a Rab5-specific GEF on early endosomes (E-a). ALS2 is a GEF for Rab5, whose function is required for motor neuron survival and maintenance. Activated Rab5 can recruit Rab7 activator onto endosomal membranes. Once Rab7-specific GEF (MCC) activates Rab7, effector proteins for Rab7, including cargo receptors mediating the association with Dynein, are recruited to endosomal membranes with concomitant maturation from early to late endosomes (E-b). Then, the Rab7 complex with cargo receptor and Dynein mediates the retrograde transport of late endosomes and autophagosomes in the axon (E-c). Ultimately, both late endosomes and autophagosomes are delivered to the cell body and fuse with lysosomes for degradation.

---

at the axonal terminal enables the retrograde transport of late endosomes along the microtubules in axon.[38] Polarity orientation of microtubules in axons is strictly organized with the plus end facing the axon terminal and the minus end facing the cell body.[39] Anterograde transport (from the cell body to the axon) is primarily mediated by the diverse family of plus-end-directed kinesin motor proteins, N-KIFs (KIFs with a motor region at the N-terminus), while retrograde transport (from the axon to the cell body) is mediated by the minus-end-directed motor proteins, cytoplasmic dynein.[40] Late endosomes are selectively transported in a retrograde direction through the interaction with dynein motor protein complex in axon.[38] Furthermore, acidification of late endosomes gradually proceeds during the retrograde transport,[5] which may contribute to the efficient degradation of cargos after fusing with lysosomes in the cell body. On the other hand, signal transduction molecules, which are endocytosed at the axonal terminal, are also transported by this retrograde transport via late endosomes so that neuronal cell bodies can continuously receive survival and/or trophic signals from postsynaptic cells.[41,42]

3.4 Autophagic pathways to the lysosome

Autophagy is an evolutionarily conserved self-degradation mechanism responsible for degradation of autophagic cargo, such as damaged organelles, malformed proteins produced during translation, and misfolded protein aggregates via the lysosome.[43] There are at least four types of autophagy, including macroautophagy, microautophagy, chaperone-mediated autophagy (CMA), and xenophagy, except for organelle-specific autophagy such as mitophagy.[44] It has originally been believed that autophagy involves lysosome-mediated nonselective bulk proteolysis. However, in some cases, unwanted cytosolic materials and damaged organelles are selectively removed by the system called "selective autophagy."[45]

Macroautophagy is the most well-studied form of autophagy, by which the cytoplasmic substrates are delivered to lysosomes via double-membraned vesicles called autophagosomes.[43] Macroautophagy takes place in the following five steps.[46] The first step is "induction." Induction of macroautophagy, which is mediated by the ULK1-containg protein complex, occurs after metabolic or other stresses. The second step is "nucleation." During this step, the formation of the phagophore (or isolation membrane) is initiated. This action is mainly triggered by a protein complex containing VPS34, a PtdIns3K of class III. The third step is "elongation." Elongation of the phagophore is mediated by two ubiquitin-like conjugation systems, the Atg12-Atg5 and the LC3-PE, both of which are indispensable for the autophagosome formation. The fourth step is "cargo recognition." LC3 on the autophagosomal membrane recruits autophagic cargos via the interaction with autophagy adaptor protein SQSTM1/p62, which recognizes cytoplasmic ubiquitinated substrates. At this step, cargos are accurately enwrapped by extended phagophore, and autophagosome formation is completed. The fifth step is "autolysosome formation." Once autophagosomes are formed, they either fuse directly with lysosomes to form autolysosomes or first fuse with late endosomes to form amphisomes followed by fusion with lysosomes. Through these processes, autophagic cargos entrapped in the autophagosome are digested by lysosomal enzymes and the degraded primary components are released from lysosomes to the cytosol for recycling. The lysosomes themselves can be regenerated so that the process can continuously proceed.

Microautophagy is the degradation process that involves the uptake of cytosolic materials by direct invagination of the lysosomal membrane.[47] However, the mechanism underling microautophagy in the lysosome remains unclear. Besides microautophagy in lysosomes, a similar phenomenon is observed in late endosomes. Proteins and organelles can also be targeted to late endosomes for degradation in mammals through what is known as endosomal microautophagy.[48] Endosomal microautophagy occurs during the formation of multivesicular bodies (MVBs). Substrates containing KFERQ(-like) motifs are first recognized by cytosolic HSC70, and then delivered to endosomes via the interaction of HSC70 with phosphatidylserine in the endosomal membrane. The endosomal sorting complex required for transport (ESCRT) machinery executes inward membrane curvature formation and scission of the endosomal membrane, generating the intraluminal vesicles, MVBs.[49] Finally, late endosomes containing MVBs fuse with lysosomes, thereby engulfed substrates and MVBs are simultaneously degraded.

CMA is the degradation process that involves the translocation of the target proteins with a specific consensus sequence into the lysosomes via lysosomal surface proteins.[47,50] First, the KFERQ-like motif-bearing cytosolic substrates are trapped by HSC70.[51] Then, the chaperone-substrate complex is recruited to lysosome-associated membrane protein type 2A (LAMP2A) on the lysosomal membrane, inducing the oligomerization of LAMP2A. Formation of LAMP2A oligomer is essential for the translocation of substrates to the lumen.[52] Lysosomal HSC70 associated with the luminal side of LAMP2A oligomer also assists the translocation of substrates. Finally, internalized substrates are degraded by lysosomal enzymes. Since approximately 45% of proteins in the mammalian proteome contain a canonical KFERQ-like motif,[53] the contribution of HSC70-mediated protein degradation system, such as CMA and endosomal microautophagy, for protein degradation could be considerable. Indeed, mutations in several genes that play an essential role in CMA and endosomal microautophagy are linked to neurodegenerative diseases,[54] suggesting that HSC70-mediated protein degradation system strongly contributes to neuronal survival and maintenance.

Xenophagy is defined as a specialized form of autophagy dedicated to eliminating pathogens in cells.[55] After infection, the surface of pathogens is rapidly polyubiquitinated in the cytosol of host cells, allowing for recognition by SQSTM1/p62.[56] The pathogen-SQSTM1/p62 complex is then captured by LC3 on the phagophores and enwrapped by autophagosomes.

Subsequently, autophagosomes containing pathogens are fused with lysosomes for digestion. In addition to such direct bacterial xenophagy, pathogens hidden in endosomes are also subjected to xenophagy. Once pathogens invade and sneak in early endosomes through endocytosis, endosomes themselves become damaged and some molecules on the cytosolic surface are ubiquitinated.[57] Such abnormal endosomes are recognized by SQSTM1/p62 and degraded by the autophagic system in due course.

3.5 Functional roles of autophagy in neurons

Autophagosome formation occurs anywhere in neurons, including in the axon, dendrite, and the cell body.[58] Autophagosomes formed in the distal axon are retrogradely transported to the cell body to fuse with lysosomes for substrate degradation. This unidirectional retrograde transport along the axon employs the trafficking system that resembles those for axonal late endosomes.[6] By contrast, dendritic autophagosomes exhibit bidirectional transport in a synaptic activity-dependent manner.[59] Of note, autophagy in neurons serves as more than recycling of energy sources and raw materials for anabolic reactions in cells. For example, autophagy in hippocampal neurons is implicated in memory formation, which is associated with the activity-dependent structural and functional synaptic plasticity, such as dynamics of dendritic spines, synaptic facilitation, and long-term potentiation. It is also noted that autophagic activity in the brain appears to decline in aged wild-type mice, which is associated with decreased autophagic core proteins, i.e., Atg5–Atg12 and Becn1, and with increased mTOR activity.[60] Thus, it is reasonable that various brain functions relying on autophagy may also decline with aging.

3.6 Rab GTPases involves in autophagosome and endolysosome transport in neurons

Rab proteins comprise the largest subfamily of the Ras superfamily small GTPases.[61] They act as regulators of the secretory and endocytic pathways in eukaryotic cells. As mentioned above, both late endosomes and autophagosomes generated in the distal axon are retrogradely transported by cytoplasmic dynein moving along the polarized microtubule in neurons. At this step, Rab proteins, namely Rab5 and Rab7, play crucial roles in the maturation from early endosomes/autophagosomes to late endosomes (Fig. 5). Rab5 is initially activated by the Rab5-specific guanine nucleotide exchange factor (Rab5-GEF) on early endosomes, facilitating homotypic as well as heterotypic endosome and/or autophagosome fusion. Rab7-GEF is also recruited to early endosomes.[62] Once Rab7 is activated by Rab7-GEF, not only effector proteins for Rab7 including cargo receptors but also cytoplasmic dynein are recruited to the endosomal membrane, allowing retrograde transport of endosomes and/or autophagosomes in axon. Furthermore, endosomal maturation associated with the Rab conversion from Rab5 to Rab7 is accompanied by progressive acidification of the lumen via endosomal V-ATPase, thereby cargo molecules are gradually degraded during transportation. The signaling pathway mediated by Rab5 and Rab7 is commonly observed in virtually all cell types; however, it is noted that mutations in the genes involving this pathway cause neurodegenerative diseases. For example, endosomal dysfunction induced by Rab5 overactivation can be linked to AD.[63] Furthermore, loss of function mutations in the *ALS2* gene, whose gene product ALS2 acts as a Rab5-GEF,[64,65] account for several forms of ALS and motor neuron diseases. Mutations in the *RAB7A* gene are also linked to Charcot-Marie-Tooth disease type 2B (CMT2B).[66,67] Although the mechanism by which neuronal cells are preferentially affected by dysregulation of the Rab5-Rab7-mediated signaling pathway remains unclear, it is reasonable to speculate that neurons primarily rely on this proteolytic and trafficking systems to maintain neuronal proteostasis.

4. Integration of cellular degradation systems

The UPS and APELS play a central role in protein quality control and the maintenance of cellular homeostasis. Recent studies support the existence of a cross-talk between the UPS and APLES.[68] Ub, a molecular tag once thought exclusive to proteasome degradation, is also utilized for cargo selection by some forms of autophagy. However, the ub-codes that determine the system-dependent substrate selection is still unknown. Nonetheless, some of the ub-recognizing molecules and shuttle factors, such as SQSTM1/p62 and ubiquilins, are responsible for the delivery of ubiquitinated substrates not only to the proteasomes but also for macroautophagy.[16,19] In addition, inhibition of the UPS activity results in the upregulation of macroautophagy, whereas persistent blockage of macroautophagy conversely increases the UPS activity, indicative of a possible reciprocal cross-talk between these two systems in cells.[69,70] Therefore, unraveling these many layers of interaction between the UPS and APELS will lead to a better understanding of neuronal proteostasis and ultimately mechanisms underlying neurodegeneration and neurodegenerative diseases.

References

1. Horton AC, Ehlers MD. Neuronal polarity and trafficking. *Neuron*. 2003;40:277–295.
2. Tas RP, Chazeau A, Cloin BMC, Lambers MLA, Hoogenraad CC, Kapitein LC. Differentiation between oppositely oriented microtubules controls polarized neuronal transport. *Neuron*. 2017;96:1264–1271.e5.
3. Hsu M, Guo C, Liou A, et al. Stage-dependent axon transport of proteasomes contributes to axon development. *Dev Cell*. 2015;35:418–431.
4. Tai H, Schuman EM. Ubiquitin, the proteasome and protein degradation in neuronal function and dysfunction. *Nat Rev Neurosci*. 2008;9:826–838.
5. Farfel-Becker T, Roney JC, Cheng X, Li S, Cuddy SR, Sheng Z. Neuronal soma-derived degradative lysosomes are continuously delivered to distal axons to maintain local degradation capacity. *Cell Rep*. 2019;28:51–64.e4.
6. Hill SE, Colón-Ramos DA. The journey of the synaptic autophagosome: a cell biological perspective. *Neuron*. 2020;105:961–973.
7. Tanaka K, Matsuda N. Proteostasis and neurodegeneration: the roles of proteasomal degradation and autophagy. *Biochim Biophys Acta*. 2014;1843:197–204.
8. Morfini GA, Burns M, Binder LI, et al. Axonal transport defects in neurodegenerative diseases. *J Neurosci*. 2009;29:12776–12786.
9. Guo T, Zhang D, Zeng Y, Huang TY, Xu H, Zhao Y. Molecular and cellular mechanisms underlying the pathogenesis of Alzheimer's disease. *Mol Neurodegener*. 2020;15:1–40.
10. Mann DMA, Snowden JS. Frontotemporal lobar degeneration: pathogenesis, pathology and pathways to phenotype. *Brain Pathol*. 2017;27:723.
11. Bi M, Du X, Jiao Q, Chen X, Jiang H. Expanding the role of proteasome homeostasis in Parkinson's disease: beyond protein breakdown. *Cell Death Dis*. 2021;12:154.
12. Cicardi ME, Marrone L, Azzouz M, Trotti D. Proteostatic imbalance and protein spreading in amyotrophic lateral sclerosis. *EMBO J*. 2021;40, e106389.
13. Glickman MH, Ciechanover A. The ubiquitin-proteasome proteolytic pathway: destruction for the sake of construction. *Physiol Rev*. 2002;82:373–428.
14. Dick LR, Mayer RJ, Bedford L, Lowe J, Brownell JE. Ubiquitin-like protein conjugation and the ubiquitin–proteasome system as drug targets. *Nat Rev Drug Discov*. 2011;10:29–46.
15. Swatek KN, Komander D. Ubiquitin modifications. *Cell Res*. 2016;26:399–422.
16. Kwon YT, Ciechanover A. The ubiquitin code in the ubiquitin-proteasome system and autophagy. *Trends Biochem Sci*. 2017;42:873–886.
17. Shabek N, Herman-Bachinsky Y, Buchsbaum S, et al. The size of the proteasomal substrate determines whether its degradation will be mediated by mono- or polyubiquitylation. *Mol Cell*. 2012;48:87–97.
18. Tsuchiya H, Burana D, Ohtake F, et al. Ub-ProT reveals global length and composition of protein ubiquitylation in cells. *Nat Commun*. 2018;9:524.
19. Zientara-Rytter K, Subramani S. The roles of ubiquitin-binding protein shuttles in the degradative fate of ubiquitinated proteins in the ubiquitin-proteasome system and autophagy. *Cell*. 2019;8:40.
20. Seok Ko H, Uehara T, Tsuruma K, Nomura Y. Ubiquilin interacts with ubiquitylated proteins and proteasome through its ubiquitin-associated and ubiquitin-like domains. *FEBS Lett*. 2004;566:110–114.
21. Reits EAJ, Benham AM, Plougastel B, Neefjes J, Trowsdale J. Dynamics of proteasome distribution in living cells. *EMBO J*. 1997;16:6087–6094.
22. Marshall RS, Vierstra RD. Dynamic regulation of the 26S proteasome: from synthesis to degradation. *Front Mol Biosci*. 2019;6:40.
23. Watts RJ, Hoopfer ED, Luo L. Axon pruning during *Drosophila metamorphosis*: evidence for local degeneration and requirement of the ubiquitin-proteasome system. *Neuron*. 2003;38:871–885.

24. Hegde AN, Haynes KA, Bach SV, Beckelman BC. Local ubiquitin-proteasome-mediated proteolysis and long-term synaptic plasticity. *Front Mol Neurosci.* 2014;7:96.
25. Liu K, Jones S, Minis A, Rodriguez J, Molina H, Steller H. PI31 is an adaptor protein for proteasome transport in axons and required for synaptic development. *Dev Cell.* 2019;50:509–524.e10.
26. Tonoki A, Kuranaga E, Tomioka T, et al. Genetic evidence linking age-dependent attenuation of the 26S proteasome with the aging process. *Mol Cell Biol.* 2009;29:1095–1106.
27. Yang C, Wang X. Lysosome biogenesis: regulation and functions. *J Cell Biol.* 2021;220.
28. Trivedi PC, Bartlett JJ, Pulinilkunnil T. Lysosomal biology and function: modern view of cellular debris bin. *Cell.* 2020;9:1131.
29. Kumari S, Mg S, Mayor S. Endocytosis unplugged: multiple ways to enter the cell. *Cell Res.* 2010;20:256–275.
30. Kaksonen M, Roux A. Mechanisms of clathrin-mediated endocytosis. *Nat Rev Mol Cell Biol.* 2018;19:313–326.
31. Jovic M, Sharma M, Rahajeng J, Caplan S. The early endosome: a busy sorting station for proteins at the crossroads. *Histol Histopathol.* 2010;25:99–112.
32. Mayor S, Pagano RE. Pathways of clathrin-independent endocytosis. *Nat Rev Mol Cell Biol.* 2007;8:603–612.
33. Rosales C, Uribe-Querol E. Phagocytosis: a fundamental process in immunity. *Biomed Res Int.* 2017;2017, 9042851-18.
34. Gordon S. Phagocytosis: an immunobiologic process. *Immunity.* 2016;44:463–475.
35. Kerr MC, Teasdale RD. Defining macropinocytosis. *Traffic.* 2009;10:364–371.
36. Lin XP, Mintern JD, Gleeson PA. Macropinocytosis in different cell types: similarities and differences. *Membranes.* 2020;10:117.
37. Su H, Yang F, Fu R, et al. Cancer cells escape autophagy inhibition via NRF2-induced macropinocytosis. *Cancer Cell.* 2021;39:1–16.
38. Cai Q, Lu L, Tian J, Zhu Y, Qiao H, Sheng Z. Snapin-regulated late endosomal transport is critical for efficient autophagy-lysosomal function in neurons. *Neuron.* 2010;68:73–86.
39. Yogev S, Cooper R, Fetter R, Horowitz M, Shen K. Microtubule organization determines axonal transport dynamics. *Neuron.* 2016;92:449–460.
40. Hirokawa N, Niwa S, Tanaka Y. Molecular motors in neurons: transport mechanisms and roles in brain function, development, and disease. *Neuron.* 2010;68:610–638.
41. Wang T, Martin S, Nguyen TH, et al. Flux of signalling endosomes undergoing axonal retrograde transport is encoded by presynaptic activity and TrkB. *Nat Commun.* 2016;7:12976.
42. Cosker KE, Segal RA. Neuronal signaling through endocytosis. *Cold Spring Harb Perspect Biol.* 2014;6, a020669.
43. Dikic I, Elazar Z. Mechanism and medical implications of mammalian autophagy. *Nat Rev Mol Cell Biol.* 2018;19:349–356.
44. Mizushima N, Komatsu M. Autophagy: renovation of cells and tissues. *Cell.* 2011;147:728–741.
45. Gatica D, Lahiri V, Klionsky DJ. Cargo recognition and degradation by selective autophagy. *Nat Cell Biol.* 2018;20:233–242.
46. Feng Y, He D, Yao Z, Klionsky DJ. The machinery of macroautophagy. *Cell Res.* 2014;24:24–41.
47. Sahu R, Kaushik S, Clement CC, et al. Microautophagy of cytosolic proteins by late endosomes. *Dev Cell.* 2011;20:131–139.
48. Tekirdag K, Cuervo AM. Chaperone-mediated autophagy and endosomal microautophagy: jointed by a chaperone. *J Biol Chem.* 2018;293:5414–5424.
49. Schöneberg J, Lee I, Iwasa JH, Hurley JH. Reverse-topology membrane scission by the ESCRT proteins. *Nat Rev Mol Cell Biol.* 2016;18:5.

50. Kaushik S, Cuervo AM. The coming of age of chaperone-mediated autophagy. *Nat Rev Mol Cell Biol.* 2018;19:365–381.
51. Agarraberes FA, Terlecky SR, Dice JF. An intralysosomal hsp70 is required for a selective pathway of lysosomal protein degradation. *J Cell Biol.* 1997;137:825–834.
52. Bandyopadhyay U, Cuervo AM. Entering the lysosome through a transient gate by chaperone-mediated autophagy. *Autophagy.* 2008;4:1101–1103.
53. Kirchner P, Bourdenxid M, et al. Proteome-wide analysis of chaperone- mediated autophagy targeting motifs. *PLoS Biol.* 2019;17:1–27.e3000301.
54. Ugbode C, West RJH. Lessons learned from CHMP2B, implications for frontotemporal dementia and amyotrophic lateral sclerosis. *Neurobiol Dis.* 2021;147, 105144.
55. Mao K, Klionsky DJ. Xenophagy: a battlefield between host and microbe, and a possible avenue for cancer treatment. *Autophagy.* 2017;13:223–224
56. Zheng YT, Shahnazari S, Brech A, Lamark T, Johansen T, Brumell JH. The adaptor protein p62/SQSTM1 targets invading bacteria to the autophagy pathway. *J Immunol.* 2009;183:5909–5916.
57. Grumati P, Dikic I. Ubiquitin signaling and autophagy. *J Biol Chem.* 2018;293:5404–5413.
58. Maday S, Holzbaur ELF. Compartment-specific regulation of autophagy in primary neurons. *J Neurosci.* 2016;36:5933–5945.
59. Kulkarni VV, Anand A, Herr JB, Miranda C, Vogel MC, Maday S. Synaptic activity controls autophagic vacuole motility and function in dendrites. *J Cell Biol.* 2021;220, e202002084.
60. Ott C, König J, Höhn A, Jung T, Grune T. Macroautophagy is impaired in old murine brain tissue as well as in senescent human fibroblasts. *Redox Biol.* 2016;10:266–273.
61. Homma Y, Hiragi S, Fukuda M. Rab family of small GTPases: an updated view on their regulation and functions. *FEBS J.* 2021;288:36–55.
62. Rink J, Ghigo E, Kalaidzidis Y, Zerial M. Rab conversion as a mechanism of progression from early to late endosomes. *Cell.* 2005;122:735–749.
63. Pensalfini A, Kim S, Subbanna S, et al. Endosomal dysfunction induced by directly overactivating Rab5 recapitulates prodromal and neurodegenerative features of Alzheimer's disease. *Cell Rep.* 2020;33, 108420.
64. Hadano S, Hand CK, Osuga H, et al. A gene encoding a putative GTPase regulator is mutated in familial amyotrophic lateral sclerosis 2. *Nat Genet.* 2001;29:166–173.
65. Otomo A, Hadano S, Yasuda M, et al. ALS2, a novel guanine nucleotide exchange factor for the small GTPase Rab5, is implicated in endosomal dynamics. *Hum Mol Genet.* 2003;12:1671–1687.
66. Houlden H, King RHM, Muddle JR, et al. A novel RAB7 mutation associated with ulcero-mutilating neuropathy. *Ann Neurol.* 2004;56:586–590.
67. Romano R, Rivellini C, De Luca M, et al. Alteration of the late endocytic pathway in Charcot-Marie-Tooth type 2B disease. *Cell Mol Life Sci.* 2021;78:351–372.
68. Pohl C, Dikic I. Cellular quality control by the ubiquitin-proteasome system and autophagy. *Science.* 2019;366:818–822.
69. Wang XJ, Yu J, Wong SH, et al. A novel crosstalk between two major protein degradation systems. *Autophagy.* 2013;9:1500–1508.
70. Wang C, Wang X. The interplay between autophagy and the ubiquitin–proteasome system in cardiac proteotoxicity. *Biochim Biophys Acta, Mol Basis Dis.* 2015;1852:188–194.

SECTION II

Lysosomes

CHAPTER 2

Lysosomes-neuronal degeneration in lysosomal storage disorders

Giulia Lunghi[a], Emma Veronica Carsana[a], Nicoletta Loberto, Sandro Sonnino, and Massimo Aureli
Department of Medical Biotechnology and Translational Medicine, University of Milan, Milan, Lombardy, Italy

1. Lysosomes

Lysosomes are cytoplasmic acidic organelles characterized by a perinuclear localization. They are present in all eukaryotic cells and are responsible for the catabolism of exogenous and endogenous macromolecules. Usually, they appear at the electron microscopy as sphere-shaped dense bodies with a diameter of 0.5–1 μm and with a tubular structure in dendritic cells and activated macrophages.[1] The lysosomal membrane is characterized by a phospholipids' bilayer poor in cholesterol (7–10 nm of thickness) presenting, at the luminal side, an intricate net of highly glycosylated proteins, the so-called lysosomal membrane proteins. Among them, the most abundant members are the lysosomal associated membrane proteins LAMP-1 and LAMP-2, both presenting more than 25 glycosylation sites. Beside a structural and protective role, these highly glycosylated proteins are clearly involved in processes ranging from lysosomal motility and transport of catabolites across the membrane to fusion of lysosomes with others intracellular organelles or the plasma membrane (PM).[2,3] Moreover, this particular membrane organization is fundamental to isolate within lysosomes more than 60 different hydrolytic enzymes, such as peptidases, glycosidases, phosphatases, sulfatases, lipases, esterase, and nucleases. These enzymes are mainly soluble, except for those involved in the lipid catabolism, which are principally associated with the inner layer of the lysosomal membrane. Lysosomal hydrolases work in acidic conditions (pH of about 5.0) guaranteed by the presence of the

[a]These authors contributed equally to this work.

Autophagy Dysfunction in Alzheimer's Disease and Dementia
https://doi.org/10.1016/B978-0-323-89906-2.00010-1

proton pump vacuolar ATPase.[4] In case of lysosomal membrane leakage, the cell undergoes the so called lysosomal-mediated cell death (LMCD), caused by the release of hydrolytic enzymes, such as cathepsins B, D, and L, into the cytosol.[5] These enzymes, which are able to work also at neutral pH, activate apoptotic/necrotic effectors, such as mitochondria proteins and/or caspases, leading to the onset of cell death.

In addition to their role in catabolic processes, lysosomes can also be considered calcium storage organelles. In fact, they present a calcium storage capacity comparable to that of the endoplasmic reticulum (ER).[6] Since calcium is an important second messenger involved in the regulation of different processes, such as organelles trafficking, recycling, and fusion, this evidence opens a new scenario in the role of lysosomes in maintaining cell homeostasis.

Lysosomes are also involved in the delivery of the cholesterol, contained in low-density lipoproteins, by the action of the acid lipase and its transport to the other cellular site by the action of two specific transporters: Niemann-Pick C1 and C2.[7]

Based on these considerations on the important role of the lysosomes in several cellular processes, it is clear that their biogenesis, trafficking, and turnover need a coordinated regulation. Recent evidences demonstrate the presence of a regulatory machinery present on lysosomal surface and triggered by the mechanistic target of rapamycin (mTOR). mTOR, in case of starvation or lysosomal engulfment, dissociates from the external leaflet of the lysosomal membrane and becomes inactive. This results in the dephosphorylation of the transcription factor EB that translocates into the nucleus promoting the expression of several lysosomal genes, representing the so-called CLEAR-network.[8] An alteration of the lysosomal biogenesis is correlated with severe pathologies, further supporting the role of the lysosomes in the cell physiology.

2. Lysosomal storage diseases

The importance of lysosomes in the regulation of cell homeostasis is clearly demonstrated by the effect evocated by their impairment. Indeed, the lack or reduced activity of only one of the lysosomal hydrolases or related proteins is responsible for the onset of the so called lysosomal storage disorders (LSDs). LSDs are a heterogeneous group of about 60 distinct monogenic disorders characterized by defects in lysosomal structure and metabolism that result in the accumulation of undegraded macromolecules.

In particular, LSDs are caused by mutations in genes coding for lysosomal proteins, such as lysosomal hydrolases, integral membrane proteins, transporters, enzyme modifiers and activators, or for proteins involved in glycohydrolases' maturation. Mutations in these genes affect the function of the encoded proteins, resulting in lysosomal malfunction and intralysosomal accumulation of undegraded metabolites that leads to the onset of cell damage.[9]

Classically, LSDs were classified based on the nature of the accumulated substrate in three groups: (i) mucopolysaccharidoses, resulting from a deficiency of lysosomal enzymes that are involved in the sequential degradation of glycosaminoglycans (or mucopolysaccharides), (ii) sphingolipidoses, in which undegraded sphingolipids accumulate due to an enzyme deficiency or to an activator protein defect, and (iii) oligosaccharidosis, also known as glycoproteinoses, characterized by defects of glycoprotein degradation (Table 1).[10] More recently, a classification on the base of the molecular defects in LSDs is emerging, including more pathologies recognized now as LSDs: (i) lysosomal defects not concerning enzymes, (ii) transmembrane protein defects (transporters and structural proteins), (iii) lysosomal enzyme protection defects, (iv) posttranslational processing defects of lysosomal enzymes, (v) trafficking defects in lysosomal enzymes, and (vi) polypeptide degradation defects.

LSDs often present multiple organs involvement with a continuum of clinical phenotypes ranging from mild to severe clinical manifestations[11] and with different ages of onset. The heterogeneous clinical phenotype depends on several factors as residual enzymatic activity, tissue and cell distribution of accumulated substrates, cell turnover rate, defective protein expression, and other mechanisms that influence the life span of affected cells.[12] The onset of symptoms can occur before birth, for the most severe forms, or during adulthood for the mild phenotypes. The principal pathological phenotypes are represented by hepatosplenomegaly, corneal or lenticular opacities, retinal dystrophy, optic nerve atrophy, glaucoma, blindness, bone dysplasia, abnormalities of bone density, and osteonecrosis.[13] Moreover, about two-thirds of patients was affected by LSDs, and almost all those affected by defects in SL catabolism (sphingolipidoses) also show an important neurological deficiency, which is extremely variable and heterogeneous ranging from progressive neurodegeneration and severe cognitive deficit to psychiatric and behavioral disorders.[13–16]

Even if a definitive cure for these pathologies is still absent, during the years some therapeutic options were developed. Starting from the 90s, the

Table 1 Lysosomal storage disorders classification.

Disease	Defective protein	Primary accumulated substrate
Mucopolysaccharidoses (MPS)		
MPS I (Hurler, Scheie, Hurler/Scheie)	α-Iduronidase	Dermatan sulfate, heparan sulfate
MPS II (Hunter)	Iduronate sulfatase	Dermatan sulfate, heparan sulfate
MPS III A (Sanfilippo A)	Heparan sulfamidase	Heparan sulfate
MPS III B (Sanfilippo B)	Acetyl α-glucosaminidase	Heparan sulfate
MPS III C (Sanfilippo C)	Acetyl CoA: α-glucosaminide *N*-acetyltransferase	Heparan sulfate
MPS III D (Sanfilippo D)	*N*-Acetyl glucosamine-6-sulfatase	Heparan sulfate
MPS IV A (Morquio A)	Acetyl galactosamine-6-sulfatase	Keratan sulfate, chondroiotin 6-sulfate
MPS IV B (Morquio B)	β-Galactosidase	Keratan sulfate
MPS VI (Maroteaux-Lamy)	Arylsulfatase B	Dermatan sulfate
MPS VII (Sly)	β-Glucuronidase	Dermatan sulfate, heparan sulfate, chondroiotin 6-sulfate
MPS IX (Natowicz)	Hyaluronidase	Hyluronan
Sphingolipidoses		
Fabry	α-Galactosidase A	Globotriasylceramide
Farber	Acid ceramidase	Ceramide
Gangliosidosis GM1 (Types I, II, III)	GM1 β-galactosidase	GM1 ganglioside, Keratan sulfate, oligosaccharides, glycolipids
Gangliosidosis GM2 (Tay-Sachs)	β-Hexosaminidase A	GM2 ganglioside, oligosaccharides, glycolipids
Gangliosidosis GM2 (Sandhoff)	β-Hexosaminidase A + B	GM2 ganglioside, oligosaccharides, glycolipids
Gaucher (Types I, II, III)	Glucosylceramidase	Glucosylceramide
Krabbe	β-Galactosylceramidase	Galactosylceramide
Metachromatic leucodystrophy	Arylsulfatase A	Sulfatides
Niemann-Pick (Types A, B)	Sphingomyelinase	Sphingomyelin

Oligosaccharidoses (glycoproteinoses)		
Aspartylglicosaminuria	Glycosylasparaginase	Aspartylglucosamine
Fucosidosis	α-Fucosidase	Glycoproteins, glycolipids, fucoside-rich oligosaccharides
α-Mannosidosis	α-Mannosidase	Mannose-rich oligosaccharides
β-Mannosidosis	β-Mannosidase	Man(β1 → 4)GlnNAc
Schindler	*N*-Acetylgalactosaminidase	Sialylated/asialoglycopeptides, glycolipids
Sialidosis	Neuraminidase	Oligosaccharides, glycopeptides
Glycogenoses		
Glycogenosis II/ Pompe	Acid α-glucosidase (Aacid maltase)	Glycogen
Lipidoses		
Wolman/CESD	Acid lipase	Cholesterol esters
Lysosomal protein defect		
Gangliosidosis GM2, activator defect	GM2 activator protein	GM2 ganglioside, oligosaccharides
Metachromatic leucodystrophy	Saposin B	Sulfatides
Krabbe	Saposin A	Galactosylceramide
Gaucher	Saposin C	Glucosylceramide
Transmembrane protein defect		
Transporters		
Sialic acid storage disease; infantile form (ISSD) and adult form (Salla)	Sialin	Sialic acid
Cystinosis	Cystinosin	Cystine
Niemann-Pick Type C1	Niemann-Pick type 1 (NPC1)	Cholesterol and sphingolipids
Niemann-Pick, Type C2	Niemann-Pick type 2 (NPC2)	Cholesterol and sphingolipids

Continued

Table 1 Lysosomal storage disorders classification—cont'd

Disease	Defective protein	Primary accumulated substrate
Structural proteins		
Danon	Lysosome-associated membrane protein 2	Cytoplasmatic debris and glycogen
Mucolipidosis IV	Mucolipin	Lipids
Lysosomal enzyme protection defect		
Galactosialidosis	Protective protein cathepsin A	Sialyloligosaccharides
Posttranslational processing defect		
Multiple sulfatase deficiency	Multiple sulfatase	Sulfatides, glycolipids, GAGs
Trafficking defect in lysosomal enzymes		
Mucolipidosis IIα/β, IIIα/β	GlcNAc-1-P transferase	Oligosaccharides, GAGs, lipids
Mucolipidosis IIIγ	GlcNAc-1-P transferase	Oligosaccharides, GAGs, lipids
Polypeptide degradation defect		
Pycnodysostosis	Cathepsin K	Bone proteins
Neuronal ceroid lipofuscinoses (NCLs)		
NCL 1	Palmitoyl protein thioesterase	Saposins A and D
NCL 2	Tripeptidyl peptidase 1	Subunit c of ATP synthase
NCL 3	CLN3, lysosomal transmembrane protein	Subunit c of ATP synthase
NCL 5	CLN5, soluble lysosomal protein	Subunit c of ATP synthase
NCL 6	CLN6, transmembrane protein of ER	Subunit c of ATP synthase
NCL 7	CLC7, lysosomal chloride channel	Subunit c of ATP synthase
NCL 8	CLN8, transmembrane protein of ER	Subunit c of ATP synthase
NCL 10	Cathepsin D	Saposins A and D

enzyme replacement therapy (ERT) has been introduced for the treatment of LSDs. The increase of enzyme activity leads to some improvement in clinical symptoms; however, the enzyme given intravenously cannot cross the blood-brain barrier (BBB) and enter the brain.[17,18] Thus, neurologic symptoms, which are present in about two-thirds of LSDs, do not ameliorate with the treatment. The first ERT approved was for Gaucher Disease (GD) type 1, in which lysosomal glucocerebrosidase deficiency causes glucosylceramide accumulation leading to damages in liver, spleen, kidney, lung, and bone. Administration of recombinant beta-glucocerebrosidase to type 1 GD patients increases the activity of this enzyme intracellularly, ameliorating the symptoms. The recombinant enzyme brings one or more mannose epitope that are recognized by mannose receptors present on the surface of macrophages and allow endocytosis of the receptor and of the linked enzyme which is internalized in the lysosomes. A similar approach has been used to engineer the α-galactosidase for the treatment of Fabry disease. Fabry disease is caused by the deficiency in the lysosomal enzyme involved in the catabolism of globotriaosylceramide that accumulates causing a progressive damage of kidney and other organs, including peripheral neuropathy. ERT is also available for Pompe disease, the glycogen storage disease type II, caused by a deficiency of the lysosomal acid α-glucosidase, and for few mucopolysaccharidoses.

Long-term treatment slows down disease progression; however, ERT presents several mild side effects such as diarrhea and back pain.[19,20] Therefore, several new treatment approaches are under development and study. Among these, the most promising are chaperone therapy, aimed at stabilizing the misfolded enzymes, thus improving their residual activity; administration of the enzyme mRNA, to stimulate the synthesis of the enzyme; substrate reduction therapies, for example, by the administration of inhibitors of glucosylceramide synthase in GD.

3. Impairment of lysosomal activity and alteration in the sphingolipid composition of cell membranes: A possible link with the onset of neuronal damage in LSD

In LSDs, beside the accumulation of the primary uncatabolized substrate caused by the loss of function mutations in a specific enzyme or related proteins, it is common to find a secondary accumulation of lysosomal components such as the sphingolipids.[21] Sphingolipids (SL) are amphiphilic membrane lipids characterized by the presence of a long-chain base called sphingosine in their hydrophobic portion.[22] They are asymmetrically distributed in the cell membrane and result enriched in the outer leaflet where SL play an important role in maintaining cell homeostasis.[22,23] Indeed, at PM level, thanks to their unique physicochemical properties, they are able to organize specific micro domains that, with the recruitment of selected proteins, generate a macromolecular complex involved in the control of signaling transduction, the so called lipids raft.[24] Despite SL are ubiquitous components of the PM of mammalian cells, they are particularly abundant in neurons.[25] In neurons, SL and especially gangliosides, a particular class of glycosphingolipids containing one or more sialic acid in their hydrophilic portion, are involved in a number of cellular processes fundamental for the correct development and function of the nervous system. Indeed, they are involved in the regulation of: (i) neuronal survival, (ii) neuronal differentiation, (iii) neurite outgrowth, (iv) synapses formation and transmission, and (v) neuronal cell adhesion, motility, and recognition.[26] On the other hand, alterations of the physiological composition of the PM sphingolipids are responsible for the onset of neuronal senescence and apoptosis.[26] The metabolic flows that participate to define the final SL pattern at the cell surface are different and not completely dissected so far. De novo biosynthesis occurs mainly in the ER and Golgi apparatus by the action of tightly regulated enzymes and vesicular transport.[27]

In particular, it *begins with the biosynthesis of ceramide in the ER. Ceramide is then transferred to the Golgi apparatus where it can be modified into sphingomyelins or glycosphingolipids*. The generation of sphingomyelin occurs through the transfer of a phosphocholine head group from phosphatidylcholine to ceramide by the action of sphingomyelin synthase. Alternatively, ceramide can be glycosylated into the cerebroside glucosylceramide (GlcCer) by GlcCer synthase or into the galactosylceramide (GalCer) by galactosylceramide synthase. GlcCer and GalCer are then transferred to the luminal side of the Golgi where the synthesis of more complex glycosphingolipids occurs by sequential activity of specific glycosyltransferases. Neo-synthesized sphingolipids reach the PM through vesicular transport, becoming components of the external leaflet.

Sphingolipid synthases and hydrolases are found both associated with the PM allowing directly in situ changes in the content of ceramide, sphingomyelin, neutral glycosphingolipids, and gangliosides.[28,29] PM sphingolipids in low amount are shed in the extracellular milieu and can partially re-associate with the same membrane or associate with the membranes of neighboring cells.[30] In addition, also the intracellular trafficking, consisting in the fusion of intracellular vesicles with the PM and the flow of sphingolipids among the membrane contact sites, plays a key role in the modification of the PM sphingolipid composition.[30]

The biosynthetic processes and those involved in the fine regulation of the cell surface SL composition are not enough to guarantee, for all the cell life, a stable SL pattern. Indeed, a crucial role in the SL turnover is played by the lysosomes, principal district of their catabolism.[16] The lysosomal catabolism of the SL can occur only upon the fine coordination of lysosomal hydrolases, saposins, and/or activator proteins and anionic lipids that facilitate the interaction between the SL and the specific hydrolases in order to obtain monomers that can leave the lysosomes and be recycled in the biosynthetic pathways.

In this scenario, lysosomes could be considered important organelles with an energetic role, since the recycling of catabolites originated in the lysosome for biosynthetic purposes is quantitatively relevant. However, as mentioned above, only a "healthy" lysosome is able to completely catabolize an SL in its monomers. The impairment of sphingolipids' catabolism at lysosomal level leads to the jamming of the overall flow of metabolites, with consequences on the sphingolipids composition in all cellular districts, including the PM.

As proof of principle, using an artificial model of lysosomal impairment, represented by human fibroblasts loaded with sucrose, it has been described that the impaired lysosomes are not able to catabolize the SL leading to their accumulation, not only in lysosomes but also at PM level.[31] Similar findings were obtained also in a neuronal model of Niemann-Pick type A disease (NPA). NPA is a pathology characterized by the loss of function mutation in the gene coding for the lysosomal acid sphingomyelinase, enzyme responsible for the catabolism of sphingomyelin. Interestingly, also in this case, the accumulation of undegraded SL is not only confined to the lysosomes but also affects the PM.[32]

The molecular mechanism involved in the transfer of uncatabolized SL from lysosomes is still debated. Interestingly, it has been found that cells react to the lysosomal accumulation of uncatabolized substrates enhancing the fusion of the lysosomes with the PM (called lysosomal exocytosis).[31] It is reasonable to speculate that the intralysosomal vesicles accumulating SL could exchange with the internal leaflet of the lysosomal membrane that, by the fusion with the cell surface, becomes component of the PM. In addition, the aberrant lysosomal exocytosis could induce the release, in the extracellular milieu, of the intralysosomal vesicles accumulating SL that, by shedding processes, could become component of the outer leaflet of the same cell PM or of that of neighboring cells.

These events are not the only ones involved in the modification of the PM sphingolipid composition occurring in LSDs.

The activation of lysosomal exocytosis in response to lysosomal stress is responsible also for driving a high amount of lysosomal hydrolases at the external leaflet of the PM. The concomitant aberrant presence of SL and their hydrolases induces conformational changes in the membrane architecture that are responsible for the onset of cells damage[31] (Fig. 1).

As mentioned above, the bioactive role of SL is mainly related to their capability to organize macromolecular complexes at the cells PM level

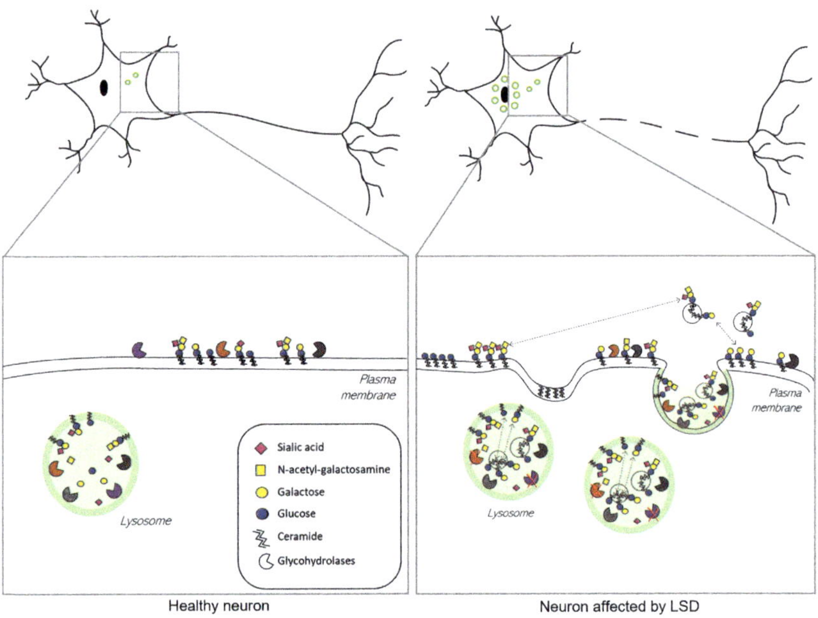

Fig. 1 Schematic representation of the effect of lysosomal impairment on the plasma membrane of neurons. Lysosomal accumulation of uncatabolized substrates promotes lysosomal biogenesis and impairs the catabolic activity of the lysosome inducing the accumulation of uncatabolized SL. Intralysosomal vesicles accumulating SL exchange with the internal leaflet of the lysosomal membrane that, by the fusion with the cell surface, becomes component of the PM. The aberrant lysosomal exocytosis also induces the release in the extracellular milieu of the intralysosomal vesicles accumulating SL that, by shedding processes, could become component of the outer leaflet of the same cell PM. In addition, the lysosomal exocytosis drives lysosomal hydrolases at the external leaflet of the PM. The concomitant aberrant presence of SL and their hydrolases induces conformational changes in the membrane architecture that are responsible for the onset of cells damage.

where they are involved in the control of the signal transduction (lipid-rafts concept). Alteration in the sphingolipid-enriched membrane domains could be responsible for impairment in the signaling events resulting in the onset of the cellular damage.

The data obtained in a cell model of GD are in line with this hypothesis. In fact, it has been demonstrated that glucosylceramide accumulation is not confined to lysosomes but occurs also in lipid rafts, possibly explaining the altered lipid and protein sorting observed in this pathological condition.[33] Moreover, it has been reported that GD is associated with insulin resistance.[34] Since insulin receptor function is regulated by its interaction in lipid rafts with glycosphingolipid[35] and in particular with GM3 ganglioside, this suggests that the altered lipid rafts organization in Gaucher cells might be responsible for altered responsiveness to insulin.

Another example is represented by psychosine (galactosylsphingosine), one of the galactohingolipids that accumulates in the brain of patients affected by Krabbe disease. This pathology is caused by the deficient activity of galactosylceramidase. Psychosine accumulates in lipid rafts leading to an altered distribution of lipid raft proteins and to inhibition of protein kinase C.[36]

ASMKO mice are an animal model for Niemann-Pick disease type A characterized by the inactivation of the gene coding for the acid sphingomyelinase. In the brain of this mice, it has been observed altered composition in terms of the fatty acid moiety in the accumulated SM. In addition, a significant increase in ganglioside content was also found, mainly due to the accumulation of monosialogangliosides GM3 and GM2 that induce the formation of "not conventional" lipid rafts.[37,38] Taken together, these observations open a new scenario on the existence of a lysosome-PM axis fundamental to maintain cell homeostasis that, in case of alterations, is responsible for the onset of neuronal damage.

4. Involvement of mitochondrial impairment in the onset of neurodegeneration in lysosomal storage diseases

Mitochondria are double-membrane organelles responsible for many metabolic functions, including ATP synthesis through oxidative phosphorylation, synthesis of biomolecules, and maintenance of calcium homeostasis.

Emerging evidences indicate an important mitochondrial dysfunction in LSDs, which seems to have an important role in the onset of neuronal damage. Indeed, neurons are postmitotic cells that are dependent on mitochondrial respiration for the production of energetic substrates. Mitochondrial

ATP production is essential for neuronal metabolism and represents the major energy supply for synaptic function, including neurotransmitter synthesis and loading in the presynaptic vesicle and their release upon the induction by action potentials.[39,40]

Since mitochondrial homeostasis critically depends on autophagy and lysosomal degradation, secondary deficits in mitochondrial function and turnover are common pathologic findings in LSDs. Upon mitochondrial damage, the autophagy machinery is activated in order to eliminate the damaged mitochondria via lysosomal degradation through a process called mitophagy. This process is limited in LSDs because of the block in lysosomal activity that leads to an increase in impaired mitochondria.[41–43] Indeed, accumulated mitochondria showed morphologic abnormalities such as fragmentation, swelling, reduced mitochondrial cristae, and inner-mitochondrial vacuolation resulting into a functional deficit in terms of mitochondrial respiration.[42–49] Mitochondrial membrane potential ($\Delta\Psi m$) was found to be decreased in several models of LSDs as Krabbe disease,[50,51] Gaucher disease,[42] and GM1-gangliosidosis.[44,52]

Many studies report that mitochondrial alterations in LSDs cannot be only due to defective autophagy, speculating on the possibility that, as described before, the accumulated macromolecules can leave the lysosomes and affect directly mitochondria interfering with their function.[44,51,53,54] For example, in the brain of $Npc^{-/-}$ mouse, a model of Niemann-Pick type C, characterized by the deletion of the gene coding for the intracellular cholesterol transporter 1, the cholesterol accumulation is not only confined to lysosomes but results also increased in the mitochondrial membrane inhibiting ATP synthase activity. The reduced availability of ATP is responsible for a decreased neuronal outgrowth and enhanced susceptibility to cellular stress.[55] In addition, in mouse models of Krabbe disease, mitochondrial transport along axons results impaired, leading to the depletion of axonal mitochondria and therefore reduced availability of ATP at the synaptic sites. The impaired axonal transport seems to be caused by psychosine accumulation, which leads to an abnormal activation of GSK3β and to an aberrant kinesin light chain phosphorylation, resulting in a premature cargo release.[54] In a Gaucher disease mouse model, amyloid precursor protein and α-synuclein accumulation have been found to inhibit complex I of mitochondrial ETC, reduce $\Delta\Psi m$, and alter calcium homeostasis.[48,56]

Mitochondrial dysfunction, besides causing a reduction in the production of ATP, is responsible also for excessive production of mitochondrial reactive oxygen species (ROS) and dysregulation of mitochondrial calcium homeostasis, common features involved in the onset of neuronal damage in LSDs.

Elevated levels of ROS production, together with a depletion in mitochondrial glutathione (GSH), which impairs mitochondrial ROS-buffering capacities, were reported in galactosylsphingosine-treated MO3.13 human oligodendrocytes, a cellular model of Krabbe disease.[50,51]

The fine regulation of calcium concentration is fundamental for synaptic functioning and neuronal survival. Nowadays, it is clear that mitochondria and the ER, the major intracellular calcium storage organelles, communicate through specific contact sites called mitochondria-associated membranes (MAMs). MAMs are rich in inositol triphosphate (IP3) and ryanodine receptors that allow the calcium flux from the ER to the mitochondrial matrix. In an in vitro model of Krabbe disease represented by a cell line of human oligodendrocyte-like cells, the loading with galactosylsphingosine induces an increase in mitochondrial calcium via the activation of the ryanodine receptor.[51] Similarly, in the β-gal$^{-/-}$ mouse, an animal model of GM1 gangliosidosis, accumulation of GM1 ganglioside in MAMs increases the open probability of IP3 receptors leading to mitochondrial calcium overload.[44] Calcium overload and subsequently prolonged mitochondrial depolarization result in the breakdown of $\Delta\Psi$m and mitochondria-induced neuronal death.[44,51,57]

On the other hand, as reported in several LSDs, the reduced capability of impaired mitochondria to help in the removal of cytosolic calcium, released for instance in response to an action potential, is considered as a proapoptotic signal.[58]

Given the importance of mitochondrial dysfunction in the development of neurodegeneration in LSDs, experimental therapeutic approaches targeting mitochondria have been explored. Promising approaches include the reduction of mitochondrial ROS through the administration of: (i) antioxidant, such as MitoQ10,[42] GSH ethyl-ester,[59] and *N*-acetylcysteine (NAC)[50,51]; (ii) direct caspase inhibition; and (iii) modulators of sirtuins, a group of proteins with important roles in mitochondrial biogenesis and mitophagy. Finally, inhibitors of glycogen synthase kinase-3 (GSK3) have been developed in recent years, since GSK3 inhibition was shown to promote mitochondrial transport along axons.[60,61] Among these, L803-mts was found to be able to restore mitochondrial trafficking along axons in a NSC34 motoneuron-like cell model of Krabbe disease.[54]

Taken together, from literature, it clearly emerges that in LSDs impaired lysosomal activity has an impact also on mitochondrial functioning, effect that contributes to the onset of the neurodegenerative phenotype.

5. Involvement of lysosomal impairment in the neuroinflammation

The term neuroinflammation is used to describe the occurrence of an immune response in the central nervous system (CNS) involving the interplay of a broad set of cellular mediators such as neurons, microglia, and astrocytes.[62]

In particular, microglia cells are the principal immune cells of the CNS that are able to sense the release of damage-associated molecular patterns (DAMPs) from damaged cells. Upon the activation, microglia switch to a phagocytic phenotype and by producing cytokines and chemokines lead to the recruitment of peripheral leukocytes to the site of inflammation. On the other hand, astrocytes are the most abundant cell population in the CNS providing neurons with neurotrophic factors. Activated by microglia, astrocytes are known to worsen the neuronal damage due to their production of chemokines, ROS, and other neurotoxic molecules. Neuroinflammation is therefore often associated with processes of astrocytosis and microgliosis, indicating the preponderance of a reactive response mediated by these cell populations.

A transient neuroinflammatory process can be protective toward impaired neurons reducing the damage by activating the innate immune system. On the contrary, persistent activation of immune cells leads to prolonged release of inflammatory mediators and neurotoxic molecules, with consequences on the blood brain barrier permeability, allowing infiltration of more peripheral macrophages and molecules, which can promote and/or worsen the neurodegeneration state.[63,64] It is still largely debated if neuroinflammation is a cause or an effect of neuronal damage.

An exemplary evidence supporting neuroinflammation's causative role toward neurodegeneration has been found concerning the aging process, where senescent astrocytes release proinflammatory cytokines and chemokines which induce the overexpression of CD38, enzyme that catalyzes the synthesis of nicotinamide from nicotinamide adenine dinucleotide (NAD^+). This leads to NAD^+ depletion and reduced activity of NAD^+-dependent enzymes, such as sirtuins and PARP-1, contributing to DNA damage, metabolic dysfunctions, and oxidative stress of neurons.[65,66]

The other way round neuronal induction of inflammation has been for instance supported by in vivo studies on a mouse model of GD characterized by the conditional knock out of the enzyme beta-glucosylcerebrosidase in neurons and astrocytes leaving the enzyme active in microglia. Surprisingly, glucosylceramide accumulates in all three cell types. These data let to speculate that the accumulation of glucosylceramide induces a damage in neurons and astrocytes which, from one side, activates microglia by the release of DAMPs and, from the other side, inactivates the native enzyme in microglial cells further enhancing the inflammatory state.[67]

The study of LSDs presenting a neuropathic phenotype has given a great contribution to shed light on the possible involvement of lysosomal impairment on both neuronal damage and microglia/astrocytes activation.

As mentioned earlier, microglia are devoted to the removal through phagocytosis of cellular debris of different origins that were degraded by lysosomes. It is clear that in case of lysosomal impairment, these cells are exposed to the aberrant accumulation of uncatabolized substrates that in turn induce their activation.[1,68] In addition, also the nature of the accumulated substrate plays an important role in defining the functioning of these cells. Indeed, accumulation of glucosylceramide in inflammatory cells and macrophages, typical feature of GD, has been associated with altered migration and antigen presentation.[69]

Patients affected by mucopolysaccharidosis and characterized by a neurodegenerative phenotype show a prolonged neuroinflammatory state mediated by microglia. In particular, the lysosomal accumulation of heparan sulfate and gangliosides alters lysosome membrane integrity, which allows the release into the cytoplasm of the storage products causing self-activation of the microglia. On the other hand, activated microglia affect neurons generating a self-propagating vicious cycle between neurodegeneration and neuroinflammation.[70]

In conclusion, dissecting the paradigm of neuroinflammation remains one of the challenges to define the complete mechanism responsible for the onset of neuronal damage in LSDs. In addition, elucidating the role of lysosomal impairment in the promotion of the neuroinflammatory state is also fundamental to understand the etiopathogenesis of several other pathologies affecting the CNS.

References

1. Hipolito VEB, Ospina-Escobar E, Botelho RJ. Lysosome remodelling and adaptation during phagocyte activation. *Cell Microbiol*. 2018;20(4):1–8.
2. Schwake M, Schroder B, Saftig P. Lysosomal membrane proteins and their central role in physiology. *Traffic*. 2013;14(7):739–748.
3. Xiong J, Zhu MX. Regulation of lysosomal ion homeostasis by channels and transporters. *Sci China Life Sci*. 2016;59(8):777–791.
4. Mego JL. The ATP-dependent proton pump in lysosome membranes: still a valid hypothesis. *FEBS Lett*. 1979;107(1):113–116.
5. Boya P, Kroemer G. Lysosomal membrane permeabilization in cell death. *Oncogene*. 2008;27(50):6434–6451.
6. Appelqvist H, Waster P, Kagedal K, Ollinger K. The lysosome: from waste bag to potential therapeutic target. *J Mol Cell Biol*. 2013;5(4):214–226.
7. Subramanian K, Balch WE. NPC1/NPC2 function as a tag team duo to mobilize cholesterol. *Proc Natl Acad Sci U S A*. 2008;105(40):15223–15224.
8. Sardiello M, Palmieri M, di Ronza A, et al. A gene network regulating lysosomal biogenesis and function. *Science*. 2009;325(5939):473–477.
9. Platt FM, Boland B, van der Spoel AC. The cell biology of disease: lysosomal storage disorders: the cellular impact of lysosomal dysfunction. *J Cell Biol*. 2012;199(5):723–734.
10. Filocamo M, Morrone A. Lysosomal storage disorders: molecular basis and laboratory testing. *Hum Genomics*. 2011;5(3):156–169.
11. Wang RY, Bodamer OA, Watson MS, Wilcox WR. Diseases AWGoDCoLS. Lysosomal storage diseases: diagnostic confirmation and management of presymptomatic individuals. *Genet Med*. 2011;13(5):457–484.
12. Jakobkiewicz-Banecka J, Gabig-Ciminska M, Banecka-Majkutewicz Z, et al. Factors and processes modulating phenotypes in neuronopathic lysosomal storage diseases. *Metab Brain Dis*. 2014;29(1):1–8.
13. Parenti G, Andria G, Ballabio A. Lysosomal storage diseases: from pathophysiology to therapy. *Annu Rev Med*. 2015;66:471–486.
14. Futerman AH, Sussman JL, Horowitz M, Silman I, Zimran A. New directions in the treatment of Gaucher disease. *Trends Pharmacol Sci*. 2004;25(3):147–151.
15. Futerman AH, van Meer G. The cell biology of lysosomal storage disorders. *Nat Rev Mol Cell Biol*. 2004;5(7):554–565.
16. Kolter T, Sandhoff K. Principles of lysosomal membrane digestion: stimulation of sphingolipid degradation by sphingolipid activator proteins and anionic lysosomal lipids. *Annu Rev Cell Dev Biol*. 2005;21:81–103.
17. Li M. Enzyme replacement therapy: a review and its role in treating lysosomal storage diseases. *Pediatr Ann*. 2018;47(5):e191–e197.
18. Concolino D, Deodato F, Parini R. Enzyme replacement therapy: efficacy and limitations. *Ital J Pediatr*. 2018;44(Suppl 2):120.
19. Revel-Vilk S, Szer J, Mehta A, Zimran A. How we manage Gaucher disease in the era of choices. *Br J Haematol*. 2018;182(4):467–480.
20. Ortiz A, Germain DP, Desnick RJ, et al. Fabry disease revisited: management and treatment recommendations for adult patients. *Mol Genet Metab*. 2018;123(4):416–427.
21. Fuller M, Futerman AH. The brain lipidome in neurodegenerative lysosomal storage disorders. *Biochem Biophys Res Commun*. 2018;504(3):623–628.
22. Sonnino S, Prinetti A, Mauri L, Chigorno V, Tettamanti G. Dynamic and structural properties of sphingolipids as driving forces for the formation of membrane domains. *Chem Rev*. 2006;106(6):2111–2125.

23. van Meer G, Hoetzl S. Sphingolipid topology and the dynamic organization and function of membrane proteins. *FEBS Lett*. 2010;584(9):1800–1805.
24. Prinetti A, Loberto N, Chigorno V, Sonnino S. Glycosphingolipid behaviour in complex membranes. *Biochim Biophys Acta*. 2009;1788(1):184–193.
25. Hansson HA, Holmgren J, Svennerholm L. Ultrastructural localization of cell membrane GM1 ganglioside by cholera toxin. *Proc Natl Acad Sci U S A*. 1977;74(9):3782–3786.
26. Piccinini M, Scandroglio F, Prioni S, et al. Deregulated sphingolipid metabolism and membrane organization in neurodegenerative disorders. *Mol Neurobiol*. 2010;41(2–3):314–340.
27. Maccioni HJ, Quiroga R, Ferrari ML. Cellular and molecular biology of glycosphingolipid glycosylation. *J Neurochem*. 2011;117(4):589–602.
28. Aureli M, Samarani M, Loberto N, et al. The glycosphingolipid hydrolases in the central nervous system. *Mol Neurobiol*. 2014;50(1):76–87.
29. Vilcaes AA, Garbarino-Pico E, Torres Demichelis V, Daniotti JL. Ganglioside synthesis by plasma membrane-associated sialyltransferase in macrophages. *Int J Mol Sci*. 2020;21(3):1–20.
30. Aureli M, Samarani M, Loberto N, et al. Neuronal membrane dynamics as fine regulator of sphingolipid composition. *Glycoconj J*. 2018;35(4):397–402.
31. Samarani M, Loberto N, Solda G, et al. A lysosome-plasma membrane-sphingolipid axis linking lysosomal storage to cell growth arrest. *FASEB J*. 2018;32(10):5685–5702.
32. Gabande-Rodriguez E, Boya P, Labrador V, Dotti CG, Ledesma MD. High sphingomyelin levels induce lysosomal damage and autophagy dysfunction in Niemann Pick disease type A. *Cell Death Differ*. 2014;21(6):864–875.
33. Hein LK, Duplock S, Hopwood JJ, Fuller M. Lipid composition of microdomains is altered in a cell model of Gaucher disease. *J Lipid Res*. 2008;49(8):1725–1734.
34. Langeveld M, Ghauharali KJ, Sauerwein HP, et al. Type I Gaucher disease, a glycosphingolipid storage disorder, is associated with insulin resistance. *J Clin Endocrinol Metab*. 2008;93(3):845–851.
35. Kabayama K, Sato T, Saito K, et al. Dissociation of the insulin receptor and caveolin-1 complex by ganglioside GM3 in the state of insulin resistance. *Proc Natl Acad Sci U S A*. 2007;104(34):13678–13683.
36. White AB, Givogri MI, Lopez-Rosas A, et al. Psychosine accumulates in membrane microdomains in the brain of krabbe patients, disrupting the raft architecture. *J Neurosci*. 2009;29(19):6068–6077.
37. Scandroglio F, Venkata JK, Loberto N, et al. Lipid content of brain, brain membrane lipid domains, and neurons from acid sphingomyelinase deficient mice. *J Neurochem*. 2008;107(2):329–338.
38. Buccinna B, Piccinini M, Prinetti A, et al. Alterations of myelin-specific proteins and sphingolipids characterize the brains of acid sphingomyelinase-deficient mice, an animal model of Niemann-Pick disease type A. *J Neurochem*. 2009;109(1):105–115.
39. de la Mata M, Cotan D, Villanueva-Paz M, et al. Mitochondrial dysfunction in lysosomal storage disorders. *Diseases*. 2016;4(4):1–10.
40. Saffari A, Kolker S, Hoffmann GF, Ebrahimi-Fakhari D. Linking mitochondrial dysfunction to neurodegeneration in lysosomal storage diseases. *J Inherit Metab Dis*. 2017;40(5):631–640.
41. Settembre C, Fraldi A, Jahreiss L, et al. A block of autophagy in lysosomal storage disorders. *Hum Mol Genet*. 2008;17(1):119–129.
42. Osellame LD, Duchen MR. Defective quality control mechanisms and accumulation of damaged mitochondria link Gaucher and Parkinson diseases. *Autophagy*. 2013;9(10):1633–1635.

43. Martins C, Hulkova H, Dridi L, et al. Neuroinflammation, mitochondrial defects and neurodegeneration in mucopolysaccharidosis III type C mouse model. *Brain.* 2015;138(Pt 2):336–355.
44. Sano R, Annunziata I, Patterson A, et al. GM1-ganglioside accumulation at the mitochondria-associated ER membranes links ER stress to Ca(2+)-dependent mitochondrial apoptosis. *Mol Cell.* 2009;36(3):500–511.
45. de Pablo-Latorre R, Saide A, Polishhuck EV, et al. Impaired parkin-mediated mitochondrial targeting to autophagosomes differentially contributes to tissue pathology in lysosomal storage diseases. *Hum Mol Genet.* 2012;21(8):1770–1781.
46. Fossale E, Wolf P, Espinola JA, et al. Membrane trafficking and mitochondrial abnormalities precede subunit c deposition in a cerebellar cell model of juvenile neuronal ceroid lipofuscinosis. *BMC Neurosci.* 2004;5:57.
47. Kennedy BE, LeBlanc VG, Mailman TM, et al. Pre-symptomatic activation of antioxidant responses and alterations in glucose and pyruvate metabolism in Niemann-Pick Type C1-deficient murine brain. *PLoS ONE.* 2013;8(12), e82685.
48. Xu YH, Xu K, Sun Y, et al. Multiple pathogenic proteins implicated in neuronopathic Gaucher disease mice. *Hum Mol Genet.* 2014;23(15):3943–3957.
49. Dasgupta N, Xu YH, Li R, et al. Neuronopathic Gaucher disease: dysregulated mRNAs and miRNAs in brain pathogenesis and effects of pharmacologic chaperone treatment in a mouse model. *Hum Mol Genet.* 2015;24(24):7031–7048.
50. Haq E, Giri S, Singh I, Singh AK. Molecular mechanism of psychosine-induced cell death in human oligodendrocyte cell line. *J Neurochem.* 2003;86(6):1428–1440.
51. Voccoli V, Tonazzini I, Signore G, Caleo M, Cecchini M. Role of extracellular calcium and mitochondrial oxygen species in psychosine-induced oligodendrocyte cell death. *Cell Death Dis.* 2014;5:e1529.
52. Takamura A, Higaki K, Kajimaki K, et al. Enhanced autophagy and mitochondrial aberrations in murine G(M1)-gangliosidosis. *Biochem Biophys Res Commun.* 2008;367(3):616–622.
53. Sanz-Blasco S, Valero RA, Rodriguez-Crespo I, Villalobos C, Nunez L. Mitochondrial Ca2+ overload underlies Abeta oligomers neurotoxicity providing an unexpected mechanism of neuroprotection by NSAIDs. *PLoS ONE.* 2008;3(7), e2718.
54. Cantuti Castelvetri L, Givogri MI, Hebert A, et al. The sphingolipid psychosine inhibits fast axonal transport in Krabbe disease by activation of GSK3beta and deregulation of molecular motors. *J Neurosci.* 2013;33(24):10048–10056.
55. Yu W, Gong JS, Ko M, et al. Altered cholesterol metabolism in Niemann-Pick type C1 mouse brains affects mitochondrial function. *J Biol Chem.* 2005;280(12):11731–11739.
56. Luth ES, Stavrovskaya IG, Bartels T, Kristal BS, Selkoe DJ. Soluble, prefibrillar alpha-synuclein oligomers promote complex I-dependent, Ca2+−induced mitochondrial dysfunction. *J Biol Chem.* 2014;289(31):21490–21507.
57. Luiro K, Kopra O, Blom T, et al. Batten disease (JNCL) is linked to disturbances in mitochondrial, cytoskeletal, and synaptic compartments. *J Neurosci Res.* 2006;84(5):1124–1138.
58. Kiselyov K, Muallem S. Mitochondrial Ca2+ homeostasis in lysosomal storage diseases. *Cell Calcium.* 2008;44(1):103–111.
59. Torres S, Matias N, Baulies A, et al. Mitochondrial GSH replenishment as a potential therapeutic approach for Niemann Pick type C disease. *Redox Biol.* 2017;11:60–72.
60. Chen S, Owens GC, Crossin KL, Edelman DB. Serotonin stimulates mitochondrial transport in hippocampal neurons. *Mol Cell Neurosci.* 2007;36(4):472–483.
61. Chen CY, Hong PW, Chen CL, et al. Ground reaction force patterns in stroke patients with various degrees of motor recovery determined by plantar dynamic analysis. *Chang Gung Med J.* 2007;30(1):62–72.

62. Lyman M, Lloyd DG, Ji X, Vizcaychipi MP, Ma D. Neuroinflammation: the role and consequences. *Neurosci Res.* 2014;79:1–12.
63. Schmid CD, Melchior B, Masek K, et al. Differential gene expression in LPS/IFNgamma activated microglia and macrophages: in vitro versus in vivo. *J Neurochem.* 2009;109(Suppl. 1):117–125.
64. Schmidt O, Rothhardt J, Roser F, et al. Millijoule pulse energy Q-switched short-length fiber laser. *Opt Lett.* 2007;32(11):1551–1553.
65. Guerreiro S, Privat AL, Bressac L, Toulorge D. CD38 in neurodegeneration and neuroinflammation. *Cell.* 2020;9(2).
66. Piedra-Quintero ZL, Wilson Z, Nava P, Guerau-de-Arellano M. CD38: an immunomodulatory molecule in inflammation and autoimmunity. *Front Immunol.* 2020;11:597959.
67. Vitner EB, Farfel-Becker T, Eilam R, Biton I, Futerman AH. Contribution of brain inflammation to neuronal cell death in neuronopathic forms of Gaucher's disease. *Brain.* 2012;135(Pt 6):1724–1735.
68. Gabande-Rodriguez E, Perez-Canamas A, Soto-Huelin B, et al. Lipid-induced lysosomal damage after demyelination corrupts microglia protective function in lysosomal storage disorders. *EMBO J.* 2019;38(2).
69. Bettman N, Avivi I, Rosenbaum H, Bisharat L, Katz T. Impaired migration capacity in monocytes derived from patients with Gaucher disease. *Blood Cells Mol Dis.* 2015;55(2):180–186.
70. Bosch ME, Kielian T. Neuroinflammatory paradigms in lysosomal storage diseases. *Front Neurosci.* 2015;9:417.

SECTION III

The autophagic pathways

CHAPTER 3

The autophagy pathway and its key regulators

Tetsushi Kataura, Nobutaka Hattori, and Shinji Saiki
Department of Neurology, Juntendo University Graduate School of Medicine, Bunkyo-ku, Tokyo, Japan

1. The autophagy machinery

Over the past two decades, autophagy has been intensively investigated using a variety of laboratory models in yeast, fly, fish, and mammals.[1–4] The molecular machinery of autophagy is evolutionarily conserved, with common autophagy-related (ATG) genes and proteins orchestrating the entire autophagy process, including autophagosome formation, autophagosome maturation, and fusion with lysosomes (Fig. 1).[5] The initiation of autophagy is regulated by the multiprotein complexes, unc-51-like autophagy activating kinase 1 (ULK1) complex, and Class III phosphatidylinositol 3-kinase (PI3K) complex.[6,7] The ULK1 complex consists of ULK1, RB1-inducible coiled-coil protein 1 (RB1CC1, also known as FIP200), ATG13, and ATG101.[6] This macromolecular complex phosphorylates ATG9 via the kinase activity of ULK1, which induces the incorporation of phospholipids from various organelles, including the endoplasmic reticulum (ER), leading to phagophore elongation.[6] The Class III PI3K complex is composed of vacuolar protein sorting 34 (VPS34), VPS15, ATG14L, and beclin-1 (BECN1), which phosphorylates phosphatidylinositol to generate phosphatidylinositol 3-phosphate (PI3P). PI3P is required for biosynthesis of the autophagosome.[7] Two ubiquitin-like conjugation systems also play a central role in the autophagy process.[8] ATG7 (E1 ubiquitin-activating enzyme-like) and ATG10 (E2 ubiquitin-conjugating enzyme-like) participate in ATG5-ATG12 conjugation, which then forms a complex with ATG16L1 that is necessary for autophagosome formation. Another complex is mediated by ATG7 and ATG3 (E2 ubiquitin-conjugating enzyme-like), which facilitates the conjugation of phosphatidylethanolamine (PE) and microtubule-associated protein 1 light chain 3 beta (MAP1LC3B or LC3B, an ATG8 subfamily member in mammals) to generate a lipidated form of LC3B (known as

https://doi.org/10.1016/B978-0-323-89906-2.00013-7

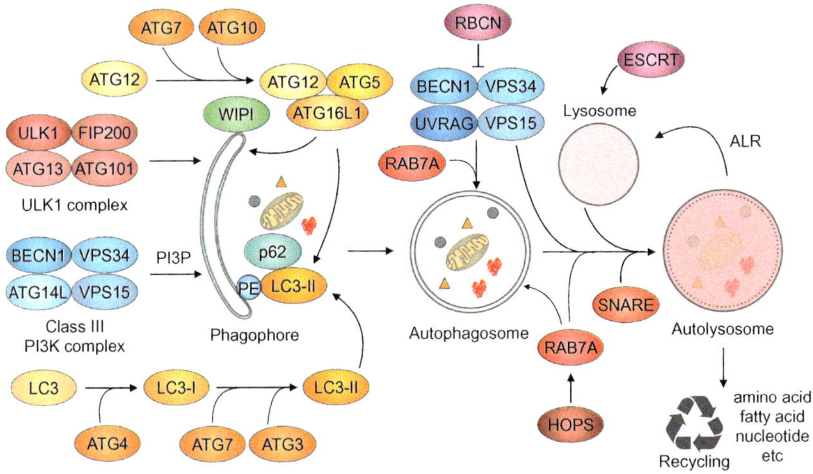

Fig. 1 The core autophagic machinery in mammalian cells. Autophagy is initiated by the ULK1 complex, and the Class III PI3K complex is involved with autophagosome formation. Two ubiquitin-like conjugation systems orchestrate phagophore expansion and LC3-II formation with ATG4 protease and WIPI. PE-conjugated LC3-II and autophagic receptors, such as p62/SQSTM1, participate in cargo selection and autophagosome maturation. The BECN1-VPS34-VPS15 complex with UVRAG and RAB7A activated by HOPS also regulate autophagosome maturation. Finally, autophagosomes fuse with lysosomes mediated by ESCRT and SNARE proteins, leading to the formation of autolysosomes where the cargo is degraded and recycled.

LC3B-II) in cooperation with ATG4 protease.[8] Both the ATG12-ATG5-ATG16L1 complex and LC3B-II have an essential role in autophagosome formation by mediating phagophore expansion, which is considered to be the major source of the membrane for autophagosomes.[9] In addition to LC3B, which is presumed to be the dominant homolog of the ATG8 family, five other homologs, LC3A, LC3C, and GABA receptor-associated protein (GABARAP) subfamily members: GABARAP, GABARAPL1, and GABARAPL2[10,11] also have individual roles in LC3-mediated phagophore expansion and GABARAP-mediated autophagosome closure.[12] LC3 and GABARAP also mediate the recognition of autophagy substrates by binding to autophagy receptors, such as sequestosome 1 (SQSTM1, also known as p62).[13] WD repeat domain phosphoinositide-interacting 1/2 (WIPI1/2) promotes the recruitment of two ubiquitin-like conjugation systems to the phagophore assembly site by interaction with PI3P, which is responsible for phagophore elongation.[14] In the next step, autophagosomes fuse with late endosomes to form amphisomes. This is followed by fusion with a lysosome to form an autolysosome in which the enclosed autophagic substrate or cargo is degraded.[15] These fusion steps are regulated by various proteins and complexes including RAS-related protein RAB7A,[16] endosomal sorting complexes required for transport (ESCRT),[17] soluble NSF attachment protein receptor (SNARE) proteins,[18] homotypic fusion and vacuole protein sorting (HOPS) complexes,[19] and the BECN1-VPS34-VPS15 complex associated with UV radiation resistance-associated gene protein (UVRAG) or Run domain Beclin-1-interacting and cysteine-rich domain-containing protein (RBCN).[20,21] RAB7A activated by UVRAG associated with the BECN1-VPS34-VPS15 complex induces autophagosome maturation, while this step is inhibited when the complex interacts with RBCN.[22] RAB7A is also activated by the HOPS complex, which mediates membrane tethering between autophagosomes and late endosomes to form the amphisomes.[23] ESCRT proteins, which are essential for multivesicular body (MVB) formation, also have a function in autophagosome maturation.[24,25] Syntaxin-17 (STX17) is a SNARE protein that is essential for autolysosome formation.[26] STX17 localizes on the autophagosomal membrane and associates with lysosomal vesicle-associated membrane protein 8 (VAMP8) and synaptosome-associated protein 29 (SNAP29), leading to the fusion of autophagosomes with lysosomes.[27] Once autolysosome formation is complete, the acidic environment maintained by vacuolar-type ATPase (V-ATPase) leads to activation of the lysosomal enzymes, such as cathepsins, that are essential to degrade the autophagic cargo.[28,29] The degradation products, including amino acids and sugars, are transported out of the autolysosome by efflux transporters.[30] In the termination stage of autophagy, autophagic lysosome reformation (ALR) occurs to recycle lysosomal membrane proteins from autolysosomes into proto-lysosomes that eventually mature into functional lysosomes.[31]

2. Key regulators and signaling pathways of autophagy

Autophagy is regulated by several intracellular signaling pathways that are activated by various stimuli, such as starvation, stress responses, and transcriptional and posttranscriptional (epigenetic) modifications.[32–34] Later, we described established autophagic pathways and their key molecules (Fig. 2).

2.1 Regulation of autophagy by MTOR

Mechanistic target of rapamycin (MTOR) is a serine/threonine protein kinase and a well-established regulator of autophagy.[35] MTOR forms two distinct functional complexes, MTOR complex 1 and 2 (MTORC1 and

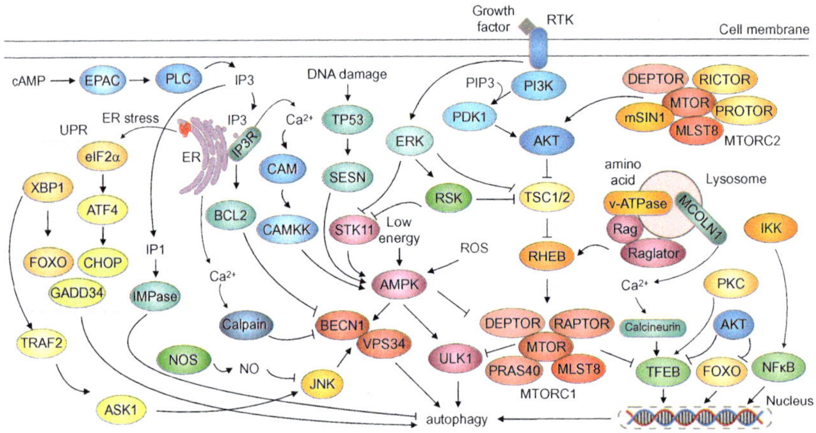

Fig. 2 The signaling pathways regulating autophagy in mammalian cells. MTORC1 negatively regulates autophagy by suppressing the ULK1 complex and TFEB transcription factor. The activity of MTORC1 is regulated by growth factor signaling, the Rag/Ragulator complex, or AMPK. MTORC2 phosphorylates AKT, which forms a positive feedback loop. AMPK positively regulates autophagy via not only MTORC1 inhibition but also direct activation of ULK1 and Class III PI3K complexes. AMPK is regulated by diverse signaling pathways, such as energy state, DNA damage, Ca^{2+}, and ROS. While ROS mainly act as an autophagy activator, nitrogen oxide inactivates autophagy via JNK inhibition. Inositol signaling negatively regulates autophagy by IP3R-mediated interaction of BCL2 with BECN. IMPase also inhibits autophagosome formation. IP3R-mediated Ca^{2+} release from the ER to the cytosol both positively and negatively regulates autophagy via Calpain and CAM/CAMKK-AMPK signaling. Ca^{2+} release from lysosomes by MCOLN1 activates the calcineurin-TFEB axis, thereby activating autophagy. The unfolded protein response caused by ER stress activates eIF2α-ATF4-CHOP/GADD34, XBP1-FOXO, and XBP1-TRAF2-ASK1-JNK signaling, resulting in autophagy induction. Transcription factors, such as TFEB, FOXO, and NFκB, regulate autophagy gene expression.

MTORC2, respectively). MTORC1 acts as a master regulator of cell growth and proliferation by regulating anabolic and catabolic pathways,[36] while MTORC2 regulates cytoskeletal structure and cell survival. MTORC1, however, has a more prominent role than MTORC2 in the regulation of autophagic activity.[37] MTORC1 consists of MTOR, regulatory-associated protein of MTOR (RAPTOR), DEP-domain containing MTOR-interacting protein (DEPTOR), proline-rich Akt substrate of 40 kDa (PRAS40) and target of rapamycin complex subunit LST8 (MLST8). MTORC1 directly phosphorylates ULK1 and ATG13, which negatively regulate the ULK1 complex, leading to the suppression of autophagy.[36] Transcription factor EB (TFEB), a master regulator of lysosomal/autophagy gene expression, is negatively regulated by MTORC1-dependent phosphorylation, which promotes the interaction of TFEB with 14-3-3, resulting in TFEB stabilization in the cytosol, thereby preventing its nuclear translocation.[38]

MTORC1 activity is controlled by nutrient conditions, energy status, and several cellular signaling pathways.[39] The V-ATPase and Ras-related GTP-binding protein (Rag)-Ragulator complex on the lysosomal surface senses amino acids and anchors MTORC1 to lysosomes.[40] Rag GTPase consists of RagA or RagB bound to RagC or RagD. The active form of the Rag complex induces the binding of RAPTOR and recruitment of MTORC1 and Ras homolog enriched in brain (RHEB) to the lysosomal surface, resulting in MTORC1 activation.[40] Growth factors, such as insulin, activate receptor tyrosine kinase (RTK), which provides a signal to PI3K (class Ia) to generate phosphatidylinositol(3,4,5)P3 (PIP3), which together with phosphoinositide-dependent kinase 1 (PDK1), activates AKT.[41] Activated AKT inhibits the activity of tuberous sclerosis complex (TSC)1/2, which acts as a GTPase-activating protein (GAP) for RHEB through phosphorylation, resulting in formation of an active GTP-bound form of RHEB.[42] This form of RHEB can directly interact with MTORC1 anchored to the lysosomal surface by the Rag-Ragulator complex.[40] AKT also provides a signal to MTORC1 in a TSC1/2-independent manner by phosphorylating PRAS40 from RAPTOR.[43] In addition, RTK-dependent RAS-mitogen-activated protein kinase (MAPK)-extracellular signal-regulated kinase (ERK) pathway activation results in phosphorylation of TSC2 by ERK and its downstream RSK, inhibiting the TSC1/2 complex and thereby triggering RHEB-mediated MTORC1 activation.[44] Depletion of amino acids or growth factors induces autophagy by disrupting these regulatory pathways. MTORC2 consists of MTOR, DEPTOR, MLST8, rapamycin-insensitive companion of mTOR (RICTOR), mammalian stress-activated Map kinase-interacting 1 (mSIN1) and protein observed with RICTOR (PROTOR). MTORC2 can regulate MTORC1 activity via phosphorylation of AKT at Ser473, which activates AKT leading to autophagy suppression.[45]

2.2 Autophagy regulation by AMPK

AMP-activated protein kinase (AMPK) is a core protein of autophagy induction and acts as a sensor for increased ATP/AMP ratios, which indicate low energy states. AMPK is regulated by energy metabolism, calcium signaling, and several protein kinases.[46] Activated AMPK suppresses MTORC1 through the inhibition of RAPTOR and the activation of

TSC1/2 by direct phosphorylation, leading to RHEB inhibition. AMPK also directly phosphorylates ULK1 and BECN1, which activates ULK1 and VPS34 complexes leading to autophagy activation in contrast to phosphorylation of ULK1 by MTORC1.[47] Under hypoxic conditions, a low ATP/AMP ratio caused by suppression of mitochondrial respiration is also sensed by AMPK.[48] A positive regulator of AMPK is serine/threonine kinase 11 (STK11, also known as LKB1), which is mainly activated by protein kinase A (PKA), a sensor protein of intracellular cyclic AMP (cAMP) although PKA can inhibit autophagy via direct phosphorylation of ULK1.[49,50] SKT11 is also inhibited by ERK and its downstream ribosomal S6 kinase (RSK).[51] Calcium-calmodulin (CAM)-dependent protein kinase kinase (CAMKK) plays a central role in calcium and nitrogen sensing. CAM is the primary intracellular calcium sensor and binding to free cytosolic Ca^{2+} facilitates its interaction with calcium/CAM-dependent protein kinase I and IV (CAMKI and CAMKIV, respectively) and CAMKK1 and CAMKK2. CAMKK2 activation permits phosphorylation of AMPK. Cells also utilize CAMKK for sensing nitrogen stress and for activating downstream AMPK.[52] Finally, DNA damage-associated activation of tumor suppressor p53 (TP53) and its downstream sestrin 1/2 (SESN1/2) activate AMPK, resulting in MTORC1 inhibition and autophagy activation.[53] TP53-associated DNA damage regulated autophagy modulator (DRAM) and TFEB activation also positively regulate autophagy.[54,55]

2.3 Regulation of autophagy by inositol and calcium signaling

Intracellular inositol levels can negatively regulate autophagy in an mTOR-independent manner. Inositol 1,4,5-triphosphate (IP3) generated from phosphatidylinositol 4,5-bisphosphate by phospholipase C (PLC) binds to IP3 receptors (IP3R) on the ER, resulting in the release of Ca^{2+} from the ER to the cytosol (and from other organelles including mitochondria and lysosomes), which mediates a range of cellular responses.[56] Intracellular Ca^{2+} can regulate autophagy (as described below). IP3R facilitates the binding of BECN1 to B-cell lymphoma 2 (BCL2), a negative regulator of autophagy, which suppresses autophagy.[57] Inositol monophosphatase (IMPase), which hydrolyzes inositol monophosphate generated from IP3 by 5′-phosphatase and inositol polyphosphate 1-phosphatase, inhibits autophagosome synthesis.[58]

Intracellular Ca^{2+} is mainly considered to be a negative regulator of autophagy, although some reports implicate Ca^{2+} in activation of autophagy.[59–61] These seemingly contradictory effects of Ca^{2+} on autophagy possibly result from the complicated crosstalk of Ca^{2+}-mediated signaling. Under basal conditions, IP3R-mediated Ca^{2+} release from the ER increases cytosolic Ca^{2+} and Ca^{2+} in mitochondria and lysosomes.[56,62] Cytosolic Ca^{2+} activates a Ca^{2+}-dependent cysteine protease, calpain, which induces the cleavage of ATG5 and BECN1 leading to the suppression of autophagy.[62,63] Calpain also activates the cAMP-rap guanine nucleotide exchange factor (RAPGEF2, also known as EPAC)-PLC pathway,[64] which forms a positive feedback loop to maintain IP3 and Ca^{2+} concentrations although cAMP may activate STK11-AMPK-mediated autophagy as described above. Ca^{2+} taken up by mitochondria through voltage-dependent anion channel (VDAC) complexes is consumed to maintain mitochondrial ATP production, which inactivates AMPK and AMPK-induced autophagy.[65] However, cytosolic Ca^{2+} can activate CAMKK and thereby AMPK-mediated autophagy.[66] The lysosomal calcium channel, mucolipin TRP cation channel 1 (MCOLN1, also known as TRPML1), mutations in which cause the lysosomal storage disorder mucolipidosis type IV (MLIV) contributes to the upregulation of autophagic genes by inducing the nuclear translocation of TFEB and activation of autophagic vesicle biogenesis through the generation of PI3P and the recruitment of essential PI3P-binding proteins to the nascent phagophore.[67] Under nutrient deprivation, MCOLN1-mediated lysosomal Ca^{2+} release regulates the activity of calmodulin-dependent calcineurin, which binds to and dephosphorylates Ser142 and Ser211 on TFEB, promoting autophagy gene expression.[68] Leucine-rich repeat kinase-2 (LRRK2) mutations, which are the most common cause of familial Parkinson's disease, elicit Ca^{2+} release from lysosomes, which triggers further Ca^{2+} release from the ER and induces autophagy via CAMKK/AMPK,[69] although various other mechanisms have been proposed for LRRK2 regulation of autophagy.[70]

2.4 Autophagy regulation by reactive oxygen species and redox signaling

Intracellular oxidative stress and excessive reactive oxygen species (ROS), which are often associated with mitochondrial dysfunction, stimulate autophagy.[71] A reduced form of ATG4 delipidates and recycles LC3, leading to the suppression of autophagy. Excessive ROS, specifically hydrogen peroxide (H_2O_2)-mediated thiol oxidation of ATG4, promotes conversion of LC3-I to LC3-II, which is essential for autophagosome formation.[72] H_2O_2 also directly oxidizes AMPK and the serine/threonine kinase, ATM, and activates STK11, which activates AMPK and thus induces autophagy.[73]

In contrast, nitric oxide (NO), synthesized by NO synthases (NOS), has negative effects on autophagy. S-nitrosylation, which is the covalent attachment of NO to a cysteine thiol in a protein, modifies and attenuates the activity of stress-activated protein kinase (SAPK, also known as JNK) and I kappa B kinase (IKK).[74] JNK inhibits the association between BCL2 and BECN1 by direct phosphorylation of BCL2,[75] and IKK activates not only AMPK but also NFκB-mediated transcriptional upregulation of autophagy genes,[76] S-nitrosylation of these proteins, therefore, suppresses autophagy. S-nitrosylation of phosphatase and tensin homolog (PTEN) and lysosomal proteins can also suppress autophagy.[77,78]

Mitochondrial ROS links calcium signaling to autophagy. Stromal interaction molecule 1 (STIM1) is activated by mitochondrial ROS and induces Ca^{2+} influx leading to CAMKK activation.[79] Mitochondrial ROS is also important for the activation of lysosomal MCOLN1.[80]

Nuclear factor erythroid 2 like 2 (NFE2L2, also known as NRF2), a master regulator of the antioxidant response, maintains the cellular redox system.[81] Kelch-like ECH associated protein 1 (KEAP1) is a negative regulator of NRF2 by mediating ubiquitination and subsequent proteasomal degradation of NRF2.[81] Under oxidative stress conditions, ROS oxidize KEAP1 and induce the dissociation of NRF2 from KEAP1, which promotes nuclear translocation of NRF2 and activates expression of its target genes, including p62/SQSTM1, an autophagy receptor protein.[82] A recent study revealed that p62/SQSTM1-droplets serve as a platform for autophagosome formation and the antioxidative stress response.[83] Unlike the association with AMPK, MTORC1, TFEB, and DRAM described earlier, TP53 negatively regulates autophagy via TP53-induced glycolysis and apoptosis regulator (TIGER). ROS-induced DNA damage can activate TP53 and its downstream TIGER, which functions to reduce intracellular fructose-2,6-bisphosphatase activity, thereby inhibiting glycolysis and activating the pentose phosphatase pathway.[54] Under these conditions, increased levels of intracellular NADPH and glutathione suppress intracellular ROS levels, resulting in suppression of autophagy.[54] ROS-mediated activation of hypoxia inducible factor 1 subunit alpha (HIF1A) can activate macroautophagic degradation of mitochondria (mitophagy) mediated by transcriptional upregulation of BCL2/adenovirus E1B 19-kDa interacting protein 3 (BNIP3) and BNIP3-like (BNIP3L/NIX).[84] HIF1A also activates cyclin dependent kinase inhibitor 1B (CDKN1B, also known as p27/KIP1)-E2F transcription factor 1 (E2F1)-mediated autophagy gene expression.[85]

2.5 Regulation of autophagy by the unfolded protein response

Accumulation of unfolded or misfolded proteins in the ER triggers ER stress. To attenuate such stress conditions, cells activate the unfolded protein response (UPR) to maintain cellular homeostasis. ER stress and UPR have been implicated in various diseases, particularly in neurodegenerative disorders including Alzheimer's disease and Parkinson's disease.[86] Under ER stress, the UPR is initiated by three transmembrane ER stress sensors, eukaryotic translation initiation factor 2A kinase 3 (EIF2AK3/PERK), endoplasmic reticulum to nucleus signaling 1 (ERN1/IRE1α), and activating transcription factor 6 (ATF6).[87] PERK activates the eukaryotic translation initiation factor 2 subunit 1A (EIF2S1/EIF2α)-ATF4-DNA damage inducible transcript 3 (DDIT3/CHOP) axis, which activates expression of autophagy genes, including ATG5, ATG7, ATG16L1, LC3, BECN1, p62/SQSTM1, and NBR1.[88] ATF6-activation is also important for the expression of DDIT3/CHOP.[89] Protein phosphatase 1 regulatory subunit 15A (PPP1R15A/GADD34) is also activated by ATF4, resulting in mTORC1 inactivation and transcriptional tuning of lysosome/autophagy gene expression.[90]

IRE1α induces the splicing of X-box binding protein 1 (XBP1) and spliced XBP1 interacts with the promotors of several genes involved in autophagy, such as BECN1.[91] IRE1α also activates the XBP1-mediated forkhead box protein O1 (FOXO1) axis and TNF receptor associated factor 2 (TRAF2)-apoptosis signal regulating kinase 1 (ASK1)-JNK signaling, which activate autophagy.[92,93]

Although ER stress and the UPR mainly activate autophagy, ER stress triggered by depletion of Ca^{2+} stores in the ER [experimentally induced by chemical treatment with thapsigargin, a sarcoendoplasmic reticulum Ca^{2+} ATPase (SERCA) blocker], clearly inhibit the fusion of autophagosomes with lysosomes. This suggests that an appropriate concentration of Ca^{2+} in the ER store is needed to form autophagosomes with sufficient luminal Ca^{2+} to enable autophagosome lysosome fusion.[94]

An increase in the abundance of unfolded proteins in mitochondria can cause mitochondrial dysfunction that activates a mitochondrial unfolded protein response (UPR^{mt}), which upregulates mitochondrial chaperone production and alleviates cytotoxicity.[95] In the UPR^{mt}, eukaryotic translation initiation factor 2-alpha kinase 4 (EIF2AK4/GCN2) phosphorylates EIF2S1/EIF2α and activates the EIF2S1/EIF2α-ATF4-CHOP axis, which is likely to activate autophagy.[96] Sirtuin 3 (SIRT3) is also activated under

UPR^{mt}, resulting in the activation of FOXO3 and thereby induction of autophagy gene expression.[97] The UPR^{mt} induces mitochondrial-selective autophagic degradation (mitophagy, described in detail later) via mitochondrial lon peptidase 1 (LONP1)-mediated translocation of PTEN-induced kinase 1 (PINK1) and E3 ubiquitin-protein ligase, parkin (PRKN), which are key mitophagy components.[98]

2.6 Regulation of autophagy by transcription factors

There are several transcription factors that regulate lysosomal/autophagy gene expression.[33] TFEB is a member of the microphthalmia/transcription factor E (MiT/TFE) family of transcription factors that also includes MITF, TFE3, and TFEC. TFEB enhances global expression of lysosomal/autophagy genes involved in the different steps of autophagy, such as BECN1, WIPI1, and ATG9 (autophagy initiation); ATG5, LC3B, and p62/SQSTM1 (autophagosome elongation and recognition of autophagic cargo); UVRAG and RAB7 (fusion with lysosomes); LAMP1 and cathepsins (lysosomal components).[38] TFEB is primarily regulated by MTORC1 (described above); however, recent studies show that protein kinase C (PKC) can directly activate TFEB,[99,100] while glycogen synthase kinase 3 (GSK3), ERK, or AKT suppress the transcriptional activity of TFEB via direct phosphorylation.[101] MITF and TFE3 are also regulators of lysosome/autophagy gene expression, and their activity is maintained via MTORC1, GSK3, or Wnt signaling.[102–104] Zinc finger with KRAB and SCAN domains 3 (ZKSCAN3) negatively regulates autophagy as a transcriptional repressor of autophagy. Under basal or nutrient-rich conditions, ZKSCAN3 localizes in the nucleus and suppresses autophagic gene expression.[105] Although the mechanism of ZKSCAN3 regulation is not fully elucidated, PKC activates downstream p38MAPK and JNK phosphorylates ZKSCAN3 leading to its translocation out of the nucleus. MTORC1 inhibition by small molecule, torin1, induces nuclear exclusion and inactivation of ZKSCAN3, indicating that MTORC1 regulates the activity of ZKSCAN3.[99,106]

FOXO1 and FOXO3 also regulate the expression of autophagic genes, such as LC3, ATG4, ATG12, BECN1, ULK1, VPS34, and RAB7.[33] Activated AKT phosphorylates FOXOs then induces the binding of FOXO proteins with 14-3-3, resulting in their cytoplasmic retention and preventing their nuclear translocation.[107] FOXO1 also regulates autophagy in a transcriptional-independent manner. While FOXO1 is deacetylated by NAD-dependent deacetylase sirtuin 2 (SIRT2) under basal conditions,

FOXO1 is acetylated possibly because of SIRT2 inhibition in response to oxidative stress or starvation, which induces the association of FOXO1 with ATG7 and autophagosome formation.[108] JNK-mediated phosphorylation of FOXOs positively regulates its transcriptional activity and thereby autophagy.[109]

The key transcription factor, NF-κB, which is mainly involved with apoptosis and inflammation, also has a function in autophagy regulation.[110] NF-κB activated by IKK can trigger autophagy by directly inducing the expression of autophagy-related genes, such as BECN1, ATG5, and LC3.[76] However, NF-κB can also inhibit autophagosome biogenesis by increasing the expression of autophagy repressors, including BCL2 and MTOR, indicating that NF-κB signaling regulates autophagy potentially in a context-dependent manner.[76]

2.7 Regulation of autophagy by microRNAs

Accumulating evidence indicates that microRNAs (miRNAs) play an important role in the regulation of autophagy.[111] miRNAs are a class of noncoding RNAs that play important roles in regulating gene expression and mainly interact with the 3′ untranslated region of target mRNAs to induce mRNA degradation and translational repression.[112] miRNA expression and abundance are strongly dependent on cell type[113]; therefore, in this section, we discuss the regulation of autophagy by miRNAs in Alzheimer's disease. In brain tissue of Alzheimer's disease patients or mouse models, decreased levels of miR-124, miR-132, miR-212, miR-214-3p, and miR-299-5p were observed.[114–117] miR-124 regulates AMPK signaling and the expression levels of p62/SQSTM1 and p38MAPK.[118,119] miR-132 and miR-212 target ATG9 and ATG5-ATG12, which are essential autophagy proteins.[116] miR-214-3p also targets ATG12,[115] and miR-299-5p represses ATG5 expression.[117] These miRNAs target positive regulators of autophagy; therefore, downregulation of these miRNAs is likely to elicit a neuroprotective effect on the Alzheimer's disease brain. Moreover, many miRNAs regulate autophagy in different cell types or species; for example, miR-101 downregulates RAB5A and ATG4,[120] miR-83 suppresses MCOLN1,[121] while miR-378 activates autophagy by downregulating PDK1.[122] Thus, many miRNAs are involved in autophagy; however, their exact mechanisms of action in the control of autophagy remain unclear. Further research will provide better understanding of the relationship between miRNAs and autophagy and also deeper insight for the development of novel therapeutic strategies that use or target miRNAs.

2.8 Regulation of autophagy by posttranslational and epigenetic modulators

Posttranslational modifications that affect the association between transcription factors and chromatin regulate autophagic gene expression.[123] The NAD^+-dependent deacetylase, SIRT1, deacetylates FOXO1 and FOXO3, resulting in the induction of autophagy gene expression.[124] SIRT1 is considered to be a regulator of autophagy by deacetylating targets, including ATG5, ATG7, and LC3B, which promotes autophagy-processing activity.[124] SIRT1 also positively controls AMPK activity by deacetylating LKB1, leading to autophagy induction.[125] Resveratrol, a polyphenolic compound, is considered to induce autophagy via SIRT1 activation.[126] The histone lysine acetyltransferase, GCN5, was recently reported as a negative regulator of autophagy. GCN5 specifically catalyzes acetylation of TFEB at K274 and K279, which disrupts TFEB dimerization and the binding of TFEB to its target gene promoters.[127] Bromodomain-containing protein 4 (BRD4) is a transcriptional repressor of lysosome/autophagy genes. In the basal state, euchromatic histone lysine methyltransferase 2 (EHMT2, also known as G9a) recruited by BRD4 catalyzes di-methylation of histone H3 at lysine residue 9 (H3K9me2) in the promoter regions of lysosome and autophagy genes, such as LC3B, p62/SQSTM1, BECN1, and ATG16L, resulting in transcriptional repression.[128] Histone H3 Arg17 dimethylation by coactivator associated arginine-methyltransferase 1 (CARM1), a key autophagy regulator, is associated with TFEB activity. CARM1 binds to the promoter region of TFEB and orchestrates lysosome/autophagy gene expression with TFEB.[129] Histone lysine acetyltransferase 8 (KAT8, known as MYST1) catalyzes H4K16ac and H4K16 deacetylation, which is associated with the downregulation of autophagy-related genes.[130] In addition, H3K4me3 and H3K56ac are downregulated, while H4K20me4 is upregulated during autophagy, although the mechanism of this action is unknown.[131]

3. Selective autophagy regulated by autophagic receptors

Autophagy is primarily considered as nonselective and a degradation process of bulky cytoplasmic proteins. However, various types of target/organelle-selective autophagy have been discovered in recent years.[132] In selective (macro)autophagy, preselected cell components recruited by type-specific autophagic receptors are degraded to remove nonfunctional organelles, foreign pathogens, or cytotoxic aggregates and maintain cellular homeostasis.[133] Selective autophagy targets specific organelles, complexes, or pathogens such as mitochondria, ER, protein aggregates, nuclei, ribosomes, peroxisomes, lysosomes, and bacteria.[134] Among them, we describe autophagic degradation of mitochondria, protein aggregates, and ER (defined as mitophagy, aggrephagy, and ER-phagy, respectively) because they are closely related to pathophysiology of neurodegenerative diseases.[135–137]

3.1 Mitophagy

Macroautophagic degradation of mitochondria (hereafter mitophagy) can remove excessive, nonfunctional, or damaged mitochondria and contribute to mitochondrial quality control.[138] The most well-known mitophagic pathway is PINK1 and E3 ubiquitin-protein ligase PRKN-dependent mitophagy. Mitophagy dysfunction is observed in familial Parkinson's disease with PINK1/PRKN mutations.[138] During mitochondrial depolarization-induced mitophagy [using a chemical mitochondrial uncoupler such as carbonyl cyanide 3-chlorophenylhydrazone (CCCP)], PINK1 accumulates on the mitochondrial surface, which induces autophosphorylation and homodimerization of PINK1.[139] Dimerized PINK1 facilitates binding to PRKN and ubiquitin. Thus, PINK1 acts as a sensor protein upon acute mitochondrial damage. PINK1 then activates PRKN via phosphorylation of ubiquitin and ubiquitin-like (UBL) domain of PRKN at S65, which recruits PRKN to the damaged mitochondria and increases E3 ubiquitin ligase activity of PRKN.[139] Ubiquitin chains formed by PRKN mediate the binding of ubiquitinated proteins of damaged mitochondria with autophagic receptors including LC3, p62/SQSTM1, calcium binding and coiled-coil domain 2 (CALCOCO2, also known as NDP52), and optineurin (OPTN), as well as RAB proteins RAB5 and RAB7A. This results in the enclosure of damaged mitochondria into autophagosomes and autophagic degradation.[133] In coordination with the PINK1/PRKN system, after mitochondrial outer membrane rapture, the mitochondrial inner membrane protein, prohibitin 2 (PHB2), binds to LC3B through its LC3 interacting region (LIR) in a PRKN-dependent manner.[140] However, many studies indicate that the PINK1/PRKN-mediated mechanism is not necessary to regulate basal mitophagy because PINK1/PRKN knockout cells are still capable of executing mitophagy.[141] Resident mitochondrial proteins, BNIP3 and BNIP3L/NIX, and the mitochondrial outer membrane protein, FUN14 domain-containing protein 1 (FUNDC1), are upregulated upon mitochondrial damage or hypoxia and act as autophagic receptor proteins.[142,143] These receptors accumulate on the damaged mitochondria and directly interact with LC3 and GABARAP through their LIR domains.[143,144] BCL2L13 (a mammalian homolog of yeast mitophagy receptor Atg32) also potentially induces mitophagy through association with LC3.[145] In addition to the autophagic/mitophagic receptors, cardiolipin and ceramide (mitochondria-specific phospholipid and sphingolipid, respectively) trigger mitophagy by directly interacting with LC3.[146]

The mitochondrial fission and fusion mechanism orchestrates mitophagy by segregating the damaged compartment from functional healthy mitochondria.[147] Mitochondrial fission-related proteins, dynamin-1 like (DNM1L, also known as DRP1) and mitochondrial fission 1 protein (FIS1), as well as mitochondrial fusion-related proteins, mitofusin1/2 (MFN1/2) and mitochondrial dynamin-like GTPase optic atrophy 1 (OPA1), regulate not only mitochondrial dynamics but also mitophagy.[148–150] Mitophagy is particularly important to maintain high-energy-requiring cells/tissue types such as neurons.[151] Compromised mitophagy causes neurodegenerative disease phenotypes. Conversely, mitophagy activation can be a novel therapeutic approach for the treatment of neurodegenerative diseases, including Alzheimer's disease and Parkinson's disease.[152]

3.2 Aggrephagy

Aggrephagy is a autophagic degradation pathway that specifically eliminates intracellular protein aggregates.[134] Cytoplasmic aggregation-prone proteins, such as amyloid plaques, Lewy bodies, and polyglutamine-expanded huntingtin (in Alzheimer's, Parkinson's, and Huntington's diseases, respectively), are hallmarks of neurodegenerative diseases; therefore, enhancing autophagy to facilitate the clearance of potentially cytotoxic aggregates using small-molecule compounds is a promising therapeutic strategy.[153]

Aggregated proteins are actively and retrogradely transported to the microtubule-organizing center (MTOC) and encapsulated into a cytoplasmic juxtanuclear structure called the aggresome. Histone deacetylase 6 (HDAC6) plays a key role in aggresome formation and aggrephagy.[154] HDAC6 recognizes the diglycine motifs of protein aggregates and induces the association between the aggregates and the dynein motor complex, a microtubule transport system, resulting in enclosure of protein aggregates into aggresomes.[154] The aggresomes are recognized by autophagic receptors and subsequently packed into autophagosomes for lysosomal degradation. OPTN, p62/SQSTM1, neighbor of BRCA1 (NBR1), toll-interacting protein (TOLLIP), and TAX1-binding protein (TAX1BP1) recognize the protein aggregates and mediate aggrephagy.[133] The ATG8 family protein, LC3C, cooperates with lysosomal tectonin beta-propeller repeat-containing protein (TECPR1) to promote aggrephagy.[155]

3.3 ER-phagy

ER-phagy (or reticulophagy) is a major cellular degradation system that is involved in clearing ER fragments and ER-resident proteins, which plays an important role in maintaining ER homeostasis.[134,156] Accumulation of unfolded proteins in the ER causes ER stress, which is implicated in various disorders, including neurodegenerative diseases, and can activate ER-phagy; therefore, enhancement of ER-phagy may be a potential therapeutic strategy for these diseases.[86,157] In mammals, eight autophagic receptors are involved with ER-phagy. The key regulators of ER-phagy are family with sequence similarity 134, member B (FAM134B), atlastin GTPase 3 (ATL3), CDK5 regulatory subunit-associated protein 3 (CDK4RAP3), cell cycle progression 1 (CCPG1), reticulon 3 (RTN3), translocation protein SEC62, testis expressed 264 (TEX264), and CALCOCO1.[133,158] In a recent study, ER-localized UFMylation, conjugation of a ubiquitin-like protein, ubiquitin-fold modifier 1 (UFM1), ribophorin1 (RPN1), and ribosomal protein L26 (RPL26) by UFM1 specific ligase 1 (UFL1) brought to the ER surface by DDRGK domain containing 1 (DDRGK1) induced ER-phagy to repress the unfolded protein response via ERN1/IRE1α signaling, which is analogous to PINK1/PRKN-mediated mitophagy.[159]

4. Concluding remarks

Autophagy is an essential catabolic process that promotes the clearance of surplus or damaged intracellular components and maintains cellular homeostasis. Given that autophagy dysfunction has been implicated in various human diseases, such as cancer, infections, and neurodegenerative diseases, stimulating autophagy by pharmacological or genetic approaches has been proposed as therapeutic intervention. Over the past two decades, understanding of autophagy at mechanistic, physiological, and pathological levels has progressed substantially. ULK1 and Class III PI3K complexes, two ubiquitin-like conjugation systems formed by several ATG proteins, are the central core of the autophagy molecular machinery. In the regulation of intracellular autophagic signaling, MTOR, a master regulator of autophagy, predominantly suppresses autophagy. In addition, AMPK, inositol, calcium, and stress response signaling can regulate autophagy (partially) in an MTOR-independent manner. Transcription factors and epigenetic modifications also affect autophagic activity via autophagy or lysosome-related gene expression. Both MTOR-dependent and MTOR-independent autophagic pathways are tightly regulated by complicated signaling pathway crosstalk, as described. Interestingly, some of the autophagy regulators, such as Ca^{2+} and

TP53, positively and negatively regulate autophagy, possibly in the context of cell conditions, cell types, organs, or the expression levels and subcellular localization of autophagic signaling sensors/receptors. This should be carefully considered when analyzing the mechanism of autophagy regulation by novel substrates or chemical compounds of interest. Notably, many of the signaling pathways involved with MTOR-independent signaling are likely to affect MTOR activity directly or indirectly via the MTOR signaling cascade. This complicates the development of autophagy-activating drugs oriented to treat diseases associated with autophagy dysfunction because in addition to autophagy, MTOR regulates diverse cellular functions, such as protein translation, cell proliferation, and metabolism, and, therefore, MTOR inhibition can cause a variety of side effects. From this point of view, small molecules that target autophagic machinery without affecting upstream signaling pathways may overcome this challenge. In recent years, organelle-specific autophagic degradation, such as mitophagy, ER-phagy, and aggrephagy, have emerged with implications for the maintenance of cell/organismal homeostasis. This selective autophagy relies on specific autophagic receptor proteins that recognize damaged organelles and cytotoxic protein aggregates. Interestingly, chemical compounds that interact with both LC3 and mutant huntingtin protein reduced mutant huntingtin levels and rescued disease phenotypes in in vivo models of Huntington's disease.[160] In line with this, "chemical linker compounds" that recognize both intracellular pathogens/dysfunctional organelles and (specific) autophagic receptors have potential as treatments against neurodegenerative diseases. Further intensive and extensive research on autophagy will continue to provide us with deeper insights to develop precise therapeutic interventions for the treatment of human diseases.

Acknowledgment

We thank Jeremy Allen, PhD, from Edanz (https://jp.edanz.com/ac) for editing a draft of this manuscript.

References

1. Dikic I, Elazar Z. Mechanism and medical implications of mammalian autophagy. *Nat Rev Mol Cell Biol*. 2018;19(6):349–364.
2. Yin Z, Pascual C, Klionsky DJ. Autophagy: machinery and regulation. *Microb Cell*. 2016;3(12):588–596.
3. Nixon RA. The role of autophagy in neurodegenerative disease. *Nat Med*. 2013;19(8):983–997.

4. Nakamura S, Yoshimori T. Autophagy and longevity. *Mol Cells.* 2018;41(1):65–72.
5. Feng Y, He D, Yao Z, Klionsky DJ. The machinery of macroautophagy. *Cell Res.* 2014;24(1):24–41.
6. Zachari M, Ganley IG. The mammalian ULK1 complex and autophagy initiation. *Essays Biochem.* 2017;61(6):585–596.
7. Mercer TJ, Gubas A, Tooze SA. A molecular perspective of mammalian autophagosome biogenesis. *J Biol Chem.* 2018;293(15):5386–5395.
8. Mizushima N. The ATG conjugation systems in autophagy. *Curr Opin Cell Biol.* 2020;63:1–10.
9. Wei Y, Liu M, Li X, Liu J, Li H. Origin of the autophagosome membrane in mammals. *Biomed Res Int.* 2018;2018:1012789.
10. Deng Z, Purtell K, Lachance V, Wold MS, Chen S, Yue Z. Autophagy receptors and neurodegenerative diseases. *Trends Cell Biol.* 2017;27(7):491–504.
11. Slobodkin MR, Elazar Z. The Atg8 family: multifunctional ubiquitin-like key regulators of autophagy. *Essays Biochem.* 2013;55:51–64.
12. Weidberg H, Shvets E, Shpilka T, Shimron F, Shinder V, Elazar Z. LC3 and GATE-16/GABARAP subfamilies are both essential yet act differently in autophagosome biogenesis. *EMBO J.* 2010;29(11):1792–1802.
13. Liu WJ, Ye L, Huang WF, et al. p62 links the autophagy pathway and the ubiqutin-proteasome system upon ubiquitinated protein degradation. *Cell Mol Biol Lett.* 2016;21:29.
14. Proikas-Cezanne T, Takacs Z, Donnes P, Kohlbacher O. WIPI proteins: essential PtdIns3P effectors at the nascent autophagosome. *J Cell Sci.* 2015;128(2):207–217.
15. Lorincz P, Juhasz G. Autophagosome-lysosome fusion. *J Mol Biol.* 2020;432(8):2462–2482.
16. Kuchitsu Y, Homma Y, Fujita N, Fukuda M. Rab7 knockout unveils regulated autolysosome maturation induced by glutamine starvation. *J Cell Sci.* 2018;131(7).
17. Lefebvre C, Legouis R, Culetto E. ESCRT and autophagies: endosomal functions and beyond. *Semin Cell Dev Biol.* 2018;74:21–28.
18. Wang Y, Li L, Hou C, et al. SNARE-mediated membrane fusion in autophagy. *Semin Cell Dev Biol.* 2016;60:97–104.
19. Jiang P, Nishimura T, Sakamaki Y, et al. The HOPS complex mediates autophagosome-lysosome fusion through interaction with syntaxin 17. *Mol Biol Cell.* 2014;25(8):1327–1337.
20. Menon MB, Dhamija S. Beclin 1 phosphorylation—at the center of autophagy regulation. *Front Cell Dev Biol.* 2018;6:137.
21. Yim WW, Mizushima N. Lysosome biology in autophagy. *Cell Discov.* 2020;6:6.
22. Wong SW, Sil P, Martinez J. Rubicon: LC3-associated phagocytosis and beyond. *FEBS J.* 2018;285(8):1379–1388.
23. Kuchitsu Y, Fukuda M. Revisiting Rab7 functions in mammalian autophagy: Rab7 knockout studies. *Cells.* 2018;7(11), 215.
24. Hurley JH. ESCRT complexes and the biogenesis of multivesicular bodies. *Curr Opin Cell Biol.* 2008;20(1):4–11.
25. Muller M, Schmidt O, Angelova M, et al. The coordinated action of the MVB pathway and autophagy ensures cell survival during starvation. *Elife.* 2015;4, e07736.
26. Itakura E, Kishi-Itakura C, Mizushima N. The hairpin-type tail-anchored SNARE syntaxin 17 targets to autophagosomes for fusion with endosomes/lysosomes. *Cell.* 2012;151(6):1256–1269.
27. Kumar S, Jain A, Farzam F, et al. Mechanism of Stx17 recruitment to autophagosomes via IRGM and mammalian Atg8 proteins. *J Cell Biol.* 2018;217(3):997–1013.
28. Man SM, Kanneganti TD. Regulation of lysosomal dynamics and autophagy by CTSB/cathepsin B. *Autophagy.* 2016;12(12):2504–2505.
29. Mauvezin C, Nagy P, Juhasz G, Neufeld TP. Autophagosome-lysosome fusion is independent of V-ATPase-mediated acidification. *Nat Commun.* 2015;6:7007.

30. Abu-Remaileh M, Wyant GA, Kim C, et al. Lysosomal metabolomics reveals V-ATPase- and mTOR-dependent regulation of amino acid efflux from lysosomes. *Science*. 2017;358(6364):807–813.
31. Chen Y, Yu L. Recent progress in autophagic lysosome reformation. *Traffic*. 2017;18(6):358–361.
32. Kroemer G, Marino G, Levine B. Autophagy and the integrated stress response. *Mol Cell*. 2010;40(2):280–293.
33. Di Malta C, Cinque L, Settembre C. Transcriptional regulation of autophagy: mechanisms and diseases. *Front Cell Dev Biol*. 2019;7:114.
34. Hu LF. Epigenetic regulation of autophagy. *Adv Exp Med Biol*. 2019;1206:221–236.
35. Dunlop EA, Tee AR. mTOR and autophagy: a dynamic relationship governed by nutrients and energy. *Semin Cell Dev Biol*. 2014;36:121–129.
36. Jung CH, Ro SH, Cao J, Otto NM, Kim DH. mTOR regulation of autophagy. *FEBS Lett*. 2010;584(7):1287–1295.
37. Saxton RA, Sabatini DM. mTOR signaling in growth, metabolism, and disease. *Cell*. 2017;168(6):960–976.
38. Napolitano G, Ballabio A. TFEB at a glance. *J Cell Sci*. 2016;129(13):2475–2481.
39. Jewell JL, Guan KL. Nutrient signaling to mTOR and cell growth. *Trends Biochem Sci*. 2013;38(5):233–242.
40. Sancak Y, Bar-Peled L, Zoncu R, Markhard AL, Nada S, Sabatini DM. Ragulator-Rag complex targets mTORC1 to the lysosomal surface and is necessary for its activation by amino acids. *Cell*. 2010;141(2):290–303.
41. Franke TF. PI3K/Akt: getting it right matters. *Oncogene*. 2008;27(50):6473–6488.
42. Dibble CC, Cantley LC. Regulation of mTORC1 by PI3K signaling. *Trends Cell Biol*. 2015;25(9):545–555.
43. Foster KG, Acosta-Jaquez HA, Romeo Y, et al. Regulation of mTOR complex 1 (mTORC1) by raptor Ser863 and multisite phosphorylation. *J Biol Chem*. 2010;285(1):80–94.
44. Mendoza MC, Er EE, Blenis J. The Ras-ERK and PI3K-mTOR pathways: cross-talk and compensation. *Trends Biochem Sci*. 2011;36(6):320–328.
45. Fu W, Hall MN. Regulation of mTORC2 signaling. *Genes (Basel)*. 2020;11(9).
46. Herzig S, Shaw RJ. AMPK: guardian of metabolism and mitochondrial homeostasis. *Nat Rev Mol Cell Biol*. 2018;19(2):121–135.
47. Li Y, Chen Y. AMPK and autophagy. *Adv Exp Med Biol*. 2019;1206:85–108.
48. Dengler F. Activation of AMPK under hypoxia: many roads leading to Rome. *Int J Mol Sci*. 2020;21(7).
49. Shackelford DB, Shaw RJ. The LKB1-AMPK pathway: metabolism and growth control in tumour suppression. *Nat Rev Cancer*. 2009;9(8):563–575.
50. Papinski D, Kraft C. Regulation of autophagy By signaling through the Atg1/ULK1 complex. *J Mol Biol*. 2016;428(9 Pt A):1725–1741.
51. Zheng B, Jeong JH, Asara JM, et al. Oncogenic B-RAF negatively regulates the tumor suppressor LKB1 to promote melanoma cell proliferation. *Mol Cell*. 2009;33(2):237–247.
52. Green MF, Anderson KA, Means AR. Characterization of the CaMKKbeta-AMPK signaling complex. *Cell Signal*. 2011;23(12):2005–2012.
53. Budanov AV. Stress-responsive sestrins link p53 with redox regulation and mammalian target of rapamycin signaling. *Antioxid Redox Signal*. 2011;15(6):1679–1690.
54. Hu W, Chen S, Thorne RF, Wu M. TP53, TP53 target genes (DRAM, TIGAR), and autophagy. *Adv Exp Med Biol*. 2019;1206:127–149.
55. Suzuki N, Johmura Y, Wang TW, et al. TP53/p53-FBXO22-TFEB controls basal autophagy to govern hormesis. *Autophagy*. 2021;1-18.
56. Berridge MJ. The inositol trisphosphate/calcium signaling pathway in health and disease. *Physiol Rev*. 2016;96(4):1261–1296.

57. Pihan P, Carreras-Sureda A, Hetz C. BCL-2 family: integrating stress responses at the ER to control cell demise. *Cell Death Differ.* 2017;24(9):1478–1487.
58. Sarkar S, Floto RA, Berger Z, et al. Lithium induces autophagy by inhibiting inositol monophosphatase. *J Cell Biol.* 2005;170(7):1101–1111.
59. Bootman MD, Chehab T, Bultynck G, Parys JB, Rietdorf K. The regulation of autophagy by calcium signals: do we have a consensus? *Cell Calcium.* 2018;70:32–46.
60. Sun F, Xu X, Wang X, Zhang B. Regulation of autophagy by Ca(2). *Tumour Biol.* 2016.
61. Hoyer-Hansen M, Bastholm L, Szyniarowski P, et al. Control of macroautophagy by calcium, calmodulin-dependent kinase kinase-beta, and Bcl-2. *Mol Cell.* 2007;25(2):193–205.
62. Yousefi S, Perozzo R, Schmid I, et al. Calpain-mediated cleavage of Atg5 switches autophagy to apoptosis. *Nat Cell Biol.* 2006;8(10):1124–1132.
63. Russo R, Berliocchi L, Adornetto A, et al. Calpain-mediated cleavage of Beclin-1 and autophagy deregulation following retinal ischemic injury in vivo. *Cell Death Dis.* 2011;2, e144.
64. Sarkar S, Ravikumar B, Floto RA, Rubinsztein DC. Rapamycin and mTOR-independent autophagy inducers ameliorate toxicity of polyglutamine-expanded huntingtin and related proteinopathies. *Cell Death Differ.* 2009;16(1):46–56.
65. Head SA, Shi W, Zhao L, et al. Antifungal drug itraconazole targets VDAC1 to modulate the AMPK/mTOR signaling axis in endothelial cells. *Proc Natl Acad Sci U S A.* 2015;112(52):E7276–E7285.
66. Nguyen TM, Combarnous Y, Praud C, Duittoz A, Blesbois E. Ca2+/calmodulin-dependent protein kinase kinases (CaMKKs) effects on AMP-activated protein kinase (AMPK) regulation of chicken sperm functions. *PLoS ONE.* 2016;11(1), e0147559.
67. Scotto Rosato A, Montefusco S, Soldati C, et al. TRPML1 links lysosomal calcium to autophagosome biogenesis through the activation of the CaMKKbeta/VPS34 pathway. *Nat Commun.* 2019;10(1):5630.
68. Medina DL, Di Paola S, Peluso I, et al. Lysosomal calcium signalling regulates autophagy through calcineurin and TFEB. *Nat Cell Biol.* 2015;17(3):288–299.
69. Gomez-Suaga P, Luzon-Toro B, Churamani D, et al. Leucine-rich repeat kinase 2 regulates autophagy through a calcium-dependent pathway involving NAADP. *Hum Mol Genet.* 2012;21(3):511–525.
70. Albanese F, Novello S, Morari M. Autophagy and LRRK2 in the aging brain. *Front Neurosci.* 2019;13:1352.
71. Yun HR, Jo YH, Kim J, Shin Y, Kim SS, Choi TG. Roles of autophagy in oxidative stress. *Int J Mol Sci.* 2020;21(9).
72. Scherz-Shouval R, Shvets E, Fass E, Shorer H, Gil L, Elazar Z. Reactive oxygen species are essential for autophagy and specifically regulate the activity of Atg4. *EMBO J.* 2007;26(7):1749–1760.
73. Shao D, Oka S, Liu T, et al. A redox-dependent mechanism for regulation of AMPK activation by Thioredoxin1 during energy starvation. *Cell Metab.* 2014;19(2):232–245.
74. Reynaert NL, Ckless K, Korn SH, et al. Nitric oxide represses inhibitory kappaB kinase through S-nitrosylation. *Proc Natl Acad Sci U S A.* 2004;101(24):8945–8950.
75. Wei Y, Pattingre S, Sinha S, Bassik M, Levine B. JNK1-mediated phosphorylation of Bcl-2 regulates starvation-induced autophagy. *Mol Cell.* 2008;30(6):678–688.
76. Salminen A, Hyttinen JM, Kauppinen A, Kaarniranta K. Context-dependent regulation of autophagy by IKK-NF-kappaB signaling: impact on the aging process. *Int J Cell Biol.* 2012;2012, 849541.
77. Kwak YD, Ma T, Diao S, et al. NO signaling and S-nitrosylation regulate PTEN inhibition in neurodegeneration. *Mol Neurodegener.* 2010;5:49.
78. Qian Q, Zhang Z, Orwig A, et al. S-Nitrosoglutathione reductase dysfunction contributes to obesity-associated hepatic insulin resistance via regulating autophagy. *Diabetes.* 2018;67(2):193–207.

79. Mungai PT, Waypa GB, Jairaman A, et al. Hypoxia triggers AMPK activation through reactive oxygen species-mediated activation of calcium release-activated calcium channels. *Mol Cell Biol*. 2011;31(17):3531–3545.
80. Zhang X, Cheng X, Yu L, et al. MCOLN1 is a ROS sensor in lysosomes that regulates autophagy. *Nat Commun*. 2016;7:12109.
81. Ma Q. Role of nrf2 in oxidative stress and toxicity. *Annu Rev Pharmacol Toxicol*. 2013;53:401–426.
82. Jiang T, Harder B, Rojo de la Vega M, Wong PK, Chapman E, Zhang DD. p62 links autophagy and Nrf2 signaling. *Free Radic Biol Med*. 2015;88(Pt B):199–204.
83. Kageyama S, Gudmundsson SR, Sou YS, et al. p62/SQSTM1-droplet serves as a platform for autophagosome formation and anti-oxidative stress response. *Nat Commun*. 2021;12(1):16.
84. Daskalaki I, Gkikas I, Tavernarakis N. Hypoxia and selective autophagy in cancer development and therapy. *Front Cell Dev Biol*. 2018;6:104.
85. Wang P, Long M, Zhang S, et al. Hypoxia inducible factor-1alpha regulates autophagy via the p27-E2F1 signaling pathway. *Mol Med Rep*. 2017;16(2):2107–2112.
86. Ghemrawi R, Khair M. Endoplasmic reticulum stress and unfolded protein response in neurodegenerative diseases. *Int J Mol Sci*. 2020;21(17).
87. Gardner BM, Pincus D, Gotthardt K, Gallagher CM, Walter P. Endoplasmic reticulum stress sensing in the unfolded protein response. *Cold Spring Harb Perspect Biol*. 2013;5(3), a013169.
88. B'Chir W, Maurin AC, Carraro V, et al. The eIF2alpha/ATF4 pathway is essential for stress-induced autophagy gene expression. *Nucleic Acids Res*. 2013;41(16):7683–7699.
89. Yang H, Niemeijer M, van de Water B, Beltman JB. ATF6 is a critical determinant of CHOP dynamics during the unfolded protein response. *iScience*. 2020;23(2):100860.
90. Gambardella G, Staiano L, Moretti MN, et al. GADD34 is a modulator of autophagy during starvation. *Sci Adv*. 2020;6(39).
91. Margariti A, Li H, Chen T, et al. XBP1 mRNA splicing triggers an autophagic response in endothelial cells through BECLIN-1 transcriptional activation. *J Biol Chem*. 2013;288(2):859–872.
92. Kishino A, Hayashi K, Hidai C, Masuda T, Nomura Y, Oshima T. XBP1-FoxO1 interaction regulates ER stress-induced autophagy in auditory cells. *Sci Rep*. 2017;7(1):4442.
93. Nishitoh H, Matsuzawa A, Tobiume K, et al. ASK1 is essential for endoplasmic reticulum stress-induced neuronal cell death triggered by expanded polyglutamine repeats. *Genes Dev*. 2002;16(11):1345–1355.
94. Ganley IG, Wong PM, Gammoh N, Jiang X. Distinct autophagosomal-lysosomal fusion mechanism revealed by thapsigargin-induced autophagy arrest. *Mol Cell*. 2011;42(6):731–743.
95. Jovaisaite V, Mouchiroud L, Auwerx J. The mitochondrial unfolded protein response, a conserved stress response pathway with implications in health and disease. *J Exp Biol*. 2014;217(Pt 1):137–143.
96. Melber A, Haynes CM. UPR(mt) regulation and output: a stress response mediated by mitochondrial-nuclear communication. *Cell Res*. 2018;28(3):281–295.
97. Audesse AJ, Dhakal S, Hassell LA, Gardell Z, Nemtsova Y, Webb AE. FOXO3 directly regulates an autophagy network to functionally regulate proteostasis in adult neural stem cells. *PLoS Genet*. 2019;15(4), e1008097.
98. Jin SM, Youle RJ. The accumulation of misfolded proteins in the mitochondrial matrix is sensed by PINK1 to induce PARK2/Parkin-mediated mitophagy of polarized mitochondria. *Autophagy*. 2013;9(11):1750–1757.
99. Li Y, Xu M, Ding X, et al. Protein kinase C controls lysosome biogenesis independently of mTORC1. *Nat Cell Biol*. 2016;18(10):1065–1077.
100. Kataura T, Tashiro E, Nishikawa S, et al. A chemical genomics-aggrephagy integrated method studying functional analysis of autophagy inducers. *Autophagy*. 2021;17(8):1856–1872.

101. Palmieri M, Pal R, Nelvagal HR, et al. mTORC1-independent TFEB activation via Akt inhibition promotes cellular clearance in neurodegenerative storage diseases. *Nat Commun*. 2017;8:14338.
102. Villegas F, Lehalle D, Mayer D, et al. Lysosomal signaling licenses embryonic stem cell differentiation via inactivation of Tfe3. *Cell Stem Cell*. 2019;24(2):257–270 e258.
103. Raben N, Puertollano R. TFEB and TFE3: linking lysosomes to cellular adaptation to stress. *Annu Rev Cell Dev Biol*. 2016;32:255–278.
104. Ploper D, De Robertis EM. The MITF family of transcription factors: role in endolysosomal biogenesis, Wnt signaling, and oncogenesis. *Pharmacol Res*. 2015;99:36–43.
105. Chauhan S, Goodwin JG, Chauhan S, et al. ZKSCAN3 is a master transcriptional repressor of autophagy. *Mol Cell*. 2013;50(1):16–28.
106. Li S, Song Y, Quach C, et al. Transcriptional regulation of autophagy-lysosomal function in BRAF-driven melanoma progression and chemoresistance. *Nat Commun*. 2019;10(1):1693.
107. Tzivion G, Dobson M, Ramakrishnan G. FoxO transcription factors; regulation by AKT and 14-3-3 proteins. *Biochim Biophys Acta*. 2011;1813(11):1938–1945.
108. Daitoku H, Sakamaki J, Fukamizu A. Regulation of FoxO transcription factors by acetylation and protein-protein interactions. *Biochim Biophys Acta*. 2011;1813(11):1954–1960.
109. Xu P, Das M, Reilly J, Davis RJ. JNK regulates FoxO-dependent autophagy in neurons. *Genes Dev*. 2011;25(4):310–322.
110. Verzella D, Pescatore A, Capece D, et al. Life, death, and autophagy in cancer: NF-kappaB turns up everywhere. *Cell Death Dis*. 2020;11(3):210.
111. Kou X, Chen D, Chen N. The regulation of microRNAs in Alzheimer's disease. *Front Neurol*. 2020;11:288.
112. O'Brien J, Hayder H, Zayed Y, Peng C. Overview of microRNA biogenesis, mechanisms of actions, and circulation. *Front Endocrinol (Lausanne)*. 2018;9:402.
113. Ludwig N, Leidinger P, Becker K, et al. Distribution of miRNA expression across human tissues. *Nucleic Acids Res*. 2016;44(8):3865–3877.
114. Du X, Huo X, Yang Y, et al. miR-124 downregulates BACE 1 and alters autophagy in APP/PS1 transgenic mice. *Toxicol Lett*. 2017;280:195–205.
115. Zhang Y, Li Q, Liu C, et al. MiR-214-3p attenuates cognition defects via the inhibition of autophagy in SAMP8 mouse model of sporadic Alzheimer's disease. *Neurotoxicology*. 2016;56:139–149.
116. Smith PY, Hernandez-Rapp J, Jolivette F, et al. miR-132/212 deficiency impairs tau metabolism and promotes pathological aggregation in vivo. *Hum Mol Genet*. 2015;24(23):6721–6735.
117. Zhang Y, Liu C, Wang J, et al. MiR-299-5p regulates apoptosis through autophagy in neurons and ameliorates cognitive capacity in APPswe/PS1dE9 mice. *Sci Rep*. 2016;6:24566.
118. Chen X, Mao R, Su W, et al. Circular RNA circHIPK3 modulates autophagy via MIR124-3p-STAT3-PRKAA/AMPKalpha signaling in STK11 mutant lung cancer. *Autophagy*. 2020;16(4):659–671.
119. Yao L, Zhu Z, Wu J, et al. MicroRNA-124 regulates the expression of p62/p38 and promotes autophagy in the inflammatory pathogenesis of Parkinson's disease. *FASEB J*. 2019;33(7):8648–8665.
120. Frankel LB, Wen J, Lees M, et al. microRNA-101 is a potent inhibitor of autophagy. *EMBO J*. 2011;30(22):4628–4641.
121. Zhou Y, Wang X, Song M, et al. A secreted microRNA disrupts autophagy in distinct tissues of *Caenorhabditis elegans* upon ageing. *Nat Commun*. 2019;10(1):4827.
122. Li Y, Jiang J, Liu W, et al. microRNA-378 promotes autophagy and inhibits apoptosis in skeletal muscle. *Proc Natl Acad Sci U S A*. 2018;115(46):E10849–E10858.
123. Wani WY, Boyer-Guittaut M, Dodson M, Chatham J, Darley-Usmar V, Zhang J. Regulation of autophagy by protein post-translational modification. *Lab Invest*. 2015;95(1):14–25.

124. Ng F, Tang BL. Sirtuins' modulation of autophagy. *J Cell Physiol.* 2013;228(12):2262–2270.
125. Lan F, Cacicedo JM, Ruderman N, Ido Y. SIRT1 modulation of the acetylation status, cytosolic localization, and activity of LKB1. Possible role in AMP-activated protein kinase activation. *J Biol Chem.* 2008;283(41):27628–27635.
126. Wu Y, Li X, Zhu JX, et al. Resveratrol-activated AMPK/SIRT1/autophagy in cellular models of Parkinson's disease. *Neurosignals.* 2011;19(3):163–174.
127. Wang Y, Huang Y, Liu J, et al. Acetyltransferase GCN5 regulates autophagy and lysosome biogenesis by targeting TFEB. *EMBO Rep.* 2020;21(1), e48335.
128. Sakamaki JI, Wilkinson S, Hahn M, et al. Bromodomain protein BRD4 is a transcriptional repressor of autophagy and lysosomal function. *Mol Cell.* 2017;66(4):517–532 e519.
129. Shin HJ, Kim H, Oh S, et al. AMPK-SKP2-CARM1 signalling cascade in transcriptional regulation of autophagy. *Nature.* 2016;534(7608):553–557.
130. Fullgrabe J, Lynch-Day MA, Heldring N, et al. The histone H4 lysine 16 acetyltransferase hMOF regulates the outcome of autophagy. *Nature.* 2013;500(7463):468–471.
131. Fullgrabe J, Klionsky DJ, Joseph B. The return of the nucleus: transcriptional and epigenetic control of autophagy. *Nat Rev Mol Cell Biol.* 2014;15(1):65–74.
132. Gatica D, Lahiri V, Klionsky DJ. Cargo recognition and degradation by selective autophagy. *Nat Cell Biol.* 2018;20(3):233–242.
133. Gubas A, Dikic I. A guide to.. the regulation of selective autophagy receptors. *FEBS J.* 2021.
134. Galluzzi L, Baehrecke EH, Ballabio A, et al. Molecular definitions of autophagy and related processes. *EMBO J.* 2017;36(13):1811–1836.
135. Lindholm D, Wootz H, Korhonen L. ER stress and neurodegenerative diseases. *Cell Death Differ.* 2006;13(3):385–392.
136. Johri A, Beal MF. Mitochondrial dysfunction in neurodegenerative diseases. *J Pharmacol Exp Ther.* 2012;342(3):619–630.
137. Olzmann JA, Li L, Chin LS. Aggresome formation and neurodegenerative diseases: therapeutic implications. *Curr Med Chem.* 2008;15(1):47–60.
138. Ashrafi G, Schwarz TL. The pathways of mitophagy for quality control and clearance of mitochondria. *Cell Death Differ.* 2013;20(1):31–42.
139. Bingol B, Sheng M. Mechanisms of mitophagy: PINK1, Parkin, USP30 and beyond. *Free Radic Biol Med.* 2016;100:210–222.
140. Yan C, Gong L, Chen L, et al. PHB2 (prohibitin 2) promotes PINK1-PRKN/Parkin-dependent mitophagy by the PARL-PGAM5-PINK1 axis. *Autophagy.* 2020;16(3):419–434.
141. McWilliams TG, Prescott AR, Montava-Garriga L, et al. Basal mitophagy occurs independently of PINK1 in mouse tissues of high metabolic Demand. *Cell Metab.* 2018;27(2):439–449 e435.
142. Ney PA. Mitochondrial autophagy: origins, significance, and role of BNIP3 and NIX. *Biochim Biophys Acta.* 2015;1853(10 Pt B):2775–2783.
143. Liu L, Feng D, Chen G, et al. Mitochondrial outer-membrane protein FUNDC1 mediates hypoxia-induced mitophagy in mammalian cells. *Nat Cell Biol.* 2012;14(2):177–185.
144. Schwarten M, Mohrluder J, Ma P, et al. Nix directly binds to GABARAP: a possible crosstalk between apoptosis and autophagy. *Autophagy.* 2009;5(5):690–698.
145. Otsu K, Murakawa T, Yamaguchi O. BCL2L13 is a mammalian homolog of the yeast mitophagy receptor Atg32. *Autophagy.* 2015;11(10):1932–1933.
146. Praharaj PP, Naik PP, Panigrahi DP, et al. Intricate role of mitochondrial lipid in mitophagy and mitochondrial apoptosis: its implication in cancer therapeutics. *Cell Mol Life Sci.* 2019;76(9):1641–1652.
147. Song M, Mihara K, Chen Y, Scorrano L, Dorn 2nd GW. Mitochondrial fission and fusion factors reciprocally orchestrate mitophagic culling in mouse hearts and cultured fibroblasts. *Cell Metab.* 2015;21(2):273–286.

148. Kageyama Y, Hoshijima M, Seo K, et al. Parkin-independent mitophagy requires Drp1 and maintains the integrity of mammalian heart and brain. *EMBO J.* 2014;33(23):2798–2813.
149. Gegg ME, Cooper JM, Chau KY, Rojo M, Schapira AH, Taanman JW. Mitofusin 1 and mitofusin 2 are ubiquitinated in a PINK1/parkin-dependent manner upon induction of mitophagy. *Hum Mol Genet.* 2010;19(24):4861–4870.
150. Liao C, Ashley N, Diot A, et al. Dysregulated mitophagy and mitochondrial organization in optic atrophy due to OPA1 mutations. *Neurology.* 2017;88(2):131–142.
151. Martinez-Vicente M. Neuronal mitophagy in neurodegenerative diseases. *Front Mol Neurosci.* 2017;10:64.
152. Wang Y, Liu N, Lu B. Mechanisms and roles of mitophagy in neurodegenerative diseases. *CNS Neurosci Ther.* 2019;25(7):859–875.
153. Rubinsztein DC, Codogno P, Levine B. Autophagy modulation as a potential therapeutic target for diverse diseases. *Nat Rev Drug Discov.* 2012;11(9):709–730.
154. Kawaguchi Y, Kovacs JJ, McLaurin A, Vance JM, Ito A, Yao TP. The deacetylase HDAC6 regulates aggresome formation and cell viability in response to misfolded protein stress. *Cell.* 2003;115(6):727–738.
155. Wetzel L, Blanchard S, Rama S, Beier V, Kaufmann A, Wollert T. TECPR1 promotes aggrephagy by direct recruitment of LC3C autophagosomes to lysosomes. *Nat Commun.* 2020;11(1):2993.
156. Bernales S, McDonald KL, Walter P. Autophagy counterbalances endoplasmic reticulum expansion during the unfolded protein response. *PLoS Biol.* 2006;4(12), e423.
157. Xu W, Ocak U, Gao L, et al. Selective autophagy as a therapeutic target for neurological diseases. *Cell Mol Life Sci.* 2021;78(4):1369–1392.
158. Mo J, Chen J, Zhang B. Critical roles of FAM134B in ER-phagy and diseases. *Cell Death Dis.* 2020;11(11):983.
159. Liang JR, Lingeman E, Luong T, et al. A genome-wide ER-phagy screen highlights key roles of mitochondrial metabolism and ER-resident UFMylation. *Cell.* 2020;180(6):1160–1177 e1120.
160. Li Z, Wang C, Wang Z, et al. Allele-selective lowering of mutant HTT protein by HTT-LC3 linker compounds. *Nature.* 2019;575(7781):203–209.

SECTION IV

Amyloid beta protein and autophagy

CHAPTER 4

Basics of amyloid β-protein in Alzheimer's disease

Azusa Sugimoto[a] and Kenjiro Ono[a,b]
[a]Division of Neurology, Department of Internal Medicine, Showa University School of Medicine, Tokyo, Japan
[b]Department of Neurology, Kanazawa University Graduate School of Medical Sciences, Kanazawa, Japan

1. What is amyloid β?

Amyloid β (Amyloid beta, abbreviated as Aβ) is the main component of amyloid plaques found in the brains of patients with Alzheimer's disease.

Aβ has been identified as a major component by biochemical analysis of cerebrovascular amyloid angiopathy and amyloid plaque accumulated in the brains of Alzheimer's disease patients and is a peptide of about 40 amino acids of 36–43 amino acids.[1] It is considered to play an important role in Alzheimer's disease as a major constituent of amyloid plaque, which is one of the pathological features of Alzheimer's disease. Amyloid plaques are composed of regularly arranged fibrillar aggregates called amyloid fibrils.

Aβ was identified by Glenner et al. In 1984, as a protein with a molecular weight of 4 kDa from cerebrovascular amyloid. Amyloid is a pathological general term for extracellular deposits that are stained orange-red by Congo red staining and show green polarization under a polarizing microscope. An electron microscope shows a fiber structure of 8–15 nm. In 1985, Master et al. prove that amyloid plaque is also composed of the same Aβ protein.

Aβ can be measured semi-quantitatively by immunostaining, and the position where Aβ is present can also be determined. Aβ is present in the brain primarily in blood vessels, as in the case of cerebral amyloid angiopathy, or in amyloid plaques in the white matter.[2] In cerebral amyloid angiopathy, Aβ forms aggregates that line blood vessels in the brain, but similar plaques also appear in some Lewy body diseases and inclusion body myositis.

Autophagy Dysfunction in Alzheimer's Disease and Dementia
https://doi.org/10.1016/B978-0-323-89906-2.00012-5

2. Production and formation

Aβ is derived from amyloid precursor protein (APP). APP is cleaved by the proteolytic enzymes α-secretase, β-secretase, and γ-secretase, while Aβ is produced by β-secretase and γ-secretase. Aβ excised by successive two-step cleavage by β-secretase and γ-secretase is secreted extracellularly.

2.1 APP

APP was revealed in 1987 that the locus is located on chromosome 21. APP is a type 1 transmembrane protein consisting of approximately 700 amino acid residues. The physiological functions of APP have been reported to include neuroprotection, neurite outgrowth, synaptogenesis promotion, cell adhesion, and synaptic vesicle axonal transport function.[3]

2.2 β-Secretase

β-Secretase activity is borne by a single transmembrane aspartic protease called beta-secretase (BACE)1, whose cleavage regulates total Aβ production.

2.3 γ-Secretase

γ-Secretase produces the C-terminus of the Aβ peptide. γ-secretase cleaves the inside of the transmembrane region of APP to produce many isoforms with a length of 30–51 amino acids.[4] The most abundant isoforms produced by γ selectors are Aβ40 and Aβ42.

2.4 α-Secretase

Most APPs are cleaved by γ-secretase after cleavage of α-secretase and released as a nonamyloid production pathway. Alpha secretase cleaves APP at position 16 of the Aβ sequence. The resulting C-terminal fragment is cleaved by γ-secretase and a short fragment called p3 is secreted. Since this pathway does not lead to Aβ production, it is protective against the development of Alzheimer's disease. ADAM9, ADAM10, and ADAM17 are considered as candidates as the main α-secretases in nerve cells.

2.5 Aβ40 and Aβ42

As the main molecular species of Aβ, Aβ40 in which the 40th amino acid is Val and Aβ42 in which the 42nd amino acid is Ala are known. Aβ40 is produced by cutting in the trans-Golgi network.[5] Aβ42 is commonly produced by cleavage in the endoplasmic reticulum.[5]

Normally, nerve cells produce nearly 10 times more Aβ40 than Aβ42.[6] Aβ42 is extremely hydrophobic and cohesive, is first deposited in the brain, and is a major component of amyloid plaque. On the other hand, deposition in cerebrovascular amyloid begins with Aβ42, but Aβ40 is predominantly deposited. However, among cerebrovascular amyloids, Aβ42 is predominant when it is deposited in capillaries.

Most of the intracellularly produced Aβ is secreted extracellularly. Aβ detected in cerebrospinal fluid and plasma is derived from secreted Aβ, exists at a constant concentration, and is determined by the balance between production and degradation.

3. Aggregation

The characteristic of Aβ is its high aggregation potential, and it aggregates to form amyloid fibrils only when it is present in a high concentration in a buffer solution. Aggregated Aβ exhibits degradation resistance. The difference in the maximum C-terminal length caused by the diversity of cleavage sites by γ-secretase during production is a factor that changes the aggregation potential of Aβ that can occur under physiological conditions.

Of these, Aβ42 is particularly critical for the disease progression because this longer form is more prone to aggregate than is the shorter Aβ40 peptide in vitro and is known to accumulate predominantly in the brains of Alzheimer's disease patients from the early stages.[7,8] Recently, it has been shown that Aβ43 is a more cohesive molecular species and is also accumulated in Alzheimer's disease brains, reaffirming the importance of the C-terminal length of Aβ.[9] Partial fraction decomposition of the N-terminal and polyglutamylation that occur after production are also considered to be important because they are highly hydrophobic. Therefore, the most important Aβ accumulated as amyloid plaque in the brains of Alzheimer's disease patients is assumed to be a molecular species in which the third glutamate is pyroglutamylated and the C-terminal ends with the 42nd alanine.

3.1 Amyloid hypothesis

In the pathophysiology of AD, the process by which Aβ is excised from APP, abnormally aggregates, and damages nerve cells is thought to play an important role, and is called amyloid hypothesis.[10]

Aβ deposition is the earliest pathologically perceived lesion, and since Aβ can aggregate and directly exhibit neuronal toxicity, abnormalities in Aβ production and accumulation are thought to be closely related to the development of Alzheimer's disease. Genetic analysis of patients with familial Alzheimer's disease (FAD) also supports this.

3.2 Oligomer hypothesis

The "amyloid hypothesis" that amyloid plaques are the cause of Alzheimer's disease has been accepted by most researchers, but has not been conclusively established. Traditionally, insoluble Aβ fibrils (fAβ) that accumulate as cerebral amyloid have been thought to exert neurotoxicity. Recently, however, attention has been focused on the study of soluble oligomers as more toxic aggregates and intermediates in the Aβ aggregation process such as protofibrils, which is called the oligomer hypothesis.[11]

The oligomeric hypothesis suggests that amyloid oligomers, rather than amyloid plaques, cause the disease. It is hypothesized that soluble Aβ oligomers cause synaptic toxicity via toxic receptors before causing neuronal cell death, and various membrane proteins have been listed as candidates for Aβ toxic receptors.

Mice that have been genetically modified to form oligomers but not amyloid plaques develop the disease. In addition, mice that have been modified to convert oligomers to amyloid plaques do not have worse symptoms than mice that express only oligomers.[12]

In AD, intracellular accumulation of tau protein is also seen, which may be involved in the disease as well as α-synuclein aggregation.

3.3 Process of Aβ aggregation

Aβ molecules aggregate to form flexible soluble oligomers in several forms. It is now believed that misfolded oligomers called "seed" induce misfolding of other Aβ molecules, triggering a chain reaction like prion infection. This oligomer is toxic to nerve cells.[13]

In the process of Aβ aggregation, structural conversion from unstructured Aβ monomer to β sheet occurs. Subsequently, oligomers are formed, protofibrils with a width of about 5 mm are formed, and fAβ, which is a mature fiber with a width of about 10 mm, is formed[11] (Fig. 1).

The process of formation of fAβ from Aβ is said to follow a polymerization nucleus-dependent polymerization model.[14] The polymerized

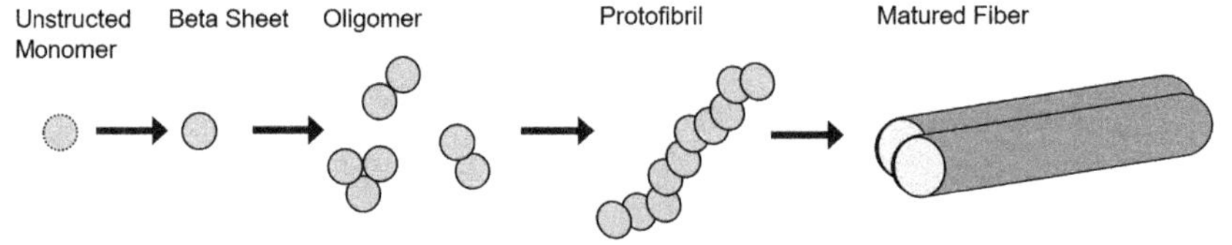

Fig. 1 Aβ aggregation model.

nucleation phase is thermodynamically less likely to occur and is the rate-determining step of the entire reaction, but once the polymerized nuclei are formed, fiber elongation proceeds according to the primary reaction kinetics model. That is, Aβ, which is a precursor protein, binds to the polymerized nucleus or the already existing fiber stump one after another while changing the three-dimensional structure, so that it proceeds rapidly. Using a high-speed atomic force microscope that can be observed in seconds, Ono et al. revealed the formation of fAβ42, which was difficult to observe in real time due to its rapid aggregation. It was shown that fAβ42 is not only stepwise formed but also has a one-way advantage. In addition, it was proved that there are two types of fAβ formed, straight and spiral, and that a switched hybrid type is also observed, and that the fibrillar structure can be changed depending on the surrounding electrolyte conditions.[15]

3.4 Soluble low-molecular-weight oligomer

Recent studies of Aβ oligomers have focused on smaller soluble oligomers as well as correlating oligomer size and biological activity.

Ono et al. succeeded in separating and extracting soluble oligomers in a stabilized state and analyzed using the extracted oligomers. As a result, it was clarified that the proportion of β-sheet structure increased in dimers, trimmers, and tetramers compared with monomers and the cytotoxicity increased accordingly.[14] In particular, it has been shown that the monomer-to-dimer process may be an important therapeutic target for AD.[16]

3.5 High-molecular-weight oligomers and off-pathways

As an intermediate in the Aβ aggregation process, not only low-molecular-weight oligomers as described above but also high-molecular-weight oligomers such as Aβ-derived diffusible ligands and protofibrils have been reported. Ono et al. compared and observed the aggregation process from the polymer oligomer protofibril and the monomer, and the protofibril was a pathway different from the pathway (on pathway) in which Aβ42 fiber is formed from the monomer, which is the final stage (off pathway) (Fig. 2). It was proposed that the monomer may be depolymerized and then enter the on-pathway.[15] In addition, the high-molecular-weight oligomer of Aβ42 exerts neurotoxicity through membrane damage mediated by active oxygen generation on the cell membrane and lipid peroxidation, resulting in decreased membrane fluidity and abnormal intracellular calcium regulation, membrane depolarization, and synaptic plasticity disorder have been shown.[17] It was also found that these effects are weak mainly in low-molecular-weight Aβ containing monomers.[17]

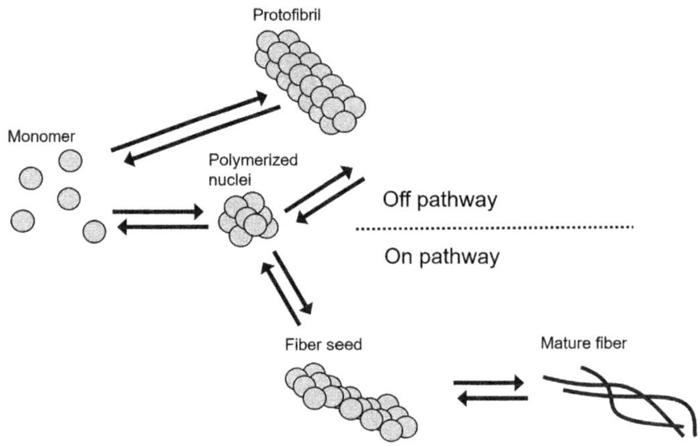

Fig. 2 Polymer oligomers and off-pathways.

4. Regulation of Aβ concentration in the brain

Aβ production in the brain is mainly carried out by neurons with the highest BACE1 expression levels.[18] The activity of BACE1, γ-secretase, and the α-secretase a disintegrin and metalloprotease 10 (ADAM10) changes according to neural activity, and the amount of Aβ produced also changes according to neural activity.[19]

The following reports have also been made in relation to increased Aβ production linked to neural activity. That is, the correlation between the level of consciousness and the amount of Aβ in the cerebrospinal fluid in coma patients,[20] the daily fluctuation of the amount of Aβ in the brain associated with sleep/wakefulness,[21] and abnormal elevation of brain activity in nondemented individuals with amyloid plaque deposits.[22]

On the other hand, experiments have shown that high concentrations of Aβ cause abnormal excitement of nerve cells, decrease of nerve activity, or death of nerve cells.

5. Decomposition and excretion

5.1 Decomposition

The half-life of Aβ in the brain is about 30 min.[23] Aβ is destroyed by several amyloid-degrading enzymes, such as neprilysin. Insulin-degrading enzymes, plasmin, endoserin-converting enzyme, cathepsin, Kallikrein-related peptidase 7 (KLK7), matrix metalloproteinase, etc. have been identified as Aβ-degrading enzymes.[24]

Aβ is also known to be phagocytosed by glial cells. Furthermore, it has been suggested that it may be excreted by transendocytosis via vascular endothelial cells. In addition, apolipoprotein E, which is the strongest genetic risk factor for Alzheimer's disease, is said to be involved in the Aβ degradation system.[25] Furthermore, it has been shown that Aβ clearance rates are significantly reduced in patients with sporadic Alzheimer's disease.[26]

5.2 Excretion

There is no lymphoid tissue in the brain. Therefore, the system corresponding to the lymphatic system involved in the excretion of waste products in the brain is called intramural Periarterial drainage (IPAD). Interstitial fluid and cerebrospinal fluid play a central role.[27]

The anatomical absorption pathway for waste products when albumin was injected into the brain gray matter was envisioned in the 1990s.[28] That is, from the basement membrane of the capillaries, it is selectively distributed in the pericapillary space of the artery, enters the nasal mucosal lymph vessels along the olfactory nerve at the base of the anterior skull, and reaches the cervical lymph nodes. Since then, the perivascular space around the cerebral arteries has come to be considered to correspond to a functional lymphatic vessel (perivascular drainage pathway).

5.3 Normal function of Aβ

Although the normal function of Aβ is not well understood, it has been discovered that it may have some activity. Examples include kinase activation,[29] protection from oxidative stress,[30] regulation of cholesterol transport,[31] function as a transcription factor,[32] and antibacterial activity (possibly associated with Aβ proinflammatory activity).[33] It is also suggested that Aβ produced in a neural activity-dependent manner may be involved in the regulation of neural activity in the brain as feedback. However, no neuroplastic abnormalities have been reported in APP knockout mice, and the physiological significance of Aβ is unknown. Studies in some animals have shown that the absence of Aβ does not cause a definite loss of physiological function.[34]

6. Relationship between Aβ and disease (including gene mutation)

6.1 Sporadic AD

Brain Aβ is elevated in patients with sporadic AD. Aβ is a major component of amyloid in the brain parenchyma and cerebrovascular disease, which contributes to cerebrovascular lesions, and is neurotoxic. It is not known how Aβ accumulates in the central nervous system and then how cell lesions begin.

6.2 FAD and genetic mutation

Autosomal dominant mutations in APP cause hereditary FAD. FAD families with duplicate mutations in the APP gene have been identified and provide a strong basis for the amyloid cascade hypothesis in Alzheimer's disease.[35]

On the other hand, autosomal dominant mutations in APP account for less than 10% of all cases, and most are not associated with such mutations.[36] FAD is thought to be the result of altered processing of proteolysis. Most of the gene mutations linked to FAD show the property of increasing Aβ production or aggregation potential.

6.2.1 Gene mutation that enhances Aβ production

The Swedish mutation and Italian mutation present near the β-secretase cleavage site increase the affinity of APP for BACE1 and increase the total Aβ production.[37,38] In addition, there is another cleavage site for β-secretase in the Aβ sequence, which is called the β′cleave site. This cleavage leads to the production of Aβ with a short N-terminus, but the Leuven mutation, which is a mutation at the β′cleavage site, suppresses β′cleavement, and as a result, has the effect of increasing total Aβ production.[39]

In addition, whole-genome sequencing analysis of the Icelandic population identified an Icelandic mutation that reduces Aβ production by 40% as a rare variant that acts protectively against Alzheimer's disease and cognitive decline associated with aging.[40] This mutation has been shown to reduce the efficiency of cleavage by β-secretase. It can be said that this mutation clarified that changes in Aβ production regulated the risk of developing Alzheimer's disease.

Recently, a functional deletion mutation of ADAM10 was found in FAD, and it was shown that suppression of the non-Aβ production pathway shifts APP metabolism to Aβ production and induces AD.[41]

6.2.2 Gene mutation that changes the production ratio of highly cohesive Aβ42

FAD mutations present near the Aβ sequence in the APP sequence do not affect the Aβ molecule itself, but change its production. The London mutation, found for the first time in the world as a gene mutation linked to FAD, has been shown to affect the C-terminal length of secreted Aβ.[42]

In addition, the Iranian mutation, Australian mutation, German mutation, French mutation, Florida mutation, Iberian mutation, London mutation, Australian mutation, Belgian mutation, etc., which are present on the C-terminal side of Aβ, all change cleavage by γ-secretase. As a result, the production ratio of Aβ42, which is particularly highly cohesive, is increased without significantly affecting the total amount of Aβ produced. Flemish mutations are also thought to reduce the effectiveness of domains that suppress γ-secretase activity and increase Aβ production.[43]

On the other hand, most FADs are linked to point mutations on the Presenilin 1 or 2 gene. It is not yet clear how these FAD mutations affect the presenilin protein. In addition to mutations that increase Aβ42 production, mutations that decrease Aβ40 production and mutations that suppress Aβ production as a whole have also been reported. However, it is common for all mutations to specifically increase the production ratio of Aβ42 compared to wild-type γ-secretase.[44] As a result, it is thought that the onset process of AD is promoted.

Mutations that suppress Aβ42 production by γ-secretase, such as the Icelandic mutation for β-secretase, have not yet been found. However, it has been reported that decreased expression of the AD-related genetic preventive factor PICALM reduces Aβ42 production by altering the intracellular transport of γ-secretase.[45]

6.2.3 Gene mutations that alter aggregation potential

Many FAD mutations are also present in the Aβ sequence and often have a significant effect on Aβ aggregation. The mutations on the N-terminal side of the Aβ sequence are the British, Tottori, and Italian mutations. Both the British and Tottori mutations enhance fAβformation.[46] For the Italian mutation, it enhances cleavage by β-secretase and at the same time enhances aggregation potential.[38]

On the other hand, as mutations located in the central part of the Aβ sequence, there are Arctic mutation, Osaka mutation, and Iowa mutation. The Dutch mutation was discovered as a mutation linked to Dutch hereditary amyloid cerebral hemorrhage. Both the Dutch and Arctic mutations have been shown to have high amyloid fibrillogenic potential in vitro.[47] In addition, Arctic mutations have been observed to enhance and stabilize the formation of protofibrils that occur during the intermediate stages of fAβ formation.[48] Aβ with the Osaka mutation does not form amyloid fibrils and remains in the form of oligomers, exhibiting synaptic toxicity.[49]

6.3 Down syndrome

The APP gene is located on chromosome 21, so people with Down syndrome have an extremely high incidence of Alzheimer's disease. Adults with Down's syndrome had an accumulation of amyloid associated with evidence of Alzheimer's disease, including memory loss, impaired executive function, and impaired visuospatial skills.[50]

7. Intervention strategy

Anti-AD therapeutic agents have been developed targeting Aβ. The development of therapeutic agents that suppress the mechanism of Aβ production by controlling secretase activity, suppress amyloid formation by inhibiting Aβ aggregation, and promote Aβ removal has been promoted.

7.1 Secretase activity regulator

Mutations in APP associated with the early onset of Alzheimer's disease have been shown to relatively increase Aβ42 production, regulating β-secretase and γ-secretase activity primarily to produce Aβ40. It has been suggested that doing so may be a cure for Alzheimer's disease.

β-secretase inhibitors prevent the initial cleavage of APP in the endoplasmic reticulum within the cell. γ-secretase inhibitors prevent the second cleavage of APP on the cell membrane and stop the subsequent formation of Aβ and toxic fragments. One of the major problems with this therapeutic approach is the need to interfere with enzymes that have functional roles other than the amyloid formation pathway.

Such examples include clinical trials approached by γ-secretase inhibitors, which faced results such as severe cognitive impairment and an increased incidence of skin cancer.[51] It is thought that γ-secretase is involved in the cleavage of many substrates, and in particular, suppression of Notch signal may have become a major problem.

Compounds that partially inhibit β-secretase and γ-secretase activities are also being sought. However, in order to initiate partial inhibition, the compound must block the large active site of aspartic protease, but at the same time, it must cross the blood-brain barrier, which is one developmental challenge.

7.2 Aβ aggregation inhibitor

Compounds targeted for Aβ aggregation that have been clinically tested are Tramiprosate (polysulfate compound), PBT2 (an analog of cryoroquinol),

ELND005 (cyloinositol, a plant sugar alcohol), and GV-971 (sodium oligomanuralate, oligosaccharide derived from the ocean).

Regarding clinical trials using Tramiprosate, it was reported in a phase II study that maintenance or improvement of cognitive function was observed in patients with mild AD,[52] but no significant effect was observed in a phase III study.[53] However, new prodrugs and biotransformers are also being developed. ALZ-801, an orally administrable prodrug of tramiprosate, significantly improved pharmacokinetic variability and gastrointestinal tolerance.[52] In addition, 3-sulfopropanoic acid, which is a metabolite of tramiprosate, has been shown to be as effective as tramiprosate in vitro.[54]

In a phase II study, PBT2 showed a significant improvement in frontal lobe function and a significant decrease in Aβ42 in cerebrospinal fluid in patients with mild to moderate AD.[55] On the other hand, in another phase II study, there was no significant effect on amyloid deposition, glucose metabolism, or cognitive function in patients with amnestic mild cognitive impairment (with amyloid deposition) and mild AD.[56]

ELND005 was conducted in a double-blind phase II study in patients with mild to moderate AD, but the high-dose group was discontinued due to serious side effects. On the other hand, although there were no significant differences in cognitive function or daily living function in the low-dose group, tolerability was confirmed and Aβ42 found in cerebrospinal fluid was significantly reduced.[57]

GV-971 is a linear molecule of manno-oligosaccharide, and a phase III study was conducted in AD patients in China, and changes were observed to improve the Alzheimer's disease evaluation scale (ADAS-cog). Phase III trials are underway and are planned for deployment in the United States, Europe, and China.[58,59]

Aggregates of pathogenic proteins, especially oligomers, are considered to be important target factors in the development of disease-modifying therapies for neurodegenerative diseases. Aβ aggregation regulators are currently being vigorously developed with the aim of preventing lesion progression in AD.

7.3 Immunotherapy

A strategy that stimulates the host's immune system to recognize and attack Aβ, or to prevent the accumulation of amyloid plaques or promote the removal of amyloid plaques and Aβ oligomers.

Oligomerization is the chemical process by which individual molecules are transformed into chains of a finite number of molecules. Prevention of Aβ oligomerization is an example of active and passive immunization to Aβ. In this process, antibodies to Aβ are used to reduce the level of amyloid plaques in the brain. Amyloid levels are reduced by removal by microglia or by facilitating the redistribution of peptides from the brain to the systemic circulation. Antibodies targeting Aβ that have been clinically tested include aducanumab, bapineuzumab, crenezumab, gantenerumab, and solanezumab.[60,61] Aβ vaccines such as CAD106 and UB-311 are also in clinical trials.[60]

Questions have been raised about the effectiveness of immunotherapy. Multiple anti-Aβ42 antibodies have been shown to have minimal cognitive protective effects in each study. However, it is also possible that the symptoms were too advanced at the time of the study. Further development is needed to apply it to asymptomatic patients and evaluate its effectiveness at an early stage of disease progression.

7.4 Other strategies

Long-term use of cholesterol-lowering drugs such as statins has been shown to be associated with a reduced incidence of Alzheimer's disease. In mice with genetically modified APP, cholesterol-lowering drugs have been shown to suppress pathology overall. Although the mechanism is not well understood, cholesterol-lowering drugs appear to have a direct impact on APP processing.[62]

Norvaline is an arginase inhibitor that easily crosses the blood-brain barrier and reduces the loss of arginine in the brain. Aβ accumulation is associated with L-arginine deficiency and neurodegeneration. Mice treated with norvaline had improved spatial memory, increased neuroplasticity-related proteins, and decreased Aβ.[63]

7.5 Summary of intervention strategies and the future

Based on the amyloid cascade hypothesis, Aβ-targeted anti-Alzheimer's disease strategies have been expected as radical treatments.

On the other hand, other than Aβ are involved in the pathophysiology of AD. There has long been a lot of debate about the relationship between Aβ accumulation and neurofibrillary tangles, which are tau lesions. While the disease specificity of amyloid plaque is high in the autopsy brain, neurofibrillary tangles are observed in various neurodegenerative diseases, and amyloid plaque accumulation is also observed in healthy nondementia subjects.

Since then, it has been found that almost all FAD gene mutations enhance Aβ accumulation. In addition, mutations that suppress Aβ production were identified as protective mutations against cognitive decline. In addition, frontotemporal lobar degeneration was found that did not show amyloid plaque accumulation due to mutations in the tau gene.

These findings suggest that Aβ is a toxic molecule that causes Alzheimer's disease and that tau lesions are a process directly involved in neuronal cell death downstream. Using model mice, it was also shown that Aβ accumulation enhances tau lesions and increases tau concentration in cerebrospinal fluid.[64,65]

However, the development of treatments for many Aβ has failed so far. In particular, in the clinical trial of Aβ vaccine therapy AN-1792, the decline in cognitive function was not suppressed, although patients with amyloid plaque accumulation disappeared.[66] Therefore, anti-Aβ therapy based on the amyloid cascade hypothesis was questioned.

On the other hand, recent large-scale clinical observational studies and biomarker analysis of FAD mutation carriers have revealed that Aβ accumulation begins more than 15–20 years before the onset of Alzheimer's disease. Tau lesions are subsequently triggered. It was also found that healthy subjects with amyloid plaque accumulation and mild cognitive impairment have a significantly higher probability of developing Alzheimer's disease.[67]

Based on the above, Aβ accumulation is considered to be the most upstream process in Alzheimer's disease lesions as brain amyloidosis. At present, it is necessary to correctly understand the risk of developing AD in individuals in the presymptomatic stage and to start anti-Aβ therapy as preemptive medicine.

References

1. Hamley IW. The amyloid beta peptide: a chemist's perspective. Role in Alzheimer's and fibrillization. *Chem Rev.* 2012;112(10):5147–5192. https://doi.org/10.1021/cr3000994.
2. Heurling K, Leuzy A, Zimmer ER, Lubberink M, Nordberg A. Imaging β-amyloid using [(18)F]flutemetamol positron emission tomography: from dosimetry to clinical diagnosis. *Eur J Nucl Med Mol Imaging.* 2016;43(2):362–373. https://doi.org/10.1007/s00259-015-3208-1.
3. Müller UC, Zheng H. Physiological functions of APP family proteins. *Cold Spring Harb Perspect Med.* 2012;2(2). https://doi.org/10.1101/cshperspect.a006288, a006288.
4. Olsson F, Schmidt S, Althoff V, et al. Characterization of intermediate steps in amyloid beta (Aβ) production under near-native conditions. *J Biol Chem.* 2014;289(3):1540–1550. https://doi.org/10.1074/jbc.M113.498246.
5. Hartmann T, Bieger SC, Brühl B, et al. Distinct sites of intracellular production for Alzheimer's disease A beta40/42 amyloid peptides. *Nat Med.* 1997;3(9):1016–1020. https://doi.org/10.1038/nm0997-1016.
6. Asami-Odaka A, Ishibashi Y, Kikuchi T, Kitada C, Suzuki N. Long amyloid beta-protein secreted from wild-type human neuroblastoma IMR-32 cells. *Biochemistry.* 1995;34(32):10272–10278. https://doi.org/10.1021/bi00032a022.

7. Jarrett JT, Berger EP, Lansbury Jr PT. The carboxy terminus of the beta amyloid protein is critical for the seeding of amyloid formation: implications for the pathogenesis of Alzheimer's disease. *Biochemistry.* 1993;32(18):4693–4697. https://doi.org/10.1021/bi00069a001.
8. Walsh DM, Selkoe DJ. A beta oligomers - a decade of discovery. *J Neurochem.* 2007;101(5):1172–1184. https://doi.org/10.1111/j.1471-4159.2006.04426.x.
9. Watkins TA, Emery B, Mulinyawe S, Barres BA. Distinct stages of myelination regulated by gamma-secretase and astrocytes in a rapidly myelinating CNS coculture system. *Neuron.* 2008;60(4):555–569. https://doi.org/10.1016/j.neuron.2008.09.011.
10. Hardy J, Selkoe DJ. The amyloid hypothesis of Alzheimer's disease: progress and problems on the road to therapeutics. *Science.* 2002;297(5580):353–356.
11. Ono K. Alzheimer's disease as oligomeropathy. *Neurochem Int.* 2018;119:57–70.
12. Gandy S, Simon AJ, Steele JW, et al. Days to criterion as an indicator of toxicity associated with human Alzheimer amyloid-beta oligomers. *Ann Neurol.* 2010;68(2):220–230. https://doi.org/10.1002/ana.22052.
13. Haass C, Selkoe DJ. Soluble protein oligomers in neurodegeneration: lessons from the Alzheimer's amyloid beta-peptide. *Nat Rev Mol Cell Biol.* 2007;8(2):101–112. https://doi.org/10.1038/nrm2101.
14. Jarrett JT, Lansbury Jr PT. Seeding "one-dimensional crystallization" of amyloid: a pathogenic mechanism in Alzheimer's disease and scrapie? *Cell.* 1993;73(6):1055–1058. https://doi.org/10.1016/0092-8674(93)90635-4.
15. Watanabe-Nakayama T, Ono K, Itami M, Takahashi R, Teplow DB, Yamada M. High-speed atomic force microscopy reveals structural dynamics of amyloid β1-42 aggregates. *Proc Natl Acad Sci U S A.* 2016;113(21):5835–5840. https://doi.org/10.1073/pnas.1524807113.
16. Ono K, Condron MM, Teplow DB. Structure–neurotoxicity relationships of amyloid β-protein oligomers. *Proc Natl Acad Sci U S A.* 2009;106(35):14745–14750.
17. Yasumoto T, Takamura Y, Tsuji M, et al. High molecular weight amyloid β 1-42 oligomers induce neurotoxicity via plasma membrane damage. *FASEB J.* 2019;33(8):9220–9234.
18. Fukumoto H, Tomita T, Matsunaga H, Ishibashi Y, Saido TC, Iwatsubo T. Primary cultures of neuronal and non-neuronal rat brain cells secrete similar proportions of amyloid beta peptides ending at A beta40 and A beta42. *Neuroreport.* 1999;10(14):2965–2969. https://doi.org/10.1097/00001756-199909290-00017.
19. Kamenetz F, Tomita T, Hsieh H, et al. APP processing and synaptic function. *Neuron.* 2003;37(6):925–937. https://doi.org/10.1016/s0896-6273(03)00124-7.
20. Brody DL, Magnoni S, Schwetye KE, et al. Amyloid-beta dynamics correlate with neurological status in the injured human brain. *Science.* 2008;321(5893):1221–1224. https://doi.org/10.1126/science.1161591.
21. Kang JE, Lim MM, Bateman RJ, et al. Amyloid-beta dynamics are regulated by orexin and the sleep-wake cycle. *Science.* 2009;326(5955):1005–1007. https://doi.org/10.1126/science.1180962.
22. Sperling RA, Laviolette PS, O'Keefe K, et al. Amyloid deposition is associated with impaired default network function in older persons without dementia. *Neuron.* 2009;63(2):178–188. https://doi.org/10.1016/j.neuron.2009.07.003.
23. Yamada K, Yabuki C, Seubert P, et al. Abeta immunotherapy: intracerebral sequestration of Abeta by an anti-Abeta monoclonal antibody 266 with high affinity to soluble Abeta. *J Neurosci.* 2009;29(36):11393–11398. https://doi.org/10.1523/JNEUROSCI.2021-09.2009.
24. Zhang S, Iwata K, Lachenmann MJ, et al. The Alzheimer's peptide a beta adopts a collapsed coil structure in water. *J Struct Biol.* 2000;130(2-3):130–141. https://doi.org/10.1006/jsbi.2000.4288.
25. Castellano JM, Kim J, Stewart FR, et al. Human apoE isoforms differentially regulate brain amyloid-β peptide clearance. *Sci Transl Med.* 2011;3(89):89ra57. https://doi.org/10.1126/scitranslmed.3002156.

26. Mawuenyega KG, Sigurdson W, Ovod V, et al. Decreased clearance of CNS beta-amyloid in Alzheimer's disease. *Science*. 2010;330(6012):1774. https://doi.org/10.1126/science.1197623.
27. Albargothy NJ, Johnston DA, MacGregor-Sharp M, et al. Convective influx/glymphatic system: tracers injected into the CSF enter and leave the brain along separate periarterial basement membrane pathways. *Acta Neuropathol*. 2018;136(1):139–152. https://doi.org/10.1007/s00401-018-1862-7.
28. Cserr HF, Knopf PM. Cervical lymphatics, the blood-brain barrier and the immunoreactivity of the brain: a new view. *Immunol Today*. 1992;13(12):507–512. https://doi.org/10.1016/0167-5699(92)90027-529.
29. Tabaton M, Zhu X, Perry G, Smith MA, Giliberto L. Signaling effect of amyloid-beta(42) on the processing of AbetaPP. *Exp Neurol*. 2010;221(1):18–25. https://doi.org/10.1016/j.expneurol.2009.09.002.
30. Baruch-Suchodolsky R, Fischer B. Abeta40, either soluble or aggregated, is a remarkably potent antioxidant in cell-free oxidative systems. *Biochemistry*. 2009;48(20):4354–4370. https://doi.org/10.1021/bi802361k.
31. Igbavboa U, Sun GY, Weisman GA, He Y, Wood WG. Amyloid beta-protein stimulates trafficking of cholesterol and caveolin-1 from the plasma membrane to the Golgi complex in mouse primary astrocytes. *Neuroscience*. 2009;162(2):328–338. https://doi.org/10.1016/j.neuroscience.2009.04.049.
32. Bailey JA, Maloney B, Ge YW, Lahiri DK. Functional activity of the novel Alzheimer's amyloid β-peptide interacting domain (AβID) in the APP and BACE1 promoter sequences and implications in activating apoptotic genes and in amyloidogenesis. *Gene*. 2011;488(1-2):13–22. https://doi.org/10.1016/j.gene.2011.06.017.
33. Li H, Liu CC, Zheng H, Huang TY. Amyloid, tau, pathogen infection and antimicrobial protection in Alzheimer's disease -conformist, nonconformist, and realistic prospects for AD pathogenesis. *Transl Neurodegener*. 2018;7:34. Published 2018 Dec 24 https://doi.org/10.1186/s40035-018-0139-3.
34. Luo Y, Bolon B, Damore MA, et al. BACE1 (beta-secretase) knockout mice do not acquire compensatory gene expression changes or develop neural lesions over time. *Neurobiol Dis*. 2003;14(1):81–88. https://doi.org/10.1016/s0969-9961(03)00104-9.
35. Rovelet-Lecrux A, Hannequin D, Raux G, et al. APP locus duplication causes autosomal dominant early-onset Alzheimer disease with cerebral amyloid angiopathy. *Nat Genet*. 2006;38(1):24–26. https://doi.org/10.1038/ng1718.
36. Alzheimer's Association. 2008 Alzheimer's disease facts and figures. *Alzheimers Dement*. 2008;4(2):110–133. https://doi.org/10.1016/j.jalz.2008.02.005.
37. Mullan M, Crawford F, Axelman K, et al. A pathogenic mutation for probable Alzheimer's disease in the APP gene at the N-terminus of beta-amyloid. *Nat Genet*. 1992;1(5):345–347. https://doi.org/10.1038/ng0892-345.
38. Di Fede G, Catania M, Morbin M, et al. A recessive mutation in the APP gene with dominant-negative effect on amyloidogenesis. *Science*. 2009;323(5920):1473–1477. https://doi.org/10.1126/science.1168979.
39. Zhou L, Brouwers N, Benilova I, et al. Amyloid precursor protein mutation E682K at the alternative β-secretase cleavage β'-site increases Aβ generation. *EMBO Mol Med*. 2011;3(5):291–302. https://doi.org/10.1002/emmm.201100138.
40. Jonsson T, Atwal JK, Steinberg S, et al. A mutation in APP protects against Alzheimer's disease and age-related cognitive decline. *Nature*. 2012;488(7409):96–99. https://doi.org/10.1038/nature11283.
41. Suh J, Choi SH, Romano DM, et al. ADAM10 missense mutations potentiate β-amyloid accumulation by impairing prodomain chaperone function. *Neuron*. 2013;80(2):385–401. https://doi.org/10.1016/j.neuron.2013.08.035.

42. Suzuki N, Cheung TT, Cai XD, et al. An increased percentage of long amyloid beta protein secreted by familial amyloid beta protein precursor (beta APP717) mutants. *Science*. 1994;264(5163):1336–1340. https://doi.org/10.1126/science.8191290.
43. Tian Y, Bassit B, Chau D, Li YM. An APP inhibitory domain containing the Flemish mutation residue modulates gamma-secretase activity for Abeta production. *Nat Struct Mol Biol*. 2010;17(2):151–158. https://doi.org/10.1038/nsmb.1743.
44. Kumar-Singh S, Theuns J, Van Broeck B, et al. Mean age-of-onset of familial alzheimer disease caused by presenilin mutations correlates with both increased Abeta42 and decreased Abeta40. *Hum Mutat*. 2006;27(7):686–695. https://doi.org/10.1002/humu.20336.
45. Lambert JC, Ibrahim-Verbaas CA, Harold D, et al. Meta-analysis of 74,046 individuals identifies 11 new susceptibility loci for Alzheimer's disease. *Nat Genet*. 2013;45(12):1452–1458. https://doi.org/10.1038/ng.2802.
46. Hori Y, Hashimoto T, Wakutani Y, et al. The Tottori (D7N) and English (H6R) familial Alzheimer disease mutations accelerate Abeta fibril formation without increasing protofibril formation. *J Biol Chem*. 2007;282(7):4916–4923. https://doi.org/10.1074/jbc.M608220200.
47. Murakami K, Irie K, Morimoto A, et al. Neurotoxicity and physicochemical properties of Abeta mutant peptides from cerebral amyloid angiopathy: implication for the pathogenesis of cerebral amyloid angiopathy and Alzheimer's disease. *J Biol Chem*. 2003;278(46):46179–46187. https://doi.org/10.1074/jbc.M301874200.
48. Nilsberth C, Westlind-Danielsson A, Eckman CB, et al. The 'Arctic' APP mutation (E693G) causes Alzheimer's disease by enhanced Abeta protofibril formation. *Nat Neurosci*. 2001;4(9):887–893. https://doi.org/10.1038/nn0901-887.
49. Tomiyama T, Nagata T, Shimada H, et al. A new amyloid beta variant favoring oligomerization in Alzheimer's-type dementia. *Ann Neurol*. 2008;63(3):377–387. https://doi.org/10.1002/ana.21321.
50. Hartley SL, Handen BL, Devenny D, et al. Cognitive decline and brain amyloid-β accumulation across 3 years in adults with Down syndrome. *Neurobiol Aging*. 2017;58:68–76. https://doi.org/10.1016/j.neurobiolaging.2017.05.019.
51. Kikuchi K, Kidana K, Tatebe T, Tomita T. Dysregulated metabolism of the amyloid-β protein and therapeutic approaches in Alzheimer disease. *J Cell Biochem*. 2017;118(12):4183–4190. https://doi.org/10.1002/jcb.26129.
52. Geerts H. NC-531 (Neurochem). *Curr Opin Investig Drugs*. 2004;5:95–100.
53. Aisen PS, Gauthier S, Ferris SH, et al. Tramiprosate in mild-to-moderate Alzheimer's disease—a randomized, double-blind, placebo-controlled, multi-centre study (the Alphase Study). *Arch Med Sci*. 2011;7(1):102–111.
54. Manzano S, Agüera L, Aguilar M, Olazarán J. A review on tramiprosate (homotaurine) in Alzheimer's disease and other neurocognitive disorders. *Front Neurol*. 2020;11:614.
55. Lannfelt L, Blennow K, Zetterberg H, et al. Safety, efficacy, and biomarker findings of PBT2 in targeting A beta as a modifying therapy for Alzheimer's disease: a phase IIa, double-blind, randomised, placebo-controlled trial. *Lancet Neurol*. 2008;7, 779786.
56. http://www.alzforum.org/news/research-news/pbt2-takesdive-phase-2-alzheimers-trial.
57. Salloway S, Sperling R, Keren R, et al. A phase 2 randomized trial of ELND005, scyllo-inositol, in mild to moderate Alzheimer disease. *Neurology*. 2011;77(13):1253–1262.
58. Lozupone M, Solfrizzi V, D'Urso F, et al. Anti-Amyloid-β Protein agents for the treatment of Alzheimer's Disease: an update on emerging drugs. *Expert Opin Emerg Drugs*. 2020;25(3):319–335.
59. https://www.alzforum.org/therapeutics/gv-971.
60. Cummings J, Lee G, Mortsdorf T, Ritter A, Zhong K. Alzheimer's disease drug development pipeline: 2017. *Alzheimers Dement*. 2017;3(3):367–384. Published 2017 May 24 https://doi.org/10.1016/j.trci.2017.05.002.

61. Schilling S, Rahfeld JU, Lues I, Lemere CA. Passive Aβ immunotherapy: current achievements and future perspectives. *Molecules.* 2018;23(5):1068. Published 2018 May 3 https://doi.org/10.3390/molecules23051068.
62. Lee JY, Cole TB, Palmiter RD, Suh SW, Koh JY. Contribution by synaptic zinc to the gender-disparate plaque formation in human Swedish mutant APP transgenic mice. *Proc Natl Acad Sci U S A.* 2002;99(11):7705–7710. https://doi.org/10.1073/pnas.092034699.
63. Polis B, Srikanth KD, Elliott E, Gil-Henn H, Samson AO. L-norvaline reverses cognitive decline and synaptic loss in a murine model of Alzheimer's disease. *Neurotherapeutics.* 2018;15(4):1036–1054. https://doi.org/10.1007/s13311-018-0669-5.
64. Götz J, Chen F, van Dorpe J, Nitsch RM. Formation of neurofibrillary tangles in P301l tau transgenic mice induced by Abeta 42 fibrils. *Science.* 2001;293(5534):1491–1495. https://doi.org/10.1126/science.1062097.
65. Maia LF, Kaeser SA, Reichwald J, et al. Changes in amyloid-β and Tau in the cerebrospinal fluid of transgenic mice overexpressing amyloid precursor protein. *Sci Transl Med.* 2013;5(194):194re2. https://doi.org/10.1126/scitranslmed.3006446.
66. Nicoll JA, Wilkinson D, Holmes C, Steart P, Markham H, Weller RO. Neuropathology of human Alzheimer disease after immunization with amyloid-beta peptide: a case report. *Nat Med.* 2003;9(4):448–452. https://doi.org/10.1038/nm840.
67. Reiman EM, Chen K, Liu X, et al. Fibrillar amyloid-beta burden in cognitively normal people at 3 levels of genetic risk for Alzheimer's disease. *Proc Natl Acad Sci U S A.* 2009;106(16):6820–6825. https://doi.org/10.1073/pnas.0900345106.

CHAPTER 5

Molecular linkages among Aβ, tau, impaired mitophagy, and mitochondrial dysfunction in Alzheimer's disease

Tomas Schmauck-Medina[a], Thale D.J.H. Patrick-Brown[a,b], Shi-qi Zhang[a], Alexandra Gilbert[c], and Evandro F. Fang[a,d]

[a]Department of Clinical Molecular Biology, University of Oslo and Akershus University Hospital, Lørenskog, Norway
[b]Division of Surgery, Inflammatory Diseases and Transplantation, Oslo University Hospital, Oslo, Norway
[c]Department of Cell and Developmental Biology, UCL, London, United Kingdom
[d]The Norwegian Centre on Healthy Ageing (NO-Age), Oslo, Norway

1. Introduction

In the pathophysiology of Alzheimer's disease (AD), the two classic hallmarks of disease are the presence of extracellular plaques, mainly made out of amyloid-β peptides (Aβ), and intracellular tangles consisting mostly of microtubule-associated, hyperphosphorylated protein tau (p-tau).[1] Aβ is a cleaved product by β-secretase (also known as BACE1) and γ-secretase from the integral membrane protein amyloid precursor protein (APP). Aβ can be produced in different sizes (from 36 to 43 amino acids in length), differing in their C-terminus, from which the 42-amino acid peptide is the least soluble and most prone to aggregation. Presenilin 1 (PSEN1) is a subunit of γ-secretase, and mutations in *PSEN1* and *APP* may increase the generation of $A\beta_{1-42}$ peptides.[2] For example, *PSEN1* mutations cause the most common type of familial AD, FAD.[3] In addition to the known extracellular aggregation of Aβ plaques in AD, intracellular Aβ peptides play both physiological and pathological functions in neurons.[4,5] In addition to pathological Aβ, accumulation of pathological tau tangles is arguably another cause of AD. While tau is mostly known to participate in microtubule assembly,[6] other roles, such as its participation in different cell signaling pathways, have also been described.[7,8] Tauopathies are a group of neurodegenerative diseases with abnormal hyperphosphorylation of microtubule-associated

Autophagy Dysfunction in Alzheimer's Disease and Dementia
https://doi.org/10.1016/B978-0-323-89906-2.00007-1

protein Tau (MAPT), leading to the formation of neurofibrillary tangles. Tauopathies include AD and many other diseases. While there is no Tau mutation in AD, Tau mutations are present in other tauopathies, such as tangle-only dementia, Pick's disease, progressive supranuclear palsy, corticobasal degeneration, argyrophilic grain disease, and globular glial tauopathy, among others.[9] One of the most well-recognized puzzles in the field has been how one protein (tau) could possibly lead to so many different types of diseases. Recent structural studies have shown that different diseases possess distinct conformers of tau filaments.[9] Additionally, the least soluble and aggregative forms of Aβ and p-tau may be considered the most pathological in AD.[10] It is interesting to note that Aβ plaques and tau tangles do not seem to act independently from each other, but rather interact to produce the characteristic phenotypes of AD. For instance, injection of $A\beta_{1-42}$ into mutant tau mice (expressing mutant human tau P301L) dramatically accelerated tau pathology as shown by increased number of tangles[11] (limitations of AD models reviewed in Ref. 12). Accumulation of brain Aβ plaques and tau tangles may be caused by increased production and reduced elimination of Aβ and tau, such as might be seen through impaired cellular clearance pathways, also termed autophagy.

Autophagy is a powerful cellular process that encompasses the different pathways that cells use to engulf and transport cellular substrates to the acidic lysosomes for degradation and recycling.[13,14] While there is an age-dependent reduction in autophagy, it is also compromised in AD.[15] Mitochondrial autophagy (termed mitophagy) is a subtype of selective macroautophagy, whereby damaged and superfluous mitochondria are recognized and selected for lysosomal degradation (Fig. 1). Compromised mitophagy is a risk factor in AD and exacerbates AD progression.[16,17] Accumulation of brain Aβ plaques and tau tangles may also be related to impaired glymphatic clearance during sleep. Indeed, an emerging biomarker for the development of AD is sleep dysfunction.[18] Circadian rhythm dysregulation, insomnia, fragmented sleep, and "sundowning" are common clinical manifestations associated with both early and late stages of AD, along with simultaneous deficits in mitochondrial function and mitophagy.[19] Both sleep and mitochondrial dysfunction are risk factors for AD that appear preclinically,[20,21] while early sleep fragmentation predicts the severity of cognitive decline in late stage AD.[18]

Here, we summarize the roles of mitochondria and mitophagy in neurons and the brain, and their linkages to Aβ plaques, tau tangles, and other AD pathologies. In addition to including molecular mechanisms of defective mitophagy in AD, novel therapeutic strategies targeting mitophagy stimulation are also included.

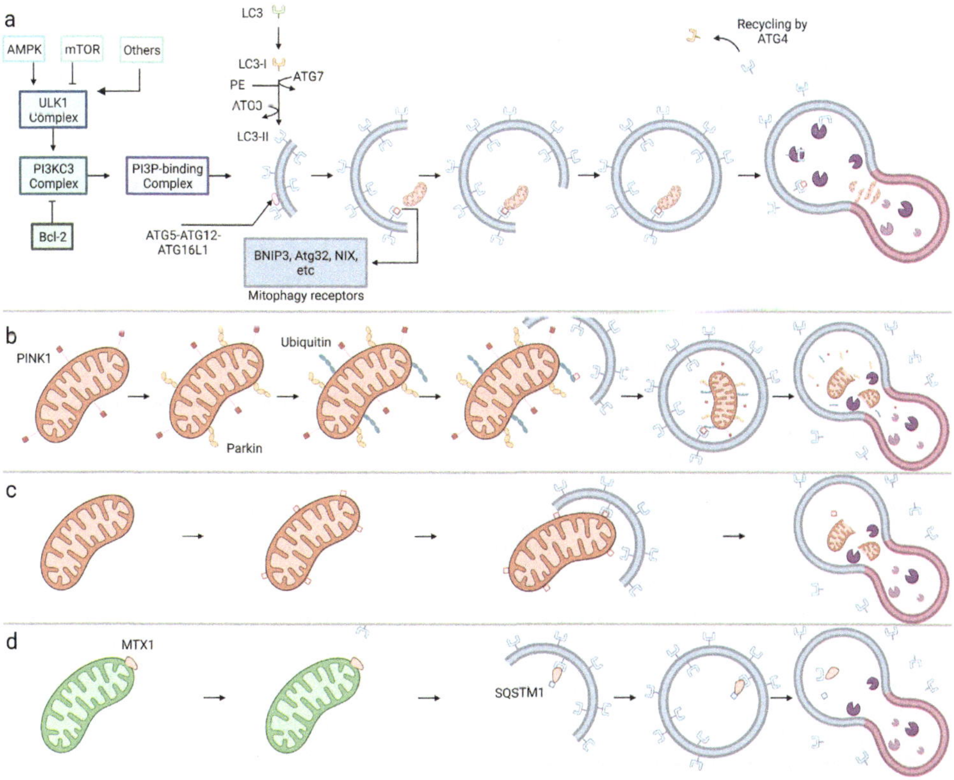

Fig. 1 Mechanisms of mitophagy. (A) In cells, the initiation of the macroautophagic machinery is at least partially regulated by mTOR and AMPK, where the ULK1 complex is a common target. This complex begins nucleation through the activation of the PI3KC3 complex, a process that can be inhibited through the interaction of Bcl-2 and Beclin1. Nucleation starts the production of phosphatidylinositol-3-phosphate (PI3P), which recruits WIPI proteins and DFCP1. LC3 is initially cleaved by ATG4, converting it into LC3-I. ATG7, phosphatidylethanolamine (PE), and ATG3 then conjugate LC3-I into LC3-II. Members of the PI3P complex can interact with ATG5-ATG12-ATG16L1 and boost conjugation of LC3. Receptors (e.g., NIX) bind both the mitochondria and to LC3-II on the phagosome. The phagosome further develops into an autophagosome, acidifies, and fuses with a lysosome, turning into an autophagolysosome, where degradation of the targets occurs. (B) One of the most common types of mitophagy is the PINK1-Parkin-dependent pathway. Initially, PINK1 targets dysfunctional mitochondria and accumulates on their surface; this in turn recruits Parkin, which ubiquitinates the mitochondria. Receptors bind to both the ubiquitin and the LC3 on the forming phagosome. The phagophore further develops into an enclosed autophagosome, which acidifies and fuses with the lysosome, further degrading the targeted mitochondrion. (C) The process of mitophagy can also occur in the absence of PINK1-Parkin, where receptors (e.g., NDP52) directly bind to the damaged mitochondria. (D) The process of mitophagy can degrade components of the mitochondria through the piecemeal mitophagy pathway. It has been shown that mitochondrial components such as MTX1 can be targeted by LC3 and SQSTM1 and further degraded through mitophagy. Selective targeting of mitochondrial proteins is relevant to maintain mitochondrial protein homeostasis.

2. Mitochondrial dysfunction

2.1 Mitochondria play fundamental roles in development and activities of neurons

Neurons form a unique cell type within the body. Transferring signals from center to center within the brain, as well as out to the body itself, neurons expend a large amount of energy in performance of their daily tasks. This energy is produced by the mitochondria within the cells. Due to their relatively higher energy expenditure, neurons often contain a large volume of mitochondria for ATP production. In addition, neurons rely on the efficient transport of energy to different subcellular regions in order to ensure appropriate levels of ATP production in more remote locations,[22,23] which itself requires increased energy production. A mitochondrion consists of an outer membrane, and inner membrane and matrix; the TCA cycle and oxidative phosphorylation (OXPHOS) take place in mitochondria and produce ATP. In order to bring the raw materials required for ATP production into the mitochondria, translocases located on the outer membrane (TOMs) mediate the movement of molecules into the mitochondria, while translocases located on the inner membrane (TIMs) mediate the movement of molecules into the aqueous mitochondrial matrix for use.[24] Mitochondria are dynamic subcellular organelles as they constantly go through fusion and fission. Common proteins involved in mitochondrial fusion are mitofusin 1 and 2 (MFN1 and MFN2) and optic atrophy 1 (OPA1), while proteins such as mitochondrial fission factor (MFF), the cytosolic GTPase dynamin-related protein 1 (DRP1), mitochondrial fission protein 1 (FIS1), mitochondrial dynamics proteins of 49 and 51 kDa (MiD49 and MiD51, respectively), and dynamin 2 (DNM2; possibly not essential in some cell types) divide one mitochondrion to smaller ones.[25–27]

In the nervous system, mitochondria play critical roles in a multitude of processes, such as neuronal growth and differentiation, synaptic plasticity, neuronal firing, and the arbitration of neuronal survival or death[16,28]; thus, their proper functioning may be key to the maintenance of brain health. Recent studies have shown that the healthy human brain is able to undertake regeneration of itself after hippocampal injury, but this process is severely lacking in the AD brain.[29] It is known that during neuronal differentiation, the number of mitochondria within in the neuron increases,[28] and if mitochondrial amplification is suppressed, neurons are not able to differentiate.[30] These findings, emphasizing the importance of mitochondria in brain development and health, suggest that the progression of AD could be impeded, at least partially, by targeting the maintenance of mitochondrial homeostasis. Further supporting this suggestion, research suggests that the accumulation of damaged mitochondria, mainly due to compromised mitophagy and the ubiquitin-proteasome system (UPS), contributes to neuronal loss in both the elderly and individuals with accelerated aging diseases, as well as in AD and other neurodegenerative diseases.[14,31,32] These findings highlight the importance of a healthy mitochondrial pool and proper functioning of the mechanisms that maintain mitochondrial homeostasis to maintain healthy aging and neuroprotection.

2.2 Mitochondrial dysfunction is an early feature of AD

Mounting evidence from experiments using human brain tissue and animal models indicates that there may be altered morphology and impaired function of mitochondria in AD. Proteins involved in mitochondrial dynamics are also altered in AD. Mitochondrial fission proteins such as DRP1, MFF (including phosphorylated MFF at S^{146}), MiD49, MiD51, and FIS1 have bene shown to be upregulated in AD, in line with the excessive mitochondrial fragmentation seen in AD.[17,33–35] In the brains of both early- and late-onset AD patients, as well as in some AD mouse models, mitochondrial abnormalities are visible, including accumulation of mitochondrial DNA (mtDNA), higher reactive oxygen species (ROS), reduced mitochondrial membrane potential, and low ATP production. Compromised ATP production in AD could be attributed to impaired axonal transport of mitochondria and reduced activities of members of OXPHOS (such as enzyme complexes I, II, and V).[17,23,35,36] Furthermore, abnormalities in mitochondrial function could be related to the impairment of other mitochondrial proteins such as cytochrome *c* oxidase (COX), α-ketoglutarate dehydrogenase complex (KGDHC), and pyruvate dehydrogenase complex (PDHC).[37–39] In line with the importance of mitochondria in metabolism, metabolic derailment is one of the earliest features in AD brains[40]; for example, there is a glial upregulation of fatty acid metabolism to compensate for neuronal glucose hypometabolism in AD,[41] while oxidative damage has been shown to precede Aβ deposition.[42]

While it is not fully clear whether mitochondrial dysfunction arises due to Aβ/tau pathologies or the other way around, age, genetic, and environmental factors cause compromised mitophagy, which results in the accumulation of damaged mitochondria during aging and especially in AD. Accumulation of damaged mitochondria could cause or exacerbate Aβ and tau pathologies via different mechanisms, including activation of the NLRP3 inflammation pathway,[17,43] and indeed it has been reported that NLRP3 activation leads to Aβ and tau pathologies.[44,45] Emerging evidence from postmortem human brain samples and animal studies suggests a high likelihood that mitochondrial dysfunction occurs ahead of detectable Aβ and tau pathologies and AD clinical phenomena (reviewed in Ref. 4). The entorhinal cortex (EC) is pivotal in memory (especially spatial memory) as it links the neocortex and the hippocampus. EC layer II (EC L-II) is the earliest brain region affected by neurofibrillary tangles (NFTs)[46]; reelin-immunoreactive neurons in EC L-II express intracellular amyloid-beta Aβ (iAβ) in early AD when no extracellular Aβ plaques are yet visible.[5] Interestingly, mitochondrial dysfunction happens in a spatiotemporal pattern with the EC being the earliest region exhibiting changes. In the human brain, when compared with cognitively healthy individuals, reduced activity of mitochondrial complexes I, II, IV, and V are already detectable in the EC at Braak stages I–II.[47] This is likely followed by a similar reduction in mitochondrial activity in the hippocampus,[17] with other areas (the parietal, prefrontal, and frontal cortices) exhibiting dysfunctional mitochondria at later stages.[48,49] The cerebellum does not show AD pathologies until a very late stage of AD, and consequently, it incurs less mitochondrial damage and oxidative stress than the neocortical areas.[50] Collectively, mitochondrial dysfunction is already present in the earliest stages of AD, much earlier than the symptomatology arises. In addition, there is a high anatomical and spatiotemporal match between tau pathology and mitochondrial dysfunction in AD.

2.3 Interactions of mitochondria with Aβ and tau

Evidence from over three decades of studies shows a convoluted and vicious cycle between mitochondrial dysfunction and pathological Aβ and tau. As mitochondrial dysfunction increases the aggregation of Aβ and tau, pathological Aβ and tau can cause and exacerbate mitochondrial dysfunction (Fig. 2). It is suggested that genes involved in mitochondrial energy metabolism are upregulated as a compensatory response to the expression of both mutant amyloid-beta precursor protein (APP) and Aβ.[51,52]

While Aβ can be found in the brain of cognitively healthy people,[53] there is evidence that a small portion of Aβ exists in the mitochondria of the human AD brain[42] and that inhibiting energy production increases the production of Aβ[51,54] (Aβ can use translocase of the outer membrane (TOM) to be transported into the mitochondria[52]). In addition, APP can block TOM40 and TIM23, arresting the entry of nuclear DNA-encoded mitochondrial proteins such as cytochrome *c* oxidase subunits 4 and 5b.[55] This results in an increase in hydrogen peroxide levels and a decrease in cytochrome *c* oxidase

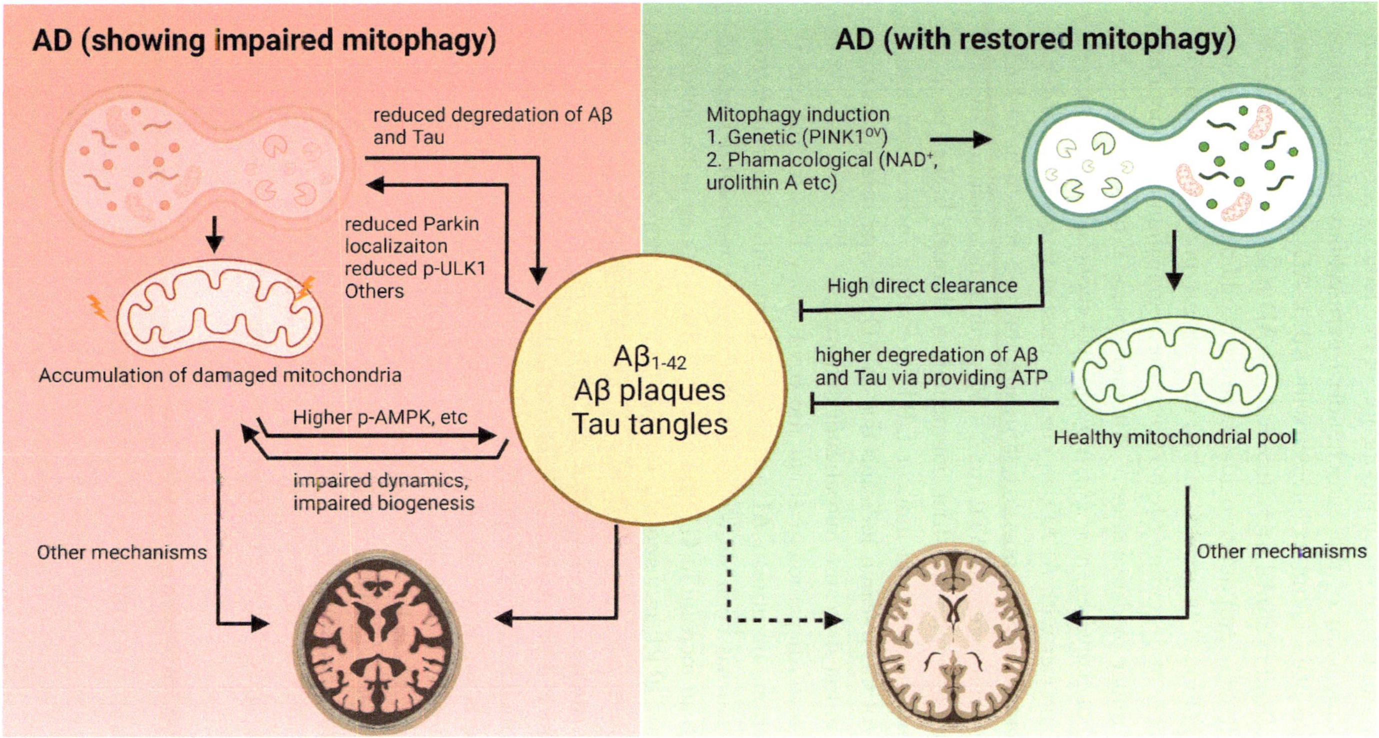

Fig. 2 Boosting mitophagy as a potential therapy for AD. Left: Impaired mitophagy results in accumulation of damaged mitochondria, contributing to Aβ and tau pathologies. Furthermore, in a vicious cycle, Aβ and tau increase ROS production while altering mitochondrial fission/fusion (dynamics) and biogenesis, exacerbating mitochondrial dysfunction. In addition, the presence of Aβ and tau decreases the levels of important mitophagy/autophagy proteins such as the mitochondrial localization of Parkin and p-ULK1. These may contribute to risk or exacerbation of the clinical appearance of AD. Right: Pharmacological or genetic restoration of defective mitophagy/autophagy allows for the clearance of dysfunctional mitochondria, as well as pathological Aβ and tau proteins, resulting in brain protection.

activity, which can then result in production of free radicals.[52,55] Cyclophilin D (CypD), a mediator of the mitochondrial permeability transition pore (PTP), has also been found to interact with Aβ in AD models, increasing mitochondrial and neuronal stress. Interestingly, AD mice lacking CypD are protected from cell death due to oxidative stress, in addition to experiencing improved cognition.[56] In mitochondria, the enzyme presequence protease (PreP) is capable of degrading Aβ; however, the degradation of Aβ by PreP is eradicated under oxidative stress.[57] Another potential target may be Aβ binding alcohol dehydrogenase (ABAD). In AD models, Aβ may directly or indirectly interact with ABAD, an enzyme present in the mitochondria, which may create mitochondrial dysfunction through reduction of COX activity, membrane potential, ATP, etc.[58,59] Regarding fission-fusion dynamics, Aβ can induce mitochondrial fragmentation through higher levels of S-nitrosylation of the fission-inducer DRP1, resulting in neuronal injury and death.[60] All this research indicates that in AD, Aβ may directly interact with mitochondria and cause mitochondrial damage.

In addition to Aβ, pathological tau also has a strong links to mitochondrial dysfunction. In human AD neurons, where mitochondrial fission dominates,[17] tau has been shown to directly interact with DRP1 and to correlate with an increase in GTPase activity.[60] By blocking mitochondrial localization of DRP1, mutated tau can produce mitochondrial elongation in neurons, together with cell death,[61–64] in addition to neurotoxicity.[65] Furthermore, studies have shown that phosphorylation at certain sites affects the movement and distribution of neuronal mitochondria.[66] In the model (rTg)4510, which is a reversible tau mouse, it was shown that the distribution of mitochondria across somata and neurites progressively worsens with age. However, when the expression of tau was inhibited, a normal distribution was recovered.[67] These localization abnormalities were also observed in human brain samples.[67] Finally, truncated forms of tau were associated with fragmented mitochondria, which in turn correlate with decreased levels of the mitochondrial fusion protein OPA1.[68] Collectively, pathological tau causes damage and distribution abnormality to mitochondria in the neurons.

3. Defective mitophagy in AD

3.1 Molecular mechanisms of mitophagy

As mentioned before, mitophagy is a type of selective autophagy through which damaged mitochondria are marked and recognized by the cellular autophagic machinery, enveloped by autophagosomes, and then degraded through fusion with lysosomes. Several mitophagy pathways have been identified, but the PTEN-induced kinase 1 (PINK1)/Parkin pathway is probably one of the most studied.[69] PINK1-Parkin-dependent mitophagy is initiated by a depolarization of the mitochondrial inner membrane caused by mitochondrial damage. This stops the normal movement of PINK1 into the mitochondria for degradation, resulting in the accumulation of PINK1 anchored on the mitochondrial outer membrane.[70] The PINK1 associated with the surface then begins to recruit the E3 ubiquitin ligase Parkin, which further ubiquitylates specific mitochondrial outer membrane proteins, eventually causing formation of a phagophore. The phagophore then associates with the lysosome that in turn degrades the damaged organelle. To note, Parkin can function as a signal amplifier through adding more ubiquitin chains to the mitochondria, enabling stronger mitophagy; in line with this, the PINK1-dependent, Parkin-independent pathway is less efficient in mitochondrial elimination compared with the PINK1-Parkin-dependent mitophagy pathway.[71–73] In PINK1/Parkin-independent mitophagy pathways, recognition of damaged (in some specific conditions, sulfurous but healthy mitochondria) is mediated by other proteins such as NIX, BNIP3, FKBP8, FUNDC1, BCL2L13, PHB2, and Cardiolipin, among others (reviewed in Ref. 16). Other mitophagy pathways and nonmitophagic mitochondrial elimination pathways are detailed in published reviews.[4,74] The mitophagy machinery is a promising druggable target as bioavailable molecules, such as nicotinamide riboside (NR), nicotinamide mononucleotide (NMN), and urolithin A (UA) induce mitophagy and increase neuronal function.[17,75]

3.2 Defective mitophagy in AD

Mitophagy is compromised in AD. The basal level of mitophagy in postmortem hippocampi from AD patients is reported to be 30%–50% reduced compared with cognitively normal individuals.[17] In line with this, mitophagy is diminished in the iPSC-derived cortical neurons of AD patients, in the neurons of *Caenorhabditis elegans*, as well as in the hippocampal tissues of both APP/PS1 and 3xTgAD mice.[17,76] Defective mitophagy leads to the accumulation of dysfunctional mitochondria, which could result in increased mitochondrial Ca^{2+} release, bioenergetic deficiency, loss of membrane potential, and oxidative stress at a cellular levels.[35,77] By extrapolation, accumulation of damaged mitochondria could also lead to Aβ and tau pathologies, with detailed mechanisms we have summarized elsewhere.[4,23]

While the molecular drivers of defective mitophagy in AD are not fully understood, there are some clues. First, impaired mitophagy in AD could be contributed to by the impaired initiation of the mitophagy machinery as the active phosphorylated form of ULK1, one of the initiative proteins of the autophagy machinery, is impeded in AD human hippocampal tissues.[17] Second, impaired mitophagy in AD could be caused by pathological tau as overexpression of human wild-type (hTau) and FTD mutant tau (hP301L) inhibit mitophagy in neuroblastoma cells and in *C. elegans*, via reduction of the mitochondrial translocation of Parkin.[78] Whether and how pathological Aβ peptides inhibit mitophagy remains to be determined. Importantly, pharmacological[17] or genetic[79] restoration of mitophagy impedes memory loss in AD animal models, indicating that compromised mitophagy plays an important role in AD progression.

Compromised mitophagy in AD could be also ascribed to aging via nicotinamide adenine dinucleotide (NAD^+) depletion. NAD^+ is a fundamental metabolite involved in a broad spectrum of cellular activities, such as cellular bioenergetics, genomic stability, mitochondrial homeostasis, adaptive stress responses, and cell survival.[80] In neurons, NAD^+ is necessary for neuronal development, synaptic plasticity, and neuronal stress resistance. Neurons are sensitive to the reduction of NAD^+ levels.[31,81,82] A deficiency in neuronal NAD^+ affects neuronal health and survival through disruption of the maintenance of balance between mitochondrial biogenesis and mitophagy.[80] NAD^+ modulates mitochondrial bioenergetics through several pathways including the NAD^+/SIRT1-PGC1α pathway, which plays a major role in mitochondrial biogenesis, neuroprotection, and longevity.[31] Furthermore, NAD^+ stimulates mitophagy/autophagy via transcriptional or translational upregulation of several mitophagy/autophagy proteins, such as NIX (DCT-1 in worms), ULK1 (UNC-51 in worms), Beclin-1, LC3, ATG-12, and BNIP3.[82,83] NAD^+ also posttranslationally activates many proteins directly or indirectly involved in mitophagy/autophagy via NAD^+-SIRTs-dependent deacetylation (or other processes), such as ATG5, ATG7, LC3, FOXO1, and FOXO3.[83–85] There is also a likelihood that autophagy maintains cellular NAD(H) pools.[86] Intriguingly, NAD^+ augmentation forestalls memory loss in both Aβ- and mutant Tau-harboring AD *C. elegans* models via PINK1 or Parkin, or in a DCT-1-dependent manner.[17] Collectively, NAD^+ and mitophagy/autophagy regulate each other to maintain cellular homeostasis, with in-depth molecular mechanisms still requiring further exploration. Since NAD^+ is reduced in the elderly, we have proposed that age-dependent impairment of the NAD^+-mitophagy-mitochondrial homeostasis axis is a risk factor for AD, as well as for Parkinson's disease (PD), Huntington's disease (HD), and amyotrophic lateral sclerosis (ALS).[80]

Upregulating mitophagy is a novel and likely also multifunctional therapeutic strategy for AD.[87] Studies in human cells and animals show that stimulation of autophagy via rapamycin reduces tau pathologies.[17] In other words, the mTOR inhibitor rapamycin could stimulate autophagy and also mitophagy as a result inhibiting tau.[88,89] Pharmacological stimulation of mitophagy, for example, via NAD^+ precursors (NR, NMN), urolithin A, and actinonin, impeded memory loss as well as both Aβ and tau pathologies in both 5xFAD and 3xTgAD mouse models, showing translational potential.[17] Through the use of machine learning and a cross specics workflow, we have identified two new robust mitophagy inducers Kaempferol and Rhapontigenin. These molecules inhibited memory loss and AD pathologies (both Aβ and pTau) in both *C. elegans* and mouse models of AD in a mitophagy-dependent manner.[76] Furthermore, clinical studies on NR, NMN, and urolithin A show that they are safe (at the doses studied) and bioavailable; clinical trials on neurodegenerative, muscle, and metabolic diseases are in progress.[80,90,91]

4. Mitochondrial dysfunction and defective mitophagy at the circuit and behavioral level

Compromised mitophagy followed by accumulation of damaged mitochondria could contribute to the synaptic loss, circuit pathologies, and disturbed behaviors (e.g., bad sleep) seen in AD patients. Synaptic loss and excitotoxicity in neurons have implications that manifest in memory loss, cognitive dysfunction, poor sleep and circadian dysregulation, among others. A well-known brain structure affected by AD is the hippocampus, specifically the CA1 excitatory pyramidal neurons. In AD, the spread of pathological Tau starts in the trans-EC and EC layer II, spreading to the hippocampus, and later to the whole cortex.[4,92] Poor spatial memory is characteristic of AD and is thought to be related to neuronal loss or dysfunction in EC layer II and hippocampus.[17,93] Although the direct mechanism mediating neuronal loss remains a debate, evidence points toward compromised mitochondrial dynamics and mitophagy as key regulators of Aβ-dependent synaptic loss.[17,64] Autophagy proteins Atg5 and mitophagy protein PINK1 are underexpressed in APP mutant mice as well as AD patient-derived pluripotent cells.[17,93] This underexpression is accompanied by fewer dendritic spines in hippocampal neurons of APP mice.[93] Correspondingly, cognitive deficits and poor spatial memory using the Morris water maze are characteristic of APP mutant mice as well as

APP/PS1 mice, which overexpress Aβ and extracellular tau tangles.[17,93] These data suggest a possible linkage between altered mitophagy dynamics and impaired neuronal circuits associated with AD.

While disrupted sleep is common in and likely a risk factor of AD, restoration of the NAD^+-mitophagy axis may improve sleep in AD. When injected with tau fibrils, mice that were chronically sleep-deprived displayed aggregates in the locus coeruleus (LC), a key population of neurons involved in the waking response through excitatory noradrenergic signaling to the hippocampus and the cortex.[94,95] The norepinephrine released from the LC at hippocampal and cortical synapses is essential for long-term potentiation of the postsynaptic targets of LC projections.[95–97] The LC also commonly shows neurodegeneration in sporadic AD.[98] The vulnerability of the LC to AD degeneration is thought to be due to its high bioenergetic load since these neurons are the primary source of norepinephrine to many brain regions (reviewed in Ref. 99). Degenerating neurons in the LC contribute to memory loss through their connections with hippocampal neurons.[100] The LC controls wakefulness and is suppressed by inhibitory GABAergic input to promote sleep-like states.[101] Similarly, ROS levels in LC neurons fluctuate over the sleep-wake cycle.[102] Circadian dysregulation, insomnia, fragmented sleep, and sundowning are common clinical manifestations associated with both early and late stages of AD.[101] Early stages of AD have characteristic fragmented sleep correlated with the presence of soluble Aβ oligomers.[103] Disrupted sleep, which affects 40% of AD patients, reduces quality of life and is a common reason for AD patient institutionalization.[19,104] Similarly, tau and Aβ proteins fluctuate with the sleep-wake cycle, where Aβ in the cerebrospinal fluid of mice is cleared during sleep, it accumulates during wakefulness and is its further agglomeration is therefore exacerbated through sleep deprivation. CSF tau is also increased by 50% after sleep deprivation compared with controls.[105] Interestingly, NAD^+ and NADH have strong circadian fluctuations along with SIRT1 and PGC1α that dampen with aging and AD, as does mitophagy.[106,107] It was reported that NAD^+ augmentation markedly improved sleep quality in old mice; mechanistically, mitochondrial respiration rhythms, and late evening activity were restored by NAD^+ augmentation to youthful levels through PER2 (clock repressor) nuclear translocation.[108] However, whether NAD^+ repletion-induced memory improvement in the AD animals could be at least partially contributed to by the improvement of sleep in AD remains to be determined. By extrapolation, targeting mitophagic pathways to improve clearance of damaged mitochondria may alleviate symptoms of sleep fragmentation and memory loss, slowing the symptomatic progression of AD. Further experiments to test this hypothesis are necessary.

5. Future research

Although much research has been done looking into the potential mechanisms behind AD, many questions remain unanswered; we do not yet know the exact mechanisms behind compromised mitophagy, for example. In order to address this question, human CSF or blood samples, matched with high-quality clinical cognitive data from longitudinal cohorts, would be ideal. Additionally, the exploration of mitophagy proteins as biomarkers of AD and prodromal AD (also named mild cognitive impairment/MCI) is a potential target for research. As well, in view of the continued failure of clinical trials of drug candidates targeting solely Aβ, p-tau, or inflammation, alternative approaches that target broader aspects of AD pathologies, such as potentially compromised mitophagy, hold promise. Finally, should the mechanisms behind AD appear to point to dysfunctional mitophagy, the screening and characterization of specific, safe, and bioavailable mitophagy inducers may be warranted. It is important to note here that there is no perfect mouse model of AD. The commonly used mouse models of AD, like those mimicking Aβ pathology (APP/PS1, 5xFAD) and Tau pathology (Tau P301S from the Goedert group and the Tau P301S (Line PS19) from the Lee group) only recapitulate some of the features of AD. Thus, mechanistic and interventional data from AD mice should be validated in tissues and stem cells from AD humans for confirmation, and before moving to any clinical studies.

Compromised mitophagy can lead to the accumulation of damaged mitochondria, potentially resulting in a risk of AD.[17] Failure to remove damaged mitochondria could release cell death signals,[109] activate inflammation via the NLRP3-Caspase 1-IL1β pathway,[17] and induce mitochondrial dysfunction-associated senescence (MiDAS),[110] all of which have been linked to the progression of AD. Additionally, there is likely a strong "mutually causative" linkage between Aβ/Tau pathologies and defective mitophagy as summarized in this chapter and elsewhere.[23]

Acknowledgments

This project was supported by the National Natural Science Foundation of China (#81971327), HELSE SØR-ØST (#2017056, #2020001, #2021021), the Research Council of Norway (#262175), Akershus University Hospital (#269901, #261973, #262960), the Civitan Norges Forskningsfond for Alzheimers sykdom (#281931), and the Czech Republic-Norway KAPPA programme (with Martin Vyhnálek, #TO01000215). S.Q. Zhang was funded by the China Scholarship Council (http://www.csc.edu.cn/); the funders had no role in study design, data collection and analysis, decision to publish, or preparation of the manuscript. Cartoon images in Figs. 1 and 2 were generated using BioRender via subscription through the Fang laboratory.

Competing interests

E.F.F. has CRADA arrangement with ChromaDex and is consultants to Aladdin Healthcare Technologies, the Vancouver Dementia Prevention Centre, Intellectual Labs, and MindRank AI Ltd. Other authors declare no competing interests.

References

1. Murphy MP, LeVine 3rd H. Alzheimer's disease and the amyloid-beta peptide. *J Alzheimers Dis.* 2010;19(1):311–323.
2. Suarez-Calvet M, Belbin O, Pera M, et al. Autosomal-dominant Alzheimer's disease mutations at the same codon of amyloid precursor protein differentially alter Abeta production. *J Neurochem.* 2014;128(2):330–339.
3. Campion D, Dumanchin C, Hannequin D, et al. Early-onset autosomal dominant Alzheimer disease: prevalence, genetic heterogeneity, and mutation spectrum. *Am J Hum Genet.* 1999;65(3):664–670.
4. Kobro-Flatmoen A, Lagartos-Donate MJ, Aman Y, Edison P, Witter MP, Fang EF. Re-emphasizing early Alzheimer's disease pathology starting in select entorhinal neurons, with a special focus on mitophagy. *Ageing Res Rev.* 2021;67, 101307.
5. Kobro-Flatmoen A, Nagelhus A, Witter MP. Reelin-immunoreactive neurons in entorhinal cortex layer II selectively express intracellular amyloid in early Alzheimer's disease. *Neurobiol Dis.* 2016;93:172–183.
6. Weingarten MD, Lockwood AH, Hwo SY, Kirschner MW. A protein factor essential for microtubule assembly. *Proc Natl Acad Sci U S A.* 1975;72(5):1858–1862.
7. Dixit R, Ross JL, Goldman YE, Holzbaur EL. Differential regulation of dynein and kinesin motor proteins by tau. *Science.* 2008;319(5866):1086–1089.
8. Li HL, Wang HH, Liu SJ, et al. Phosphorylation of tau antagonizes apoptosis by stabilizing beta-catenin, a mechanism involved in Alzheimer's neurodegeneration. *Proc Natl Acad Sci U S A.* 2007;104(9):3591–3596.
9. Goedert M. Tau proteinopathies and the prion concept. *Prog Mol Biol Transl Sci.* 2020;175:239–259.
10. Jarrett JT, Berger EP, Lansbury Jr PT. The carboxy terminus of the beta amyloid protein is critical for the seeding of amyloid formation: implications for the pathogenesis of Alzheimer's disease. *Biochemistry.* 1993;32(18):4693–4697.
11. Gotz J, Chen F, van Dorpe J, Nitsch RM. Formation of neurofibrillary tangles in P301l tau transgenic mice induced by Abeta 42 fibrils. *Science.* 2001;293(5534):1491–1495.
12. Drummond E, Wisniewski T. Alzheimer's disease: experimental models and reality. *Acta Neuropathol.* 2017;133(2):155–175.
13. Rubinsztein DC, Marino G, Kroemer G. Autophagy and aging. *Cell.* 2011;146(5):682–695.
14. Aman Y, Schmauck-Medina T, Hansen M, et al. Autophagy in healthy ageing and disease. *Nat Aging.* 2021;1:634–650.

15. Menzies FM, Fleming A, Rubinsztein DC. Compromised autophagy and neurodegenerative diseases. *Nat Rev Neurosci*. 2015;16(6):345–357.
16. Lou G, Palikaras K, Lautrup S, Scheibye-Knudsen M, Tavernarakis N, Fang EF. Mitophagy and neuroprotection. *Trends Mol Med*. 2019;26(1):8–20.
17. Fang EF, Hou Y, Palikaras K, et al. Mitophagy inhibits amyloid-beta and tau pathology and reverses cognitive deficits in models of Alzheimer's disease. *Nat Neurosci*. 2019;22(3):401–412.
18. Winer JR, Mander BA, Helfrich RF, et al. Sleep as a potential biomarker of tau and beta-amyloid burden in the human brain. *J Neurosci*. 2019;39(32):6315–6324.
19. Wang C, Holtzman DM. Bidirectional relationship between sleep and Alzheimer's disease: role of amyloid, tau, and other factors. *Neuropsychopharmacology*. 2020;45(1):104–120.
20. Wang W, Zhao F, Ma X, Perry G, Zhu X. Mitochondria dysfunction in the pathogenesis of Alzheimer's disease: recent advances. *Mol Neurodegener*. 2020;15(1):30.
21. Wu H, Dunnett S, Ho YS, Chang RC. The role of sleep deprivation and circadian rhythm disruption as risk factors of Alzheimer's disease. *Front Neuroendocrinol*. 2019;54, 100764.
22. Cheng A, Wan R, Yang JL, et al. Involvement of PGC-1alpha in the formation and maintenance of neuronal dendritic spines. *Nat Commun*. 2012;3:1250.
23. Kerr JS, Adriaanse BA, Greig NH, et al. Mitophagy and Alzheimer's disease: cellular and molecular mechanisms. *Trends Neurosci*. 2017;40(3):151–166.
24. Neupert W, Herrmann JM. Translocation of proteins into mitochondria. *Annu Rev Biochem*. 2007;76:723–749.
25. Fonseca TB, Sanchez-Guerrero A, Milosevic I, Raimundo N. Mitochondrial fission requires DRP1 but not dynamins. *Nature*. 2019;570(7761):E34–E42.
26. Loson OC, Song Z, Chen H, Chan DC. Fis1, Mff, MiD49, and MiD51 mediate Drp1 recruitment in mitochondrial fission. *Mol Biol Cell*. 2013;24(5):659–667.
27. Leal NS, Martins LM. Mind the gap: mitochondria and the endoplasmic reticulum in neurodegenerative diseases. *Biomedicine*. 2021;9(2):227.
28. Mattson MP, Gleichmann M, Cheng A. Mitochondria in neuroplasticity and neurological disorders. *Neuron*. 2008;60(5):748–766.
29. Moreno-Jimenez EP, Flor-Garcia M, Terreros-Roncal J, et al. Adult hippocampal neurogenesis is abundant in neurologically healthy subjects and drops sharply in patients with Alzheimer's disease. *Nat Med*. 2019;25(4):554–560.
30. Vayssiere JL, Cordeau-Lossouarn L, Larcher JC, Basseville M, Gros F, Croizat B. Participation of the mitochondrial genome in the differentiation of neuroblastoma cells. *In Vitro Cell Dev Biol*. 1992;28A(11–12):763–772.
31. Fang EF, Scheibye-Knudsen M, Brace LE, et al. Defective mitophagy in XPA via PARP-1 hyperactivation and NAD(+)/SIRT1 reduction. *Cell*. 2014;157(4):882–896.
32. Fang EF, Scheibye-Knudsen M, Chua KF, Mattson MP, Croteau DL, Bohr VA. Nuclear DNA damage signalling to mitochondria in ageing. *Nat Rev Mol Cell Biol*. 2016;17(5):308–321.
33. Cai Q, Tammineni P. Alterations in mitochondrial quality control in Alzheimer's disease. *Front Cell Neurosci*. 2016;10:24.
34. Chakravorty A, Jetto CT, Manjithaya R. Dysfunctional mitochondria and mitophagy as drivers of Alzheimer's disease pathogenesis. *Front Aging Neurosci*. 2019;11:311.
35. Pradeepkiran JA, Reddy PH. Defective mitophagy in Alzheimer's disease. *Ageing Res Rev*. 2020;64, 101191.
36. Hou Y, Lautrup S, Cordonnier S, et al. NAD(+) supplementation normalizes key Alzheimer's features and DNA damage responses in a new AD mouse model with introduced DNA repair deficiency. *Proc Natl Acad Sci U S A*. 2018;115(8):E1876–E1885.
37. Casley CS, Canevari L, Land JM, Clark JB, Sharpe MA. Beta-amyloid inhibits integrated mitochondrial respiration and key enzyme activities. *J Neurochem*. 2002;80(1):91–100.

38. Sorbi S, Bird ED, Blass JP. Decreased pyruvate dehydrogenase complex activity in Huntington and Alzheimer brain. *Ann Neurol*. 1983;13(1):72–78.
39. Gibson GE, Sheu KF, Blass JP, et al. Reduced activities of thiamine-dependent enzymes in the brains and peripheral tissues of patients with Alzheimer's disease. *Arch Neurol*. 1988;45(8):836–840.
40. Mosconi L, Tsui WH, De Santi S, et al. Reduced hippocampal metabolism in MCI and AD: automated FDG-PET image analysis. *Neurology*. 2005;64(11):1860–1867.
41. Demarest TG, Varma VR, Estrada D, et al. Biological sex and DNA repair deficiency drive Alzheimer's disease via systemic metabolic remodeling and brain mitochondrial dysfunction. *Acta Neuropathol*. 2020;140(1):25–47.
42. Pratico D, Uryu K, Leight S, Trojanoswki JQ, Lee VM. Increased lipid peroxidation precedes amyloid plaque formation in an animal model of Alzheimer amyloidosis. *J Neurosci*. 2001;21(12):4183–4187.
43. Lautrup S, Lou G, Aman Y, Nilsen H, Tao J, Fang EF. Microglial mitophagy mitigates neuroinflammation in Alzheimer's disease. *Neurochem Int*. 2019;129, 104469.
44. Ising C, Venegas C, Zhang S, et al. NLRP3 inflammasome activation drives tau pathology. *Nature*. 2019;575(7784):669–673.
45. Heneka MT, Kummer MP, Stutz A, et al. NLRP3 is activated in Alzheimer's disease and contributes to pathology in APP/PS1 mice. *Nature*. 2013;493(7434):674–678.
46. Braak H, Braak E. Neuropathological stageing of Alzheimer-related changes. *Acta Neuropathol*. 1991;82(4):239–259.
47. Armand-Ugon M, Ansoleaga B, Berjaoui S, Ferrer I. Reduced mitochondrial activity is early and steady in the entorhinal cortex but it is mainly unmodified in the frontal cortex in Alzheimer's disease. *Curr Alzheimer Res*. 2017;14(12):1327–1334.
48. Wang J, Xiong S, Xie C, Markesbery WR, Lovell MA. Increased oxidative damage in nuclear and mitochondrial DNA in Alzheimer's disease. *J Neurochem*. 2005;93(4):953–962.
49. Jadiya P, Kolmetzky DW, Tomar D, et al. Impaired mitochondrial calcium efflux contributes to disease progression in models of Alzheimer's disease. *Nat Commun*. 2019;10(1):3885.
50. Thubron EB, Rosa HS, Hodges A, et al. Regional mitochondrial DNA and cell-type changes in post-mortem brains of non-diabetic Alzheimer's disease are not present in diabetic Alzheimer's disease. *Sci Rep*. 2019;9(1):11386.
51. Reddy PH, McWeeney S, Park BS, et al. Gene expression profiles of transcripts in amyloid precursor protein transgenic mice: up-regulation of mitochondrial metabolism and apoptotic genes is an early cellular change in Alzheimer's disease. *Hum Mol Genet*. 2004;13(12):1225–1240.
52. Hansson Petersen CA, Alikhani N, Behbahani H, et al. The amyloid beta-peptide is imported into mitochondria via the TOM import machinery and localized to mitochondrial cristae. *Proc Natl Acad Sci U S A*. 2008;105(35):13145–13150.
53. Jack Jr CR, Knopman DS, Chetelat G, et al. Suspected non-Alzheimer disease pathophysiology—concept and controversy. *Nat Rev Neurol*. 2016;12(2):117–124.
54. Velliquette RA, O'Connor T, Vassar R. Energy inhibition elevates beta-secretase levels and activity and is potentially amyloidogenic in APP transgenic mice: possible early events in Alzheimer's disease pathogenesis. *J Neurosci*. 2005;25(47):10874–10883.
55. Devi L, Prabhu BM, Galati DF, Avadhani NG, Anandatheerthavarada HK. Accumulation of amyloid precursor protein in the mitochondrial import channels of human Alzheimer's disease brain is associated with mitochondrial dysfunction. *J Neurosci*. 2006;26(35):9057–9068.
56. Du H, Guo L, Fang F, et al. Cyclophilin D deficiency attenuates mitochondrial and neuronal perturbation and ameliorates learning and memory in Alzheimer's disease. *Nat Med*. 2008;14(10):1097–1105.

57. Alikhani N, Ankarcrona M, Glaser E. Mitochondria and Alzheimer's disease: amyloid-beta peptide uptake and degradation by the presequence protease, hPreP. *J Bioenerg Biomembr.* 2009;41(5):447–451.
58. Takuma K, Yao J, Huang J, et al. ABAD enhances Abeta-induced cell stress via mitochondrial dysfunction. *FASEB J.* 2005;19(6):597–598.
59. Lustbader JW, Cirilli M, Lin C, et al. ABAD directly links Abeta to mitochondrial toxicity in Alzheimer's disease. *Science.* 2004;304(5669):448–452.
60. Cho DH, Nakamura T, Fang J, et al. S-nitrosylation of Drp1 mediates beta-amyloid-related mitochondrial fission and neuronal injury. *Science.* 2009;324(5923):102–105.
61. Wang X, Su B, Siedlak SL, et al. Amyloid-beta overproduction causes abnormal mitochondrial dynamics via differential modulation of mitochondrial fission/fusion proteins. *Proc Natl Acad Sci U S A.* 2008;105(49):19318–19323.
62. Wang X, Su B, Lee HG, et al. Impaired balance of mitochondrial fission and fusion in Alzheimer's disease. *J Neurosci.* 2009;29(28):9090–9103.
63. Wang X, Su B, Fujioka H, Zhu X. Dynamin-like protein 1 reduction underlies mitochondrial morphology and distribution abnormalities in fibroblasts from sporadic Alzheimer's disease patients. *Am J Pathol.* 2008;173(2):470–482.
64. Manczak M, Calkins MJ, Reddy PH. Impaired mitochondrial dynamics and abnormal interaction of amyloid beta with mitochondrial protein Drp1 in neurons from patients with Alzheimer's disease: implications for neuronal damage. *Hum Mol Genet.* 2011;20(13):2495–2509.
65. DuBoff B, Gotz J, Feany MB. Tau promotes neurodegeneration via DRP1 mislocalization in vivo. *Neuron.* 2012;75(4):618–632.
66. Shahpasand K, Uemura I, Saito T, et al. Regulation of mitochondrial transport and inter-microtubule spacing by tau phosphorylation at the sites hyperphosphorylated in Alzheimer's disease. *J Neurosci.* 2012;32(7):2430–2441.
67. Kopeikina KJ, Carlson GA, Pitstick R, et al. Tau accumulation causes mitochondrial distribution deficits in neurons in a mouse model of tauopathy and in human Alzheimer's disease brain. *Am J Pathol.* 2011;179(4):2071–2082.
68. Perez MJ, Vergara-Pulgar K, Jara C, Cabezas-Opazo F, Quintanilla RA. Caspase-cleaved tau impairs mitochondrial dynamics in Alzheimer's disease. *Mol Neurobiol.* 2018;55(2):1004–1018.
69. Cai Q, Jeong YY. Mitophagy in Alzheimer's disease and other age-related neurodegenerative diseases. *Cells.* 2020;9(1):150.
70. Pickrell AM, Youle RJ. The roles of PINK1, parkin, and mitochondrial fidelity in Parkinson's disease. *Neuron.* 2015;85(2):257–273.
71. Lazarou M, Sliter DA, Kane LA, et al. The ubiquitin kinase PINK1 recruits autophagy receptors to induce mitophagy. *Nature.* 2015;524(7565):309–314.
72. Chen RH, Chen YH, Huang TY. Ubiquitin-mediated regulation of autophagy. *J Biomed Sci.* 2019;26(1):80.
73. Fivenson EM, Lautrup S, Sun N, et al. Mitophagy in neurodegeneration and aging. *Neurochem Int.* 2017;109:202–209.
74. Palikaras K, Lionaki E, Tavernarakis N. Mechanisms of mitophagy in cellular homeostasis, physiology and pathology. *Nat Cell Biol.* 2018;20(9):1013–1022.
75. Ryu D, Mouchiroud L, Andreux PA, et al. Urolithin a induces mitophagy and prolongs lifespan in *C. elegans* and increases muscle function in rodents. *Nat Med.* 2016;22(8):879–888.
76. Xie C, Zhuang XX, Niu Z, et al. Amelioration of Alzheimer’s disease pathology by mitophagy inducers identified via machine learning and a cross-species workflow. *Nat Biomed Eng.* 2022;6:76–93.
77. Reddy PH, Oliver DM. Amyloid beta and phosphorylated tau-induced defective autophagy and mitophagy in Alzheimer's disease. *Cells.* 2019;8(5):488.

78. Cummins N, Tweedie A, Zuryn S, Bertran-Gonzalez J, Gotz J. Disease-associated tau impairs mitophagy by inhibiting Parkin translocation to mitochondria. *EMBO J.* 2019;38(3):e99360.
79. Du F, Yu Q, Yan S, et al. PINK1 signalling rescues amyloid pathology and mitochondrial dysfunction in Alzheimer's disease. *Brain.* 2017;140(12):3233–3251.
80. Lautrup S, Sinclair DA, Mattson MP, Fang EF. NAD(+) in brain aging and neurodegenerative disorders. *Cell Metab.* 2019;30(4):630–655.
81. Fang EF, Kassahun H, Croteau DL, et al. NAD(+) replenishment improves lifespan and healthspan in ataxia telangiectasia models via mitophagy and DNA repair. *Cell Metab.* 2016;24(4):566–581.
82. Fang EF, Hou Y, Lautrup S, et al. NAD(+) augmentation restores mitophagy and limits accelerated aging in Werner syndrome. *Nat Commun.* 2019;10(1):5284.
83. Fang EF. Mitophagy and NAD(+) inhibit Alzheimer disease. *Autophagy.* 2019;15(6):1112–1114.
84. Lee IH, Cao L, Mostoslavsky R, et al. A role for the NAD-dependent deacetylase Sirt1 in the regulation of autophagy. *Proc Natl Acad Sci U S A.* 2008;105(9):3374–3379.
85. Huang R, Xu Y, Wan W, et al. Deacetylation of nuclear LC3 drives autophagy initiation under starvation. *Mol Cell.* 2015;57(3):456–466.
86. Sedlackova L, Otten EG, Scialo F, I. KV. Autophagy promotes cell and organismal survival by maintaining NAD(H) pools. *bioRxiv.* 2020. https://doi.org/10.1101/2020.01.31.928424.
87. Kingwell K. Turning up mitophagy in Alzheimer disease. *Nat Rev Drug Discov.* 2019;18:252.
88. Silva MC, Nandi GA, Tentarelli S, et al. Prolonged tau clearance and stress vulnerability rescue by pharmacological activation of autophagy in tauopathy neurons. *Nat Commun.* 2020;11(1):3258.
89. Li Q, Gao S, Kang Z, et al. Rapamycin enhances mitophagy and attenuates apoptosis after spinal ischemia-reperfusion injury. *Front Neurosci.* 2018;12:865.
90. Andreux PA, Blanco-Bose W, Ryu D, et al. The mitophagy activator urolithin A is safe and induces a molecular signature of improved mitochondrial and cellular health in humans. *Nat Metab.* 2019;1:593–603.
91. Yoshino M, Yoshino J, Kayser BD, et al. Nicotinamide mononucleotide increases muscle insulin sensitivity in prediabetic women. *Science.* 2021;372:1224–1229.
92. Price JL, Davis PB, Morris JC, White DL. The distribution of tangles, plaques and related immunohistochemical markers in healthy aging and Alzheimer's disease. *Neurobiol Aging.* 1991;12(4):295–312.
93. Manczak M, Kandimalla R, Yin X, Reddy PH. Hippocampal mutant APP and amyloid beta-induced cognitive decline, dendritic spine loss, defective autophagy, mitophagy and mitochondrial abnormalities in a mouse model of Alzheimer's disease. *Hum Mol Genet.* 2018;27(8):1332–1342.
94. Iba M, Guo JL, McBride JD, Zhang B, Trojanowski JQ, Lee VM. Synthetic tau fibrils mediate transmission of neurofibrillary tangles in a transgenic mouse model of Alzheimer's-like tauopathy. *J Neurosci.* 2013;33(3):1024–1037.
95. Sara SJ. The locus coeruleus and noradrenergic modulation of cognition. *Nat Rev Neurosci.* 2009;10(3):211–223.
96. Bacon TJ, Pickering AE, Mellor JR. Noradrenaline release from locus coeruleus terminals in the hippocampus enhances excitation-spike coupling in CA1 pyramidal neurons via beta-adrenoceptors. *Cereb Cortex.* 2020;30(12):6135–6151.
97. Palacios-Filardo J, Mellor JR. Neuromodulation of hippocampal long-term synaptic plasticity. *Curr Opin Neurobiol.* 2019;54:37–43.

98. Braak H, Del Tredici K. The preclinical phase of the pathological process underlying sporadic Alzheimer's disease. *Brain.* 2015;138(Pt 10):2814–2833.
99. Mather M, Harley CW. The locus coeruleus: essential for maintaining cognitive function and the aging brain. *Trends Cogn Sci.* 2016;20(3):214–226.
100. Eschenko O, Mello-Carpes PB, Hansen N. New insights into the role of the locus coeruleus-noradrenergic system in memory and perception dysfunction. *Neural Plast.* 2017;2017:4624171.
101. Gervasoni D, Darracq L, Fort P, Souliere F, Chouvet G, Luppi PH. Electrophysiological evidence that noradrenergic neurons of the rat locus coeruleus are tonically inhibited by GABA during sleep. *Eur J Neurosci.* 1998;10(3):964–970.
102. Sanchez-Padilla J, Guzman JN, Ilijic E, et al. Mitochondrial oxidant stress in locus coeruleus is regulated by activity and nitric oxide synthase. *Nat Neurosci.* 2014;17(6):832–840.
103. Prinz PN, Vitaliano PP, Vitiello MV, et al. Sleep, EEG and mental function changes in senile dementia of the Alzheimer's type. *Neurobiol Aging.* 1982;3(4):361–370.
104. Hope T, Keene J, Gedling K, Fairburn CG, Jacoby R. Predictors of institutionalization for people with dementia living at home with a carer. *Int J Geriatr Psychiatry.* 1998;13(10):682–690.
105. Holth JK, Fritschi SK, Wang C, et al. The sleep-wake cycle regulates brain interstitial fluid tau in mice and CSF tau in humans. *Science.* 2019;363(6429):880–884.
106. Nakahata Y, Sahar S, Astarita G, Kaluzova M, Sassone-Corsi P. Circadian control of the NAD+ salvage pathway by CLOCK-SIRT1. *Science.* 2009;324(5927):654–657.
107. Chang HC, Guarente L. SIRT1 mediates central circadian control in the SCN by a mechanism that decays with aging. *Cell.* 2013;153(7):1448–1460.
108. Levine DC, Hong H, Weidemann BJ, et al. NAD(+) controls circadian reprogramming through PER2 nuclear translocation to counter aging. *Mol Cell.* 2020;78(5):835–849 e837.
109. Yuan J, Amin P, Ofengeim D. Necroptosis and RIPK1-mediated neuroinflammation in CNS diseases. *Nat Rev Neurosci.* 2019;20(1):19–33.
110. Wiley CD, Velarde MC, Lecot P, et al. Mitochondrial dysfunction induces senescence with a distinct secretory phenotype. *Cell Metab.* 2016;23(2):303–314.

CHAPTER 6

Endocytosis in β-amyloid biology and Alzheimer's disease

Bradlee L. Heckmann[a,b,c,*] **and Douglas R. Green**[d,*]
[a]USF Health Neuroscience Institute, Tampa, FL, United States
[b]Department of Molecular Medicine, University of South Florida Morsani College of Medicine, Tampa, FL, United States
[c]Byrd Alzheimer's Center, Tampa, FL, United States
[d]Department of Immunology, St. Jude Children's Research Hospital, Memphis, TN, United States

1. Introduction

Alzheimer's disease (AD) is one of the most prevalent and devasting neurodegenerative conditions, characterized by progressive pathological neuronal loss leading to severe impairments in memory and cognition.[1] From an overly simplistic view, AD is often held as a spontaneous age-associated pathology with the emergence of symptomatic profiles measurable in elderly individuals of 65+ years of age (late-onset AD). However, highly aggressive inherited or genetically linked AD, referred to as early-onset AD, presents as early as 40–50 years of age and is often accompanied by an exaggerated time course of disease progression compared to late-onset disease. While the factors leading to disease establishment and ultimately disease progression differ between early- and late-onset AD, both forms of disease are characterized by the deposition of the neurotoxic peptide β-amyloid (Aβ) into amyloid plaques, hallmark features of AD.[1,2]

Neurotoxic Aβ has long been considered as the primary driving factor for AD pathology, often referred to as the amyloid hypothesis.[3] Aβ found in the AD brain varies between individuals but is most commonly composed of two predominant forms, $A\beta_{1-40}$ and $A\beta_{1-42}$. In addition, other Aβ peptides are present in AD and have been shown to contribute to disease pathology.[1,4–6] Rigorous studies over the past two decades have provided convincing evidence illustrating the processing of amyloid precursor protein (APP) into various Aβ peptide states. APP is a critically important type-I transmembrane glycoprotein in neurons that has a variety of proposed

*These authors contributed equally to this work.

Autophagy Dysfunction in Alzheimer's Disease and Dementia
https://doi.org/10.1016/B978-0-323-89906-2.00009-5

functions ranging from regulation of neuronal development and signaling to the maintenance of neuronal homeostasis in the adult organism and in development.[7,8] Aβ is produced by the proteolytic cleavage of transmembrane APP by β- and γ-secretases leading to release of cleaved Aβ intracellularly.[9,10] The amyloidogenic pathway of APP processing is complex and results in a variety of Aβ peptides that serve a variety of both known and unknown functions and has been well reviewed previously.[11]

How Aβ becomes neurotoxic and leads to the establishment of AD is poorly understood. Novel in vivo imaging techniques developed within the past 10 years have increased the complexity surrounding Aβ production and secretion by neurons, as mounting evidence demonstrating that while all AD patients have Aβ deposition, Aβ accumulation alone does not necessarily lead to AD pathogenesis as nonsymptomatic individuals with extraneuronal Aβ deposition have now been well documented.[12,13] These data bring into question the validity of the amyloid hypothesis.

Although debate regarding physiological versus pathological functions of Aβ remain, it is clear that accumulation of Aβ is ubiquitously present in the AD brain and is a major contributor to disease progression and pathology. An effort has been underway for the better part of the new millennium to target Aβ utilizing small molecule inhibition and more recently monoclonal antibody neutralization,[14–16] all with little success. Mounting evidence is demonstrating the importance of pathways such as phagocytosis and endocytosis in the recognition and clearance of extraneuronal Aβ and genetic evidence reveals alterations and impairments in these and associated pathways in the human AD brain.[17–19] Moreover, endocytic processes are known to function not only in the uptake of extra-neuronal Aβ but also play distinct roles in the processing and production of Aβ. Herein, we focus our efforts on detailing the endocytic mechanisms responsible for the processing, production, and clearance of neurotoxic Aβ and the effects of modulating these pathways on AD pathology.

2. The flavors of endocytosis in the brain

As with other cell and tissue types, endocytosis in the brain refers to a variety of pathways that result in the internalization of extracellular cargoes and receptors into the cell. The most predominant of these pathways are clathrin-mediated endocytosis (often referred to as receptor-mediated endocytosis and hereafter endocytosis) and phagocytosis, both are extensively described in this chapter. As with other cell types, neuronal and glial endocytosis is regulated by a variety of multiprotein complexes and occurs in a stepwise fashion. While these regulatory pathways have been well reviewed in full detail,[20–23] in brief receptor ligation on the plasma membrane results in signaling leading to the recruitment of intracellular clathrin and the formation of a clathrin-coated vesicle or pit. This clathrin cage is rapidly shed following vesicle internalization and the vesicle subsequently fuses with early endosomes for downstream trafficking within the cell. Although ubiquitous across cell types, the machinery that regulates retrieval, degradation, or recycling of cargo at the early endosome is of particular importance in neurons and microglia. Sorting machinery including the retromer-complex, retriever-complex, ESCRT machinery, and the WASH-complex play defined roles in cargo processing at the early endosome and direct downstream end points for endosomal cargoes and internalized receptors (Fig. 1).

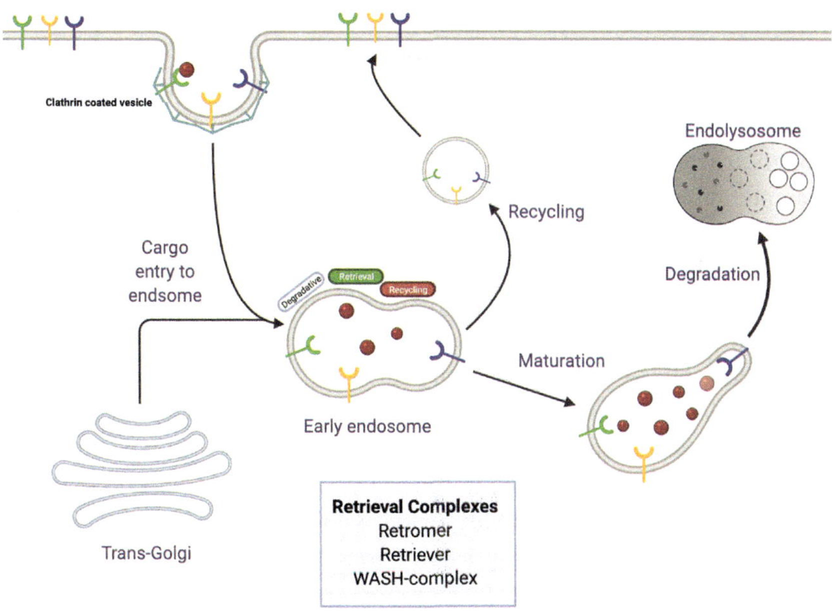

Fig. 1 End points of endocytosis. Endocytic cargoes are derived from extracellular uptake following receptor ligation or from transport from the trans-Golgi network. Machinery at the level of the early endosome is responsible for the retrieval and sorting of cargoes in an end point-dependent manner. Complexes involved in sorting include retromer, retriever for endosomal maturation and/or recycling, and the WASH-complex primarily for cargo degradation. Subsets of cargo are sorted for endosomal maturation and eventual lysosomal fusion and degradation. Receptors or receptor-cargo complexes can likewise be shuttled for recycling to the plasma membrane. All three major endosomal end points are observed in Aβ processing or clearance as described in the main text.

In the context of Aβ production, components of the retromer-complex are of particular importance as further discussed herein. At its core, the retromer-complex contains the endocytic proteins VPS26, VPS29, sorting nexins (SNXs), and anchored by VPS35, the master conductor of endosomal sorting.[24]

Mutations in the retromer-complex, in particular within VPS35, are known risk factors for neurodegenerative conditions including Parkinson's disease and perhaps roles in AD.[25,26] Furthermore, it is believed that other endosomal processes downstream of cargo sorting, including recruitment of WASH-complex components for lysosomal targeting, are dependent on the assembly of at least certain components of the retromer machinery.[27] Therefore, alterations in retromer function have a broad impact on endocytic function beyond simple cargo recognition at the early endosome. As endosomal maturation occurs, distinct roles for the Rab-GTPase family of proteins are required for proper trafficking of vesicles. Although Rab proteins including Rab5 and Rab7 are shared between a variety of vesicular pathways, including crosstalk between endosomes, autophagosomes, and lysosomes, mutations in Rab7, have been identified in the AD brain and thought to contribute to not only aberrant Aβ production but impaired Aβ clearance, discussed in more detail later.[28]

Falling under the umbrella of endocytosis is the process of phagocytosis. Long thought to be the primary mechanism of Aβ recognition and clearance by innate immune cells of the brain, the microglia, phagocytosis is now known to simply be one mechanism of clearance. Unlike receptor-mediated endocytosis, phagocytosis is typically restricted to professional phagocytic cells of the immune system, which include microglia. These macrophage-like cells are the brain's primary defense against any immunological perturbation ranging from a pathogen or dying cell to Aβ and other protein aggregates.[29] Although microglia possess many of the same properties as their peripheral counterparts, microglia are distinct from macrophages and arise from a unique developmental lineage.[30] Moreover, brain region-specific microglia can differ dramatically with respect to their genetic signatures and abilities to phagocytose or endocytose cargoes. This heterogeneity within microglial populations is further exacerbated under pathological conditions including those found in AD. Regulated by similar machinery that is found in the endocytic pathway, phagocytosis is a delineated process targeting the uptake or internalization of cargoes greater than 0.5 μm in size.[31] Unlike endocytosis, phagocytic uptake is independent of clathrin-pit formation and does not have a variety of intracellular end points, rather is defined as the directed delivery of cargoes to lysosomes for cargo degradation.[31] The phagocytic pathway plays no role in the production of Aβ, but is relevant for the recognition and clearance of Aβ. Moreover, unlike endocytosis, phagocytosis requires actin polymerization and actin re-organization, illustrating another level of distinction between these molecular entities.[31]

A concerted effort between canonical endocytic processes and phagocytosis exists, and deregulation of these pathways is a key contributor to the development of Aβ deposition and ultimately AD pathology.

3. The MO(F) of Aβ

To understand the endocytic processes that recognize and govern Aβ processing and eventual clearance, it is important to first understand the various forms Aβ can assume. As a nascently formed peptide, Aβ is found as a soluble monomeric (M) species. However, human Aβ monomers rapidly aggregate into a variety of assemblies ranging from oligomers and protofibrils, to amyloid fibrils found in amyloid plaques in AD. Due to the intrinsically unstructured and disordered nature of Aβ, common methods for peptide crystallization have proven difficult. Recent advances in NMR technology have revealed slight nuisances in the structure of monomeric Aβ species with the hydrophobic C-terminus of Aβ being critical in conformational reorganization from alpha-helical to β-sheet structure that plays a central role in determining peptide aggregation state in AD.[32,33]

The primary intermediate form of Aβ aggregation is oligomer (O) formation. These oligomers typically appear as spherical aggregates that subsequently elongate into mature Aβ fibrils (f).[32,33] Like monomeric Aβ, oligomeric Aβ is mostly soluble in nature and has been suggested to be a major neurotoxic form of the peptide.[10,34] While early studies suggested that Aβ plaques are the causative agents in AD pathology, newer data have suggested that the formation of insoluble fibrils and plaques might serve as a protective mechanism to sequester aberrantly cleaved Aβ once extricated from neurons into the extraneuronal space.[10,34] Evidence demonstrating the high toxicity of both soluble monomeric and oligomeric Aβ on cell survival as well as inflammatory activation in the brain lends support to this idea.[35,36] Soluble Aβ has been shown to bind to a host of other molecules as well as surface receptors (see below) on microglia.[37–40] Additionally, soluble Aβ binds a variety of receptors found on neurons including β-adrenergic receptor, Aβ-binding p75 neurotrophin receptor, low-density lipoprotein receptor-related protein 1 (LRP1), NMDAR, and others.[17] Binding of Aβ to these receptors is thought to be fundamental in the transduction of neurotoxic signals into neurons leading to events such as mitochondrial dysfunction, ER stress, and ultimately apoptotic death.

In its native state, APP is a single-pass transmembrane protein that takes residence in the plasma membrane of neuronal axons following synthesis in the Golgi.[41] The critical steps in APP processing occur at the plasma membrane and within the trans-Golgi network (TGN). Newly synthesized APP can either be transported to the plasma membrane or directly shuttled into the endosomal compartment. Both of these events are mediated by clathrin-coated vesicles. Plasma membrane APP can be directly proteolyzed by α-secretase and then γ-secretase that does not result in the production of Aβ, or can be reinternalized in clathrin-coated pits. This latter reinternalization event facilitates the exposure of APP in an endosomal compartment containing the proteases β-secretase and γ-secretase.[42,43] Cleavage by these secretases results in the production of Aβ peptides that are either degraded through the lysosome or deposited into the extracellular space via the endocytic recycling pathway.[10] This deposition of Aβ into the extracellular brain interstitial fluid (ISF) is the event that leads to the hallmark plaque formation found in AD.

The intimate need for the endocytic machinery in the processing of mature APP into Aβ illustrates a potential therapeutic target for limiting the aberrant cleavage of APP. Pharmacological or genetic modulation of key endocytic regulators has been shown to alter APP processing. Interestingly, when and where endocytosis is interrupted results in diverse outcomes with respect to Aβ production and extraneuronal deposition. Inhibition of early stages of endocytosis, by targeting multivesicular body (MVB) production via depletion of Tsg101, leads to retention of APP and limits Aβ production.[41] In contrast, inhibiting later stages of the MVB pathway through ablation of VPS4 or CHMP6 shuttles APP to the TGN where increased processing of APP into Aβ occurs.[41] Moving further down the process, inhibition of retromer-mediated recycling of APP through ablation of VPS35 essentially halts the production of Aβ suggesting that VPS35 and likely the retromer complex as a whole is required for Aβ production.[41] These findings suggest that "stalling or interrupting" the bidirectional recycling process results in a net intracellular accumulation of APP in early endosomes and prevents its ability to be cleaved into Aβ peptides.

It is clear that endocytic processes play a direct role in the production of Aβ; however, how this necessarily applies to human AD pathology remains a question, as most individuals do not have aberrant Aβ deposition within their brains. This is likely due to two facets, (1) variances in Aβ production and/or (2) differences in Aβ clearance. It is evident from human studies that Aβ deposition is most often correlated with age. Similarly, genome wide analyses have revealed that components of the endocytic machinery have increased expression with age and are further elevated in the human AD brain. Furthermore, this upregulation of endocytic genes is not limited to minor regulators, but rather is highlighted by the major foundational components of the machinery, genes including clathrin, dynamin-1, Rab5, and caveolin-2.[19,21] In comparison, APP expression in aged and AD brains remained statistically equivalent. Aβ deposition, in contrast, is highly correlated with the upregulation of the endocytic regulators.[17] Although upregulation of neuronal endocytic pathways is not the only contributing factor to Aβ production, changes in secretase expression and activity occur as well and are definitive contributing components to Aβ processing and aberrant deposition.

Recent evidence demonstrates further correlations between elevated endocytic regulation, Aβ production, and neuronal signaling. The role for the endocytic machinery in neuronal signaling and synaptic transmission is well defined. It is tempting to hypothesize that increased endocytic function in neurons with age and further in AD brains is a compensatory mechanism to maintain homeostatic signaling processes and that this heightened endocytic function in synaptic activity may have a deleterious effect on APP homeostasis, resulting in changes to Aβ production. Interestingly, supporting this idea is the observation that approximately 70% of Aβ arises from endocytic-associated mechanisms, with a significant majority of that Aβ pool being dependent on synaptic activity.[44] Together, this suggests that changes in endocytic signaling and increased synaptic activity with age is sufficient to "prime" neurons for aberrant Aβ production.

4. Endocytosis in a phagocytic world

While clathrin-mediated endocytosis is the major endocytic process in neurons and in the processing, cleavage, and deposition of Aβ, the roles of endocytosis in the recognition and clearance extra-cellular Aβ are less clear. Early studies investigating the recognition and uptake of amyloids were originally performed in the 1960s using models of peripheral amyloidosis and defined roles for phagocytosis in clearing and degrading amyloid fibrils by macrophages.[45,46] However, it was not until the early 1990s that evidence demonstrating recognition and phagocytic uptake of Aβ by microglia was first characterized.[47–49] Due to limitations in imaging technology and the early thought that all detrimental aspects of Aβ stem from the fibrillary plaque structures that are comparatively large in size, continued investigation of phagocytosis ensued.

It has now become widely accepted that microglia recognize and phagocytose large fibrillary Aβ deposits, and this process is robustly increased when fibrillary Aβ is bound to the complement fragment, C3b.[50] Furthermore, elegant studies have revealed multiple surface receptors that are now implicated in the recognition phase of Aβ by microglial cells (Fig. 2). Of particular importance is Triggering Receptor Expressed on Myeloid Cells 2 (TREM2), wherein polymorphisms in the coding region, associated with late onset AD, reduce protein maturation and reduce protein maturation and ultimately impair the phagocytic capacity of TREM2-expressing microglia.[51,52] More recent evidence has illustrated that TREM2 may function to direct microglial association with Aβ plaques, effectively surrounding the plaque, limiting Aβ effects, and preventing neuritic damage.[51] The role of TREM2 in recognizing Aβ goes beyond simply identifying Aβ as a cargo and has more intrinsic roles within immune regulation. In response to Aβ, microglia often upregulate proinflammatory immune cytokines including the activation of interleukin-1 signaling and TNFa signaling pathways, resulting in feedback signaling which itself can alter the phagocytic and

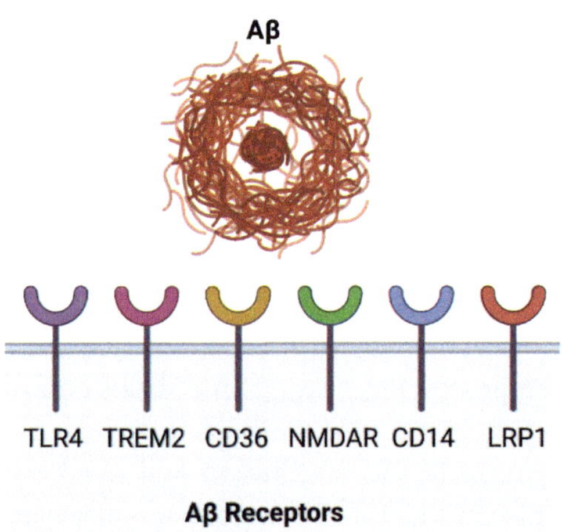

Fig. 2 Major Aβ receptors found primarily on microglia as well as neurons and astrocytes. Heterodimerization of these receptors has been shown, and they likely function in concert in the recognition of Aβ.

endocytic capacity of microglia.[53,54] Deletion or mutation of microglial TREM2 drastically impacts TREM2-dependent Aβ recognition and its ability to antagonize NFkB activation and signaling, resulting in dysregulated inflammation.[51]

Toll-like receptors have likewise been shown to be involved in the recognition of Aβ across all peptide forms including monomeric, oligomeric, and fibrillary. Of particular importance are TLR2 and TLR4, both of which are directly involved in Aβ uptake along with TREM2 and the immune scavenger receptor, Leukocyte Differentiation Antigen CD36 (CD36), as well as immune receptor cluster of differentiation 14 (CD14).[55,56] Knockout of TLR2 or TLR4 in mice leads to impaired Aβ clearance and increased Aβ deposition in brain.[57–59] Signaling through TLRs has long been shown to activate inflammatory pathways in response to microbial infection. Similarly, activation of TLRs by Aβ peptides contributes to immune polarization of microglia.[57–59]

The consequences of immune activation following Aβ recognition are multimodal in nature. Direct effects stemming from inflammatory cytokine production include neuronal impairment and eventual induction of neurodegeneration. Similarly detrimental is the autocrine impact of inflammation on microglial function. Inflammatory mediators and NFkB pathway activation have the potential to greatly reduce the ability of microglia to internalize cargoes including Aβ as described earlier. Taken together, these facets result in the establishment of a progressive reactive feedback loop that often is unresolvable in AD pathology.

Interestingly, most of the work regarding microglial dynamics and activation was in the context of Aβ phagocytosis. However, improved mouse modeling of AD and new imaging techniques have revealed that Aβ phagocytosis is only one aspect of Aβ recognition and clearance. Over the past few years, significant evidence has emerged demonstrating a primary role for canonical clathrin-mediated endocytic pathways in Aβ clearance.[56,60–63] Unlike phagocytic recognition of fibrillary Aβ, endocytic uptake of monomeric and oligomeric species of Aβ appears to be of critical importance to brain health. This area has been spurred by the realization that monomeric and oligomeric Aβ species are often more neurotoxic than their fibrillary, plaque associated counterparts that appear to be mostly inert.[64] Intraneuronal accumulation and neurotoxicity associated with oligomeric Aβ are dependent on clathrin-mediated endocytosis. Blocking dynamin-mediated vesicle trafficking limits intraneuronal oligomeric Aβ deposition and reduces neurotoxicity.[63]

Recognition and uptake of oligomeric Aβ by microglia occurs in a similar manner. Following receptor ligation, oligomeric Aβ internalization occurs through a clathrin-dependent process. Interestingly, internalization of monomeric Aβ can occur through both clathrin-dependent and -independent endocytic pathways.[17] Consistent with these nonfibriallary forms of Aβ, endocytosis of fibrillary Aβ peptides has also been observed. TLR4 and CD14 recognition of Aβ can promote clathrin-mediated endocytosis of Aβ fibrils.[62] The presence of clathrin on these Aβ fibrillary containing vesicles provides definitive evidence of delineation from canonical phagocytosis. More importantly, CD14-positive microglia have been shown to increase in response to aging, providing more credence to global endocytic pathway upregulation with age.[42]

Adding more complexity to the recognition and clearance of Aβ peptides by microglia is the observation that Aβ itself, specifically the 1–40 and 1–42 species can directly impair phagocytic and endocytic capacity in the microglia and neurons. One way in which Aβ itself interrupts the clearance of Aβ peptides is through altering the trafficking of master transcription factors such as TFEB that are required for lysosomal biogenesis and homeostasis.[65] Aβ 1–42 peptide reduces the nuclear translocation of TFEB and decreases downstream expression of TFEB targets including osteoporosis-associated transmembrane protein 1 (OSTM1).[65] OSTM1 is crucial for regulating lysosomal acidification in microglial cells, and reduction in OSTM1 functional capacity restricts lysosomal function.[65] This Aβ-induced lysosomal dysfunction subsequently leads to intravesical accumulation of Aβ and a propagation and exacerbation in inflammatory signaling paired with continual extracellular Aβ accumulation.

Defects in both the phagocytic and endocytic recognition and clearance of Aβ by microglia are now bonafide contributors to AD pathology. Both pathways likely work in tandem to limit an amyloid burden due to increase processing and production in neurons. By having the ability to recognize and internalize varying forms of Aβ monomers, oligomers, fibrils, and plaques, this multifaceted endocytic system endows microglia with means to prevent aberrant Aβ deposition. In the context of AD, this system is essentially overrun due to effects of Aβ on processes such as inflammation in addition to direct consequences and organelle impairment stemming from direct effects of Aβ.

5. Adding a modifier, a new world for endocytosis in the AD brain

From studies over the better part of the past half-century, it has become apparent that endocytic processes are central components in Aβ production and AD pathology. Nuances in phagocytosis versus endocytosis that have recently been revealed and recognition of varying Aβ peptides has strengthened our understanding of how neurons produce Aβ and how microglia attempt to deal with increased Aβ burdens. Over the past decade, a unique modification to membrane vesicles was identified in the context of phagosomes. A well-described mediator of the autophagy pathway is the small microtubule-associated protein light-chain 3 (LC3) family of small proteins.[66] LC3 is processed by multiple complexes referred to as the autophagy lipidation machinery that ultimately conjugates ATG4-cleaved LC3 to phosphatidylethanolamine (PE) in the autophagosomal membrane.[67] This lipidation of LC3 to autophagosomes functions in intracellular cargo recognition and facilitates vesicle trafficking and lysosomal fusion, resulting in cargo degradation.[67] This canonical process of LC3 lipidation in autophagy is required for the completion of autophagy and degradation of autophagosomal cargoes.

While crosstalk between the autophagy and endosomal pathways was identified and characterized previously, it was not until the discovery of LC3-associated phagocytosis (LAP) that distinct LC3 lipidation events on single membranes were observed.[68–70] In LAP, a subset of the autophagy machinery lipidates LC3 directly on phagosomes in a process that is distinct from canonical autophagy and requires unique initiating components including the protein Rubicon.[68–70] Abrogation of this LC3 lipidation in LAP proved to have severe consequences in peripheral macrophages. Inhibition of LAP via deletion of the primary regulatory protein Rubicon resulted in a significant impairment in cargo degradation and promoted robust inflammatory activation, even in response to typically immune silent cargoes such as apoptotic cell corpses.[71] Unlike autophagy, activation of LAP in macrophages is not dependent on the AMPK-mTORC1-ULK1 signaling axis and does not require the FIP200 anchored autophagy initiation complex.[68,69] A majority of other autophagy genes (ATGs) are however required (ATG5, 7, 12, 16, and others) for LAP.[68,69]

The effects of LAP abrogation, including those associated with alterations in immune polarization mentioned above, likely stem from variations in pathway cargo. Unlike autophagy, LAP cargo are extracellular in nature. A variety of ligands that stimulate the recruitment of Rubicon to the developing phagophore have been identified and include a diverse cohort ranging from pathogen moieties and dying cells to unique immune complexes.[68,69] To date, LAP activation has best been characterized by the ligation of dying cells to receptors including TIM4 as well as activation through binding of TLR2/6 and TLR4 for pathogen recognition.[68,69] Based on the involvement of the TLR family of immune receptors, one may hypothesize that LAP functions similarly in microglial cells for the recognition of extracellular cargoes in the brain.

Initial studies evaluated the role of LAP in microglia in the recognition and uptake of Aβ.[56,72,73] Abrogation of the LAP pathway via ablation of Rubicon in the 5xFAD mouse model of Aβ deposition and AD resulted in a robust exacerbation in both the timeframe and the amount of Aβ accumulation.[56] The loss of LAP further increases inflammatory cytokine production by microglia and microglial-specific deletion of the LAP pathway, which substantiated these results.[56] Commensurate increases in the hyperphosphorylation of tau, a major AD hallmark and severe behavioral and memory impairments, are observed in LAP-deficient, 5xFAD mice.[56] However, discrepancies exist between Aβ-mediated LAP activation and stimulation with other cargoes such as yeast particles. Mainly, Aβ induces LC3 lipidation on vesicles, but abrogation of LAP does not alter the delivery of Aβ to the lysosome or the subsequent degradation of that Aβ by lysosomal hydrolases (Fig. 3A and B).[56] Further investigation shows that what were presumed to be $LC3^+$ "LAPosomes" containing Aβ are in fact distinct molecular entities that are resistant to phagosomal inhibition and are positive for endocytic markers including clathrin (Fig. 3C).[56]

These distinct clathrin-coated, $LC3^+$ vesicles containing Aβ are endosomal in nature, and impairment of this novel pathway leads to severe restrictions in the recycling of a variety of receptors, including major Aβ receptors TREM2, TLR4, and CD36 (Fig. 4).[56] Although this pathway shares the molecular components that regulate LAP, it is functionally divergent and is referred to as LC3-associated endocytosis (LANDO). The discovery and characterization of LANDO mark a paradigm shift in our understanding of how Aβ is recognized and has increased our understanding of how Aβ functions to alter immune pathways in the brain. The physical process of recognition and uptake of Aβ by microglial is no longer the sole prime determinant of Aβ flux, rather modifications to the canonical endocytic pathway, including LANDO activation have key biological relevance. Furthermore, it demonstrates that simply clearing an aberrant aggregate may not be sufficient to prevent detrimental consequences. LANDO is protective by not only mitigating neuroinflammatory activation of microglia but is also critical to maintain a pool of Aβ receptors at the plasma membrane for the continued recognition and internalization of Aβ. The mechanisms by which LANDO is activated in response to particular forms of Aβ remain to be elucidated and delineated experiments evaluating the

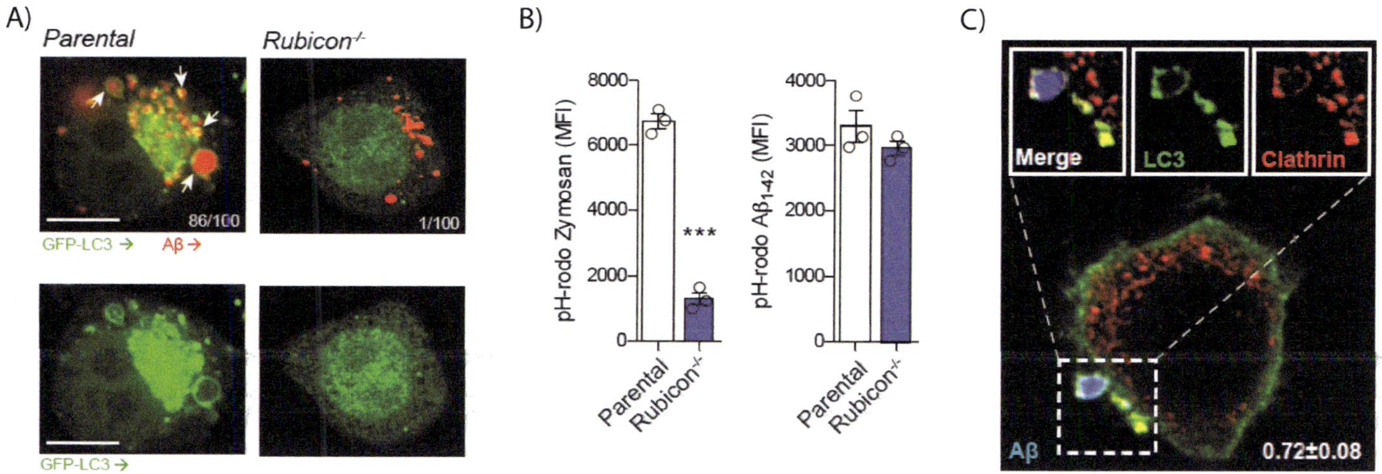

Fig. 3 (A) Representative fluorescence imaging showing LC3-recruitment (*green*) to vesicles containing Aβ (*red*) in LANDO-sufficient (parental) and LANDO-deficient (Rubicon$^{-/-}$) microglia. (B) Analysis of phagocytic zymosan or endocytic Aβ delivery to lysosomes using pH-sensitive, pH-rodo labeling in LANDO-sufficient versus LANDO-deficient microglia. (C) Immunofluorescence imaging for LC + Aβ vesicles copositive for endosomal clarthin. (Reproduced with permission from Elsevier: Heckmann BL, Teubner BJW, Tummers B, et al. LC3-associated endocytosis facilitates beta-amyloid clearance and mitigates neurodegeneration in murine Alzheimer's disease. *Cell.* 2019;178(3):536–551 e514.)

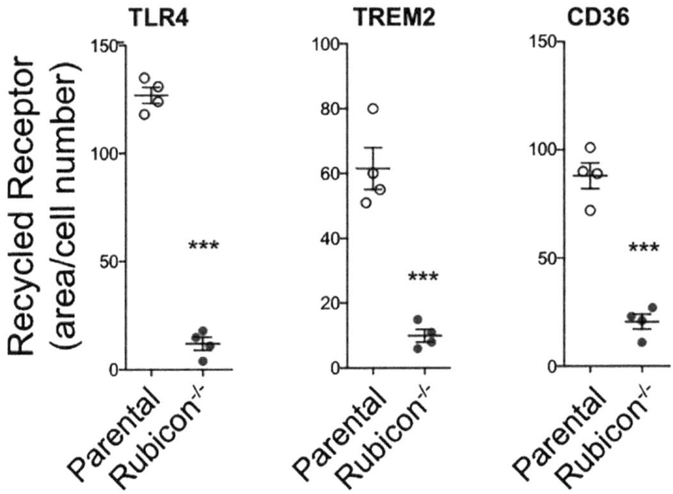

Fig. 4 Representative analysis of recycling for major Aβ receptors; TLR4, TREM2, and CD36 in LANDO-sufficient (parental) and LANDO-deficient (Rubicon$^{-/-}$) microglia. Impaired LANDO reduces receptor recycling.

activating potential of all forms of Aβ are needed. Nonetheless suppression of this key protective pathway through the downregulation of genes including Rubicon or mutation of the WD-domain of ATG16L1 with age is likely to contribute to the development and ultimately worsening of AD pathology. In mice, LANDO deficiency alone is sufficient for driving an age associated, spontaneous AD-like pathology with Aβ accumulation, tau hyperphosphorylation, neurodegeneration, and memory loss as observed in aged ATG16L1 WD-domain-deficient mice.[74] These findings further illustrate the true importance of endocytic regulation in AD.

6. Endocytosis in a starry world

The majority of focus on endocytic processes has been directed toward Aβ production in neurons or clearance by microglia. Endocytosis in other brain cells such as astrocytes has largely gone unexplored. Recent evidence illustrating the importance of astrocytes in AD has begun to emerge. More importantly, endocytosis of Aβ by astrocytes has been well documented, and multiple studies have been focused on understanding the mechanisms that govern Aβ recognition and endocytosis regulation. Silencing of LRP1 in astrocytes leads to the reduction in not only Aβ internalization but also results in the downregulation of multiple Aβ-degrading enzymes including the metalloproteases MMP2 and MMP3.[75] Furthermore, disruption of actin polymerization in astrocytes by depletion of metallothionein 3 (Mt3) demonstrated an obligatory need for actin homeostasis.[76] Mt3-deficient astrocytes have marked reduction in both Aβ 1–40 and 1–42 endocytosis. Consistent with these in vitro studies, Mt3 is downregulated in the human AD brain.[77] More recently, direct links between AD risk genes such as PICALM and APOE4 and Aβ endocytosis in astrocytes were identified. Studies using iPSC-derived human astrocytes show that APOE4 expressing astrocytes have fewer early endosomes but do not have alterations in recycling endosomes compared to healthy donor APOE3 expressing cells.[78] This results in a decrease in endocytic uptake of cargo. Interestingly, PICALM overexpression is able to reverse the APOE4-induced endocytic defect. This functional interaction between two AD genetic risk factors demonstrates unique regulation of conserved biological processes in AD and raises more questions into defining the relationship between endocytosis across varying cell types in the AD brain and the contribution of varying genetics on these processes.

7. Exploiting endocytosis for therapeutic gain

AD is one of the remaining conditions without a viable therapeutic intervention. Targeting endocytosis has been posed as a potential avenue to either limit the production of Aβ or bolster its degradation. However, all of these proposals are predicated on the assumption that the amyloid hypothesis is correct and that simply limiting Aβ accumulation will resolve the pathology. The complexity of the roles of endocytosis in the AD brain between neurons and microglia as well as the added layer of signaling alterations by pathways such as LANDO, inherently stymies enthusiasm for targeting endocytic processes. An additional layer is the critical necessity for endocytosis in synaptic transmission and neuronal function, and therefore simply inhibiting neuronal endocytosis would likely create more problems than it would solve. Other approaches to improve the function of endocytosis and possibly pathways such as LANDO in microglia are attractive, yet have their own drawbacks. One factor is the ability to restore the normal expression of a full set of regulatory proteins required for pathways including LANDO to function. Although challenges exist in targeting endocytic pathways in the AD brain, some of which are detailed above, it should not detract from the serious discussion of targeting endocytosis in AD.

8. Recycling full circle, a summary

Roles for the endocytic machinery in Aβ biology and AD have long been acknowledged, and for the better part of the past 30 years well characterized. However, the true breadth of the role played by endocytosis in AD with regard to being "simply a process in the disease" or a party to pathology has long been debated. As reviewed herein, there is mounting evidence demonstrating that endocytosis is indeed a key contributor to the establishment and progression of AD pathology. Neuronal endocytosis is a required entity in the processing of APP and aberrant production of Aβ peptides. The recycling endosome is critical for the delivery and ultimate secretion of Aβ into the brain interstitium. Moreover, distinct roles for endocytic mechanisms ranging from phagocytosis to receptor-mediated endocytosis have been illustrated in the recognition of Aβ by microglia. Newer evidence has revealed unique modifications to canonical forms of endocytosis in processes including LANDO and how pathways such as LANDO participate in mitigating aberrant immune activation, inflammation, and neurodegeneration.[79]

Over the years, elegant genetic studies and RNA sequencing analyses have corroborated genome wide changes in the machinery that regulates endocytosis in both murine models of AD as well as in human AD patients. Pairing these observations with new frontiers of investigation studying the role of endocytosis in other brain cells, including astrocytes, has provided data that parallels endocytic function in other glia and alterations in endocytic pathways consistent to those observed in neurons and microglia.

While this chapter is dedicated to the role of endocytosis with respect to Aβ, it is pertinent to illustrate the role which endocytosis plays in other aspects of AD, in particular in the spreading and aggregation of tau. Under physiological conditions, extracellular tau is internalized through endocytosis and mostly degraded through the lysosome.[80] While the receptor for tau has not been determined, newer data suggest that tau, at least in its monomeric state, binds LRP1 leading to internalization.[81] Hyperphosphorylation of tau dramatically reduces the affinity for LRP1-tau binding and is likely significant in the development of tau neurofibrillary tangles.[81] Furthermore, ApoE4 reduces the endocytic uptake of tau, leading to enhanced extracellular accumulation, consistent with what is observed with Aβ.[82] The loss of this physiological pathway of tau catabolism promotes tau uptake by other mechanisms that can result in the exacerbation of tau spreading and ultimately tau-mediated cognitive impairment.

When taken together, this surmounting evidence demonstrating the critical role of endocytic pathways at essentially every step of Aβ biology and furthermore many aspects of AD pathology highlight the importance for continued investigation into not only direct mechanisms which govern the cell-specific endocytic pathways, but moreover those that regulate the crosstalk with immune signaling, neuronal function, and cell death regulation. No matter how you traffic or recycle it, endocytosis is indeed intertwined in every aspect of Aβ and AD pathology from start to finish.

Acknowledgments

We would like to thank all members of the Heckmann and Green labs for their continued dedication to advancing science. This work was supported in part by the funding from the National Institute of Allergy and Infectious Disease and the National Cancer Institute under award numbers AI138492 and CA231423 to B.L.H. and AI40646 and CA231620 to D.R.G. The content herein is solely the responsibility of the authors and does not necessarily represent the official views of the National Institutes of Health.

Disclosures

B.L.H. and D.R.G. are coinventors on United States patent applications; 62/795,217 filed January 22, 2019, 62/797,564 filed January 28, 2019, and worldwide patent application PCT/IB2020/050504 filed January 22, 2020, through the United States Patent and Trademark Office. This work was supported by the National Institute of Allergy and Infectious Disease, the National Cancer Institute, and the National Institute of Neurological Diseases and Stroke under award numbers, R01NS124783 and L40CA231423 to B.L.H. and R01AI40646 and R35CA231620 to D.R.G. The content herein is solely the responsibility of the authors and does not necessarily represent the official views of the National Institutes of Health. B.L.H consults for Ventus Therapeutics and USA Prime Biotech and is the co-founder and CSO for Asha Therapeutics. D.R.G. consults for Ventus Therapeutics and Inzen Therapeutics.

References

1. Murphy MP, LeVine 3rd H. Alzheimer's disease and the amyloid-beta peptide. *J Alzheimers Dis.* 2010;19(1):311–323.
2. Breijyeh Z, Karaman R. Comprehensive review on Alzheimer's disease: causes and treatment. *Molecules.* 2020;25(24):5789.

3. Hardy J, Selkoe DJ. The amyloid hypothesis of Alzheimer's disease: progress and problems on the road to therapeutics. *Science*. 2002;297(5580):353–356.
4. Kwak SS, Washicosky KJ, Brand E, et al. Amyloid-beta42/40 ratio drives tau pathology in 3D human neural cell culture models of Alzheimer's disease. *Nat Commun*. 2020;11(1):1377.
5. Hammond TC, Xing X, Wang C, et al. beta-amyloid and tau drive early Alzheimer's disease decline while glucose hypometabolism drives late decline. *Commun Biol*. 2020;3(1):352.
6. Bloom GS. Amyloid-beta and tau: the trigger and bullet in Alzheimer disease pathogenesis. *JAMA Neurol*. 2014;71(4):505–508.
7. Sosa LJ, Caceres A, Dupraz S, Oksdath M, Quiroga S, Lorenzo A. The physiological role of the amyloid precursor protein as an adhesion molecule in the developing nervous system. *J Neurochem*. 2017;143(1):11–29.
8. Muller UC, Zheng H. Physiological functions of APP family proteins. *Cold Spring Harb Perspect Med*. 2012;2(2), a006288.
9. Zhang YW, Thompson R, Zhang H, Xu H. APP processing in Alzheimer's disease. *Mol Brain*. 2011;4:3.
10. O'Brien RJ, Wong PC. Amyloid precursor protein processing and Alzheimer's disease. *Annu Rev Neurosci*. 2011;34:185–204.
11. Chen GF, Xu TH, Yan Y, et al. Amyloid beta: structure, biology and structure-based therapeutic development. *Acta Pharmacol Sin*. 2017;38(9):1205–1235.
12. Selkoe DJ, Hardy J. The amyloid hypothesis of Alzheimer's disease at 25 years. *EMBO Mol Med*. 2016;8(6):595–608.
13. Kametani F, Hasegawa M. Reconsideration of amyloid hypothesis and tau hypothesis in Alzheimer's disease. *Front Neurosci*. 2018;12:25.
14. Rafii MS, Aisen PS. The search for Alzheimer disease therapeutics—same targets, better trials? *Nat Rev Neurol*. 2020;16(11):597–598.
15. Rafii MS, Aisen PS. Recent developments in Alzheimer's disease therapeutics. *BMC Med*. 2009;7:7.
16. Alexander GC, Emerson S, Kesselheim AS. Evaluation of Aducanumab for Alzheimer disease: scientific evidence and regulatory review involving efficacy, safety, and futility. *JAMA*. 2021;325(17):1717–1718.
17. Ries M, Sastre M. Mechanisms of Abeta clearance and degradation by glial cells. *Front Aging Neurosci*. 2016;8:160.
18. Cataldo AM, Peterhoff CM, Troncoso JC, Gomez-Isla T, Hyman BT, Nixon RA. Endocytic pathway abnormalities precede amyloid beta deposition in sporadic Alzheimer's disease and down syndrome: differential effects of APOE genotype and presenilin mutations. *Am J Pathol*. 2000;157(1):277–286.
19. Alsaqati M, Thomas RS, Kidd EJ. Proteins involved in endocytosis are upregulated by ageing in the normal human brain: implications for the development of Alzheimer's disease. *J Gerontol A Biol Sci Med Sci*. 2018;73(3):289–298.
20. Sole-Domenech S, Cruz DL, Capetillo-Zarate E, Maxfield FR. The endocytic pathway in microglia during health, aging and Alzheimer's disease. *Ageing Res Rev*. 2016;32:89–103.
21. Kumari S, Mg S, Mayor S. Endocytosis unplugged: multiple ways to enter the cell. *Cell Res*. 2010;20(3):256–275.
22. Galloway DA, Phillips AEM, Owen DRJ, Moore CS. Phagocytosis in the brain: homeostasis and disease. *Front Immunol*. 2019;10:790.
23. Cosker KE, Segal RA. Neuronal signaling through endocytosis. *Cold Spring Harb Perspect Biol*. 2014;6(2).
24. Seaman MN. The retromer complex—endosomal protein recycling and beyond. *J Cell Sci*. 2012;125(Pt 20):4693–4702.

25. Wen L, Tang FL, Hong Y, et al. VPS35 haploinsufficiency increases Alzheimer's disease neuropathology. *J Cell Biol.* 2011;195(5):765–779.
26. Vilarino-Guell C, Wider C, Ross OA, et al. VPS35 mutations in Parkinson disease. *Am J Hum Genet.* 2011;89(1):162–167.
27. McNally KE, Cullen PJ. Endosomal retrieval of cargo: retromer is not alone. *Trends Cell Biol.* 2018;28(10):807–822.
28. Ba L, Chen XH, Chen YL, et al. Distinct Rab7-related endosomal-autophagic-lysosomal dysregulation observed in cortex and hippocampus in APPswe/PSEN1dE9 mouse model of Alzheimer's disease. *Chin Med J (Engl).* 2017;130(24):2941–2950.
29. Li Q, Barres BA. Microglia and macrophages in brain homeostasis and disease. *Nat Rev Immunol.* 2018;18(4):225–242.
30. Chan WY, Kohsaka S, Rezaie P. The origin and cell lineage of microglia: new concepts. *Brain Res Rev.* 2007;53(2):344–354.
31. Uribe-Querol E, Rosales C. Phagocytosis: our current understanding of a universal biological process. *Front Immunol.* 2020;11:1066.
32. Xiao Y, Matsuda I, Inoue M, Sasahara T, Hoshi M, Ishii Y. NMR-based site-resolved profiling of beta-amyloid misfolding reveals structural transitions from pathologically relevant spherical oligomer to fibril. *J Biol Chem.* 2020;295(2):458–467.
33. Kotler SA, Brender JR, Vivekanandan S, et al. High-resolution NMR characterization of low abundance oligomers of amyloid-beta without purification. *Sci Rep.* 2015;5:11811.
34. Lublin AL, Gandy S. Amyloid-beta oligomers: possible roles as key neurotoxins in Alzheimer's disease. *Mt Sinai J Med.* 2010;77(1):43–49.
35. Fernandez A, Llacuna L, Fernandez-Checa JC, Colell A. Mitochondrial cholesterol loading exacerbates amyloid beta peptide-induced inflammation and neurotoxicity. *J Neurosci.* 2009;29(20):6394–6405.
36. Currais A, Quehenberger O, Aaron MA, Daugherty D, Maher P, Schubert D. Amyloid proteotoxicity initiates an inflammatory response blocked by cannabinoids. *NPJ Aging Mech Dis.* 2016;2:16012.
37. Zhao Y, Wu X, Li X, et al. TREM2 is a receptor for beta-amyloid that mediates microglial function. *Neuron.* 2018;97(5). 1023–1031.e1027.
38. Mroczko B, Groblewska M, Litman-Zawadzka A, Kornhuber J, Lewczuk P. Cellular receptors of amyloid beta oligomers (AbetaOs) in Alzheimer's disease. *Int J Mol Sci.* 2018;19(7).
39. Jarosz-Griffiths HH, Noble E, Rushworth JV, Hooper NM. Amyloid-beta receptors: the good, the bad, and the prion protein. *J Biol Chem.* 2016;291(7):3174–3183.
40. Doens D, Fernandez PL. Microglia receptors and their implications in the response to amyloid beta for Alzheimer's disease pathogenesis. *J Neuroinflammation.* 2014;11:48.
41. Choy RW, Cheng Z, Schekman R. Amyloid precursor protein (APP) traffics from the cell surface via endosomes for amyloid beta (Abeta) production in the trans-Golgi network. *Proc Natl Acad Sci U S A.* 2012;109(30):E2077–E2082.
42. Tan JZA, Gleeson PA. The role of membrane trafficking in the processing of amyloid precursor protein and production of amyloid peptides in Alzheimer's disease. *Biochim Biophys Acta Biomembr.* 2019;1861(4):697–712.
43. Evrard C, Kienlen-Campard P, Coevoet M, et al. Contribution of the endosomal-lysosomal and proteasomal systems in amyloid-beta precursor protein derived fragments processing. *Front Cell Neurosci.* 2018;12:435.
44. Cirrito JR, Kang JE, Lee J, et al. Endocytosis is required for synaptic activity-dependent release of amyloid-beta in vivo. *Neuron.* 2008;58(1):42–51.
45. Shirhama T, Cohen AS, Rodgers OG. Phagocytosis of amyloid: invitro interaction of mouse peritoneal macrphages with human amyloid fibrils and their accelerated uptake after dye binding. *Exp Mol Pathol.* 1971;14(1):10–23.

46. Lavie G, Zucker-Franklin D, Franklin EC. Degradation of serum amyloid A protein by surface-associated enzymes of human blood monocytes. *J Exp Med*. 1978;148(4):1020–1031.
47. Weldon DT, Rogers SD, Ghilardi JR, et al. Fibrillar beta-amyloid induces microglial phagocytosis, expression of inducible nitric oxide synthase, and loss of a select population of neurons in the rat CNS in vivo. *J Neurosci*. 1998;18(6):2161–2173.
48. Kopec KK, Carroll RT. Alzheimer's beta-amyloid peptide 1-42 induces a phagocytic response in murine microglia. *J Neurochem*. 1998;71(5):2123–2131.
49. Frackowiak J, Wisniewski HM, Wegiel J, Merz GS, Iqbal K, Wang KC. Ultrastructure of the microglia that phagocytose amyloid and the microglia that produce beta-amyloid fibrils. *Acta Neuropathol*. 1992;84(3):225–233.
50. Rogers J, Li R, Mastroeni D, et al. Peripheral clearance of amyloid beta peptide by complement C3-dependent adherence to erythrocytes. *Neurobiol Aging*. 2006;27(12):1733–1739.
51. Wang Y, Ulland TK, Ulrich JD, et al. TREM2-mediated early microglial response limits diffusion and toxicity of amyloid plaques. *J Exp Med*. 2016;213(5):667–675.
52. Fan KH, Feingold E, Rosenthal SL, et al. Whole-exome sequencing analysis of Alzheimer's disease in non-APOE*4 carriers. *J Alzheimers Dis*. 2020;76(4):1553–1565.
53. Cherry JD, Olschowka JA, O'Banion MK. Neuroinflammation and M2 microglia: the good, the bad, and the inflamed. *J Neuroinflammation*. 2014;11:98.
54. Akhter R, Shao Y, Formica S, Khrestian M, Bekris LM. TREM2 alters the phagocytic, apoptotic and inflammatory response to Abeta42 in HMC3 cells. *Mol Immunol*. 2021;131:171–179.
55. Doens D, Valiente PA, Mfuh AM, et al. Identification of inhibitors of CD36-amyloid beta binding as potential agents for Alzheimer's disease. *ACS Chem Nerosci*. 2017;8(6):1232–1241.
56. Heckmann BL, Teubner BJW, Tummers B, et al. LC3-associated endocytosis facilitates beta-amyloid clearance and mitigates neurodegeneration in murine Alzheimer's disease. *Cell*. 2019;178(3). 536–551.e514.
57. Vollmar P, Kullmann JS, Thilo B, et al. Active immunization with amyloid-beta 1-42 impairs memory performance through TLR2/4-dependent activation of the innate immune system. *J Immunol*. 2010;185(10):6338–6347.
58. Costello DA, Carney DG, Lynch MA. Alpha-TLR2 antibody attenuates the Abeta-mediated inflammatory response in microglia through enhanced expression of SIGIRR. *Brain Behav Immun*. 2015;46:70–79.
59. Balducci C, Frasca A, Zotti M, et al. Toll-like receptor 4-dependent glial cell activation mediates the impairment in memory establishment induced by beta-amyloid oligomers in an acute mouse model of Alzheimer's disease. *Brain Behav Immun*. 2017;60:188–197.
60. Wesen E, Lundmark R, Esbjorner EK. Role of membrane tension sensitive endocytosis and rho GTPases in the uptake of the Alzheimer's disease peptide Abeta(1-42). *ACS Chem Nerosci*. 2020;11(13):1925–1936.
61. Shi JM, Zhu L, Lan X, et al. Endocytosis is a key mode of interaction between extracellular beta-amyloid and the cell membrane. *Biophys J*. 2020;119(6):1078–1090.
62. Fujikura M, Iwahara N, Hisahara S, et al. CD14 and toll-like receptor 4 promote fibrillar Abeta42 uptake by microglia through a clathrin-mediated pathway. *J Alzheimers Dis*. 2019;68(1):323–337.
63. Du SS, Sun X, Cen J, Shi JX, An MX, Zhao WD. Dynamin-2 mediates clathrin-dependent endocytosis for amyloid-beta internalization in brain microvascular endothelial cells. *Microvasc Res*. 2021;138, 104219.
64. Martins IC, Kuperstein I, Wilkinson H, et al. Lipids revert inert Abeta amyloid fibrils to neurotoxic protofibrils that affect learning in mice. *EMBO J*. 2008;27(1):224–233.
65. Guo X, Tang P, Chen L, et al. Amyloid beta-induced redistribution of transcriptional factor EB and lysosomal dysfunction in primary microglial cells. *Front Aging Neurosci*. 2017;9:228.

66. Cheon SY, Kim H, Rubinsztein DC, Lee JE. Autophagy, cellular aging and age-related human diseases. *Exp Neurobiol.* 2019;28(6):643–657.
67. Martens S, Fracchiolla D. Activation and targeting of ATG8 protein lipidation. *Cell Discov.* 2020;6:23.
68. Heckmann BL, Green DR. LC3-associated phagocytosis at a glance. *J Cell Sci.* 2019;132(5).
69. Heckmann BL, Boada-Romero E, Cunha LD, Magne J, Green DR. LC3-associated phagocytosis and inflammation. *J Mol Biol.* 2017;429(23):3561–3576.
70. Boada-Romero E, Martinez J, Heckmann BL, Green DR. The clearance of dead cells by efferocytosis. *Nat Rev Mol Cell Biol.* 2020;21(7):398–414.
71. Cunha LD, Yang M, Carter R, et al. LC3-associated phagocytosis in myeloid cells promotes tumor immune tolerance. *Cell.* 2018;175(2). 429–441.e416.
72. Rickman AD, Hilyard A, Heckmann BL. Dying by fire: noncanonical functions of autophagy proteins in neuroinflammation and neurodegeneration. *Neural Regen Res.* 2022;17(2):246–250.
73. Heckmann BL, Teubner BJW, Tummers B, et al. LC3-associated endocytosis facilitates beta-amyloid clearance and mitigates neurodegeneration in murine Alzheimer's disease. *Cell.* 2020;183(6):1733–1734.
74. Heckmann BL, Teubner BJW, Boada-Romero E, et al. Noncanonical function of an autophagy protein prevents spontaneous Alzheimer's disease. *Sci Adv.* 2020;6(33):eabb9036.
75. Liu CC, Hu J, Zhao N, et al. Astrocytic LRP1 mediates brain Abeta clearance and impacts amyloid deposition. *J Neurosci.* 2017;37(15):4023–4031.
76. Lee SJ, Seo BR, Koh JY. Metallothionein-3 modulates the amyloid beta endocytosis of astrocytes through its effects on actin polymerization. *Mol Brain.* 2015;8(1):84.
77. Yu WH, Lukiw WJ, Bergeron C, Niznik HB, Fraser PE. Metallothionein III is reduced in Alzheimer's disease. *Brain Res.* 2001;894(1):37–45.
78. Narayan P, Sienski G, Bonner JM, et al. PICALM rescues endocytic defects caused by the Alzheimer's disease risk factor APOE4. *Cell Rep.* 2020;33(1), 108224.
79. Heckmann BL, Tummers B, Green DR. Crashing the computer: apoptosis vs. necroptosis in neuroinflammation. *Cell Death Differ.* 2019;26(1):41–52.
80. Evans LD, Wassmer T, Fraser G, et al. Extracellular monomeric and aggregated tau efficiently enter human neurons through overlapping but distinct pathways. *Cell Rep.* 2018;22(13):3612–3624.
81. Rauch JN, Luna G, Guzman E, et al. LRP1 is a master regulator of tau uptake and spread. *Nature.* 2020;580(7803):381–385.
82. Cooper JM, Lathuiliere A, Migliorini M, et al. Regulation of tau internalization, degradation, and seeding by LRP1 reveals multiple pathways for tau catabolism. *J Biol Chem.* 2021;296, 100715.

SECTION V

Autophagy and tau protein

CHAPTER 7

Autophagy and tau protein

Tadanori Hamano[a,b,c,d] **and Yoshinori Endo**[a,b]
[a]Second Department of Internal Medicine, Faculty of Medical Sciences, University of Fukui, Fukui, Japan
[b]Department of Neurology, University of Fukui Hospital, Fukui, Japan
[c]Department of Aging and Dementia (DAD), University of Fukui, Fukui, Japan
[d]Life Science Innovation Center, University of Fukui, Fukui, Japan

1. Introduction

Neurofibrillary tangles (NFTs) consisting of aberrantly phosphorylated tau and senile plaques (SPs) consisting of amyloid-β protein (Aβ) are pathological hallmarks of Alzheimer's disease (AD). Tau protein stabilizes microtubules[1] and has an important role in axonal transport.[2,3] However, once tau is extensively phosphorylated, it changes conformation and is unable to bind to microtubules. The free tau forms aggregates, called oligomers (Fig. 1). Further aggregation of tau oligomers forms NFTs. Tau oligomers are more toxic to neurons than NFTs.[4] AD, progressive supranuclear palsy (PSP), corticobasal degeneration (CBD), Argyrophilic grain disease, Pick's disease, frontotemporal dementia with parkinsonism linked to chromosome 17 (FTDP-17), Niemann-Pick disease type C, and chronic traumatic encephalopathy are collectively known as tauopathies due to the shared feature of pathological tau accumulation.[4–6]

There are several factors that cause tau aggregation. The first is the activation of tau kinases, glycogen synthase kinase 3 (GSK3β), cyclin-dependent kinase 5 (cdk5), p38 mitogen-activated protein kinase (p38MAPK), or stress-activated protein kinase (SAPK)/Jun amino terminal kinase (JNK) activation, or inactivation of the tau phosphatase, protein phosphatase 2A (PP2A). Second, as the C-terminus of tau inhibits tau aggregation, C terminal truncation of tau is a factor that promotes tau aggregation (Fig. 1A).[7] Another important factor causing tau aggregation is the impaired clearance of tau.

The ubiquitin-proteasome system (UPS) and the autophagy-lysosome system have been proposed as mechanisms of tau clearance. UPS mainly breaks down short-lived proteins, and autophagy is involved in the digestion of long-lived proteins and damaged organelles.

Autophagy Dysfunction in Alzheimer's Disease and Dementia
https://doi.org/10.1016/B978-0-323-89906-2.00006-X

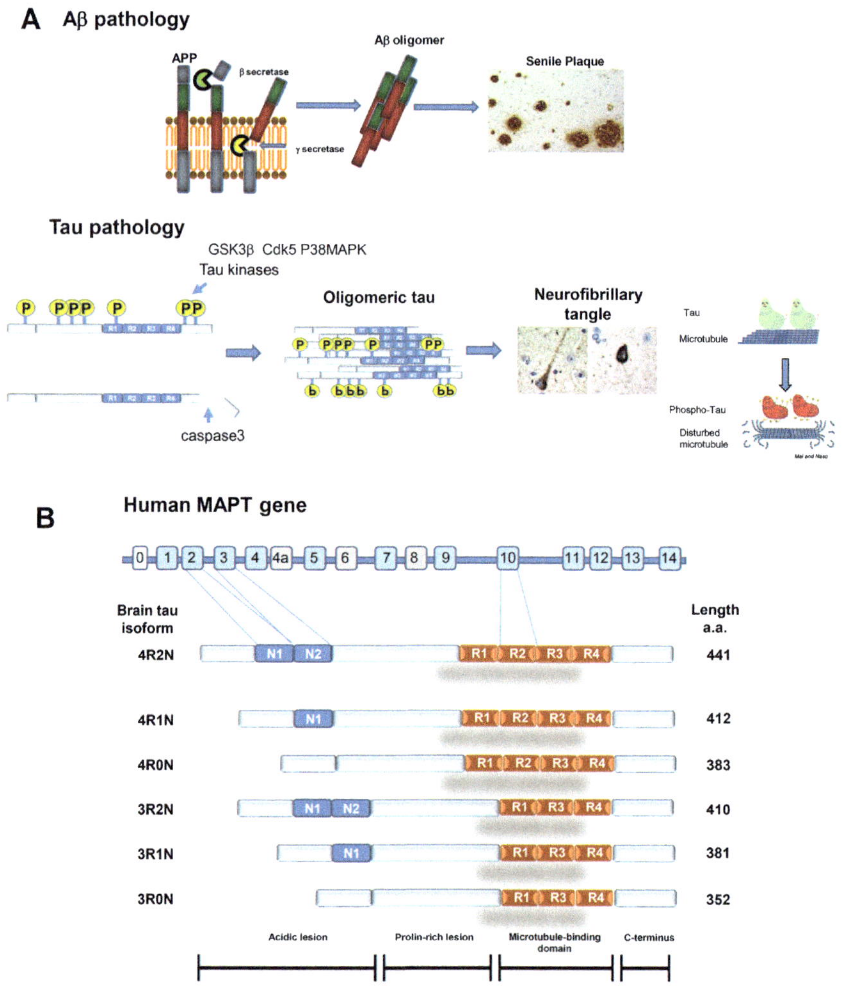

Fig. 1 Pathological hallmarks of Alzheimer's disease. (A) Amyloid β protein (Aβ) was generated by cleaving amyloid precursor protein (APP) with β secretase and γ secretase. Aβ forms oligomers and finally forms senile plaques. Once tau protein is hyperphosphorylated by tau kinases, including GSK3β or cdk5, it detaches from the microtubule and form aggregates, called tau oligomers. C terminal-truncated tau by caspase3 and N-terminal truncation accelerates tau aggregation. Tau oligomers are neurotoxic and further aggregation leads to neurofibrillary tangles (NFTs). (B) Tau isoforms and phosphorylation epitopes. There are 6 tau isoforms according to the number of N-terminal insertions and repeats of the microtubule binding domain: 4 repeat 2N (4R2N), 4 repeat 1N (4R1N), 4 repeat 0N (4R0N), 3 repeat 2N (3R2N), 3 repeat 1N (3R1N), and 3 repeat 0N (3R0N). (Revised from Fig. 1, Hamano T, Enomoto S, Shirafuji N, et al. Autophagy and tau protein. *Int J Mol Sci*. 2021;22(14):7475.)

In this chapter, we describe the process of degradation of tau proteins including polymerized tau by the autophagy-lysosome system and the possibility of autophagy regulation as a therapeutic strategy for tauopathies including AD.

2. Tau protein

2.1 Physiological functions of tau

The tau gene is located on chromosome 17 and has 6 different isoforms due to alternative splicing of exons 2, 3, and 10. The isoforms differ in the number of N-terminal insertions (0, 1, or 2) and the number of microtubule-binding repeat domains (3R or 4R), resulting in six isoforms: 4-repeat 2N (4R2N), 4-repeat 1N (4R1N), 4-repeat 0N (4R0N), 3-repeat 2N (3R2N), 3 repeat 1N (3R1N), and 3R0N (3R0N) (Fig. 1B).[3]

The main function of tau is considered to be the stabilization of microtubules.[1] Tau also participates in promoting the axonal transport of cellular cargo on microtubule tracks of neurons by specifically controlling two motor proteins, dynein, and kinesin.[2] However, the interaction of tau with microtubules is intermittent and short, metaphorically referred to as the "kiss and hop process".[8] Many previous studies demonstrated that tau regulates numerous signaling pathways. For example, tau promotes insulin-induced tyrosine (Tyr) phosphorylation of insulin receptor substrate 1 (IRS-1) and inactivates phosphatase and tensin homolog (PTEN). In conclusion, the essential physiological function of tau in the brain remains unclear.[8]

Experimental reduction or removal of tau in vivo does not alter a variety of neural properties and processes thought to be dependent on microtubules, such as neuronal health, axonal transport, synaptogenesis, and complicated brain functions. Moreover, it has been emphasized that a decrease in endogenous tau is essential for the treatment of tauopathies.[8,9]

2.2 Posttranslational modifications of tau

2.2.1 Phosphorylation of tau protein

Tau is phosphorylated at several sites by multiple tau kinases.[8,10] The phosphorylation sites in normal and AD brains are shown in Fig. 2A.[3] The phosphorylation levels of tau Serine (Ser)199, Ser202/Threonine (Thr)205, and Ser422 were comparable in normal and Braak stage I–IV brains. However, in Braak stage V/VI, they increased in the cerebral cortex and lateral cortex and further increased with increasing Braak stage. Furthermore, the largest increases in phosphorylation compared with controls were observed in

Tyr18, Thr231, and Ser199. Conversely, Ser396 tau and Ser262 tau were only weakly phosphorylated, with only a slight progression in all brain regions.[11]

2.2.2 Ubiquitination of tau protein

Ubiquitination is one of the essential posttranslational modifications that has important functions in eukaryotic cells, including protein homeostasis, signal transduction, and subcellular localization. Ubiquitin is covalently bound to lysine (Lys) residues of target proteins and functions as a degradation signal by the 26S proteasome. NFTs are also immunostained with antibodies against ubiquitin, suggesting that paired helical filament (PHF)-tau is ubiquitinated. Analysis of tau from AD brains revealed that most of the ubiquitinated tau is monoubiquitinated and only small portion was polyubiquitinated. In addition, the tau was N-terminally truncated. This implies that N-terminal cleavage of tau precedes its ubiqutination. The C-terminus of Hsc70-interacting protein (CHIP), a ubiquitin E3 ligase, cooperates with Hsp70/90 and indirectly induces tau ubiquitination.[12] The ubiquitination site of tau is shown in Fig. 2B.[3]

2.2.3 Acetylation of tau protein

Acetylation is another posttranslational modification. Lys residues are characterized by their involvement in electrostatic and hydrophobic interactions and play an essential role in tau polymerization and toxicity. Therefore, it is of interest whether acetylation at Lys residues of tau regulates tau polymerization. Cohen et al. found that CREB (cAMP-response element-binding protein)-binding protein (CBP), an acetylase, acetylates Lys residues in tau fragments containing microtubule-binding domains. Acetylation promotes the aggregation of tau.[13] The acetylation site of tau is shown in Fig. 2B.[3,14] Recently, it has been suggested that acetylated tau inhibits chaperone-mediated autophagy (CMA)-mediated degradation. Therefore, acetylated tau is thought to be preferentially degraded by macroautophagy and microautophagy. However, other reports stated that acetylation of full-length tau by acetyltransferase p300 inhibits tau aggregation, which is restored when the deacetylase histone deacetylase (HDAC6) is added.[15]

2.2.4 Nitration

Nitration is the covalent addition of a nitro group to the aromatic ring of a Tyr residue and nitrated tau exists in NFTs,[16] which is thought to cause the loss of tau binding to microtubules.[17] In particular, Tyr29-nitrated tau has been observed specifically in severe AD brains (Fig. 2D).[3]

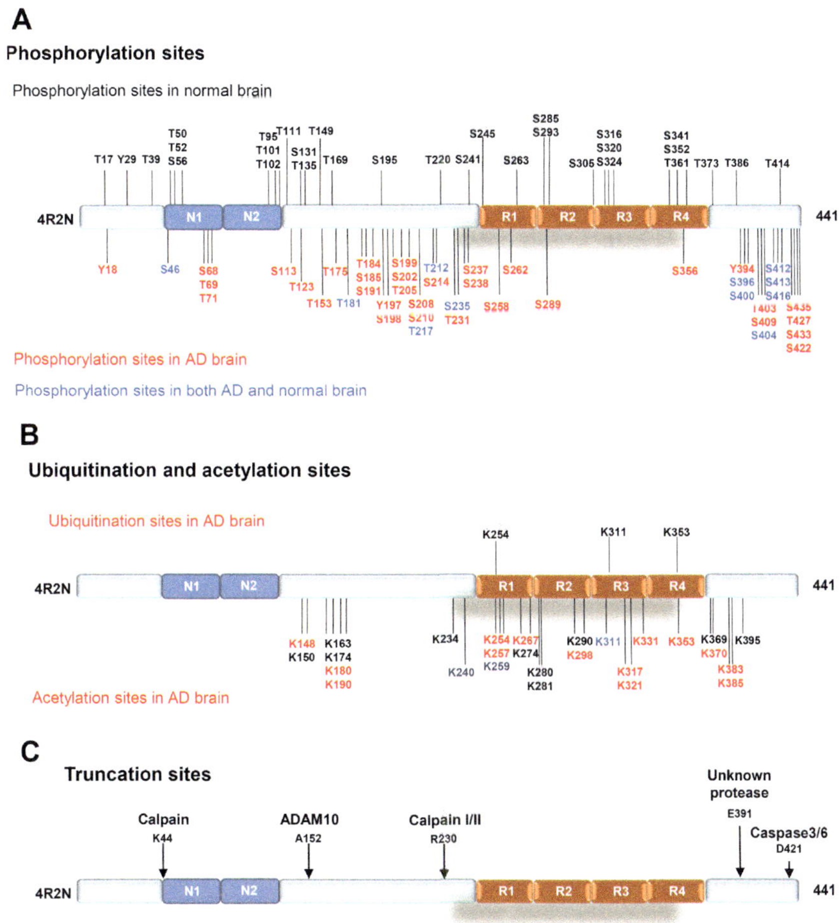

Fig. 2 Posttranslational modification of tau protein. (A) Putative sites of tau protein phosphorylation. Amino acids phosphorylated in brain with Alzheimer's disease (AD) are shown in red, those in AD and normal brains are shown in blue, and those in normal brains are shown in black. (B) Ubiquitination sites (upper) observed in AD brain, and acetylation by p300 and CREB (cAMP-response element-binding protein)-binding protein (black), only by p300 (red), or only by CREB-binding protein (blue) is also shown. (C) Truncation sites of tau protein. The sequential truncation of tau can produce tau fragments. The cleavage of tau is more likely when tau is detached from microtubules, either forming aggregates or associating with other proteins than tubulin. (D) Nitration sites of tau protein. (Revised from Fig. 2, Hamano T, Enomoto S, Shirafuji N, et al. Autophagy and tau protein. *Int J Mol Sci*. 2021;22(14):7475.)

2.2.5 Glycosylation of tau

O-linked glycosylation is a common type of glycosylation in which *O*-GlcNAc is added to Ser or Thr adjacent to Proline (Pro). Tau also undergoes this modification. However, the number of glycosylation sites is smaller than that of phosphorylation sites. When tau undergoes this modification, it becomes less phosphorylated.[18] Furthermore, when the protein *O*-GlcNAc transferase (OGT), an *O*-GlcNAc-adding enzyme, was knocked out in the forebrain, neurodegeneration was observed, and the amount of *O*-Glu-Nac tau in the AD brain decreased as the disease progressed. On the other hand, *N*-glycosylation of abnormally phosphorylated tau in AD brain functions in the maintenance of PHF structure.

2.2.6 Truncation of tau protein

As mentioned previously, tau proteins purified from the core of PHF have their N- or C-terminal portions truncated. It has been demonstrated that cleavage of tau promotes tau aggregation.[19,20] Tau is readily digested by proteases. Many studies have confirmed that tau is digested by calpain, caspases,[6,20–23] and A disintegrin and metalloproteinase 10 (ADAM10). Tau is also a substrate for aspartate endopeptidase. These studies suggest that tau cleavage is important for tau aggregation and neurodegeneration.[19,20] Several specific tau cleavages have been discovered in AD brains, including cleavage of Asp421 (D421) and Glu391 (E391), which promote tau aggregation[3,8,19] (Fig. 2C). In addition, it has been reported that tau fragments propagate between neurons via synapses and expand neurofibrillary degeneration to postsynaptic neurons.[10]

2.2.7 Tau toxicity

Aberrant phosphorylation, proteolysis, and aggregation of tau protein in the early stages of neurofibrillary degeneration (Figs. 1 and 2)[3] are neuropathologically important in the development of AD. However, it is unclear which type of tau is most toxic (misfolded fibrous aggregates, soluble hyperaggregates, fibrous aggregates, soluble hyperphosphorylated aggregates, or mislocalized) and whether the toxicity is due to a gain or loss of function.

As there is limited precise proof that tau fibers are toxic themselves, it is becoming more certain that the soluble oligomeric form of tau is more toxic to neurons and synapses.[9] In the normal brain, tau is observed primarily in axons.[2] However, in the AD brain, NFTs consisting of highly phosphorylated tau are observed not only in axons but also in the cytoplasm and dendrites.

3. Autophagy

3.1 Concept of autophagy

Autophagy (Greek for "self-eating") is a highly conserved pathway and a unique mechanism for the degradation of large organelles and proteins via lysosomes. Three types of autophagy exist: macroautophagy, CMA, and microautophagy (Fig. 3).[3,24] Usually, the term autophagy means macroautophagy. Autophagy is mainly considered as a mechanism for survival. Under physiological conditions, cells carry out low levels of autophagy to maintain homeostasis, which termed basal autophagy. However, various life-threatening events, such as hypoxia, nutrient deprivation, exposure to reactive oxygen species (ROS), microbial invasion, organelle damage, and excessive accumulation of aggregation proteins, for example, Aβ and tau, activate autophagy all at once.

Previous studies demonstrated that autophagy is highly selective.[25] Selective autophagy includes pexophagy (peroxisomes), mitophagy (mitochondria), xenophagy (bacteria), and aggrephagy (misfolded proteins with aggregation) (Fig. 4).[3]

The discovery of selective autophagy receptors in mammalian cells and the study of cytoplasmic to vacuolar targeting pathways revealed some of the mechanisms of efficient autophagy. Autophagy is highly dynamic and consists of the processes of initiation, nucleation, and elongation.

3.2 Initiation of autophagy

The ULK1 (unc-51-like kinase 1) complex is considered to be the leading initiator of mammalian autophagy.[26] ULK1 and Atg (autophagy-related protein)13 are the core of the ULK1 complex, which is further supported by scaffolding proteins Atg101, and focal adhesion kinase (FAK) interacting protein (FIP). The ULK1 complex is inhibited by stress signaling pathways, including the mechanistic target of rapamycin complex 1 (mTORC1) pathway (Fig. 4).[3] In a nutrient-rich environment, mTORC1 binds to and inhibits ULK1. On the other hand, when intracellular amino acids are depleted during starvation, the energy level-dependent sensor 5′ adenosine monophosphate-activated protein kinase (AMPK) is activated. AMPK phosphorylates mTORC1 and helps dissociate mTORC1 from ULK1. When mTORC1 is downregulated, transcription factor EB (TFEB) is translocated to the nucleus and activated, resulting in the translation of autophagy and lysosome-related genes. ULK1 activates Atg13, FIP200, and Atg10.[26] The fully bound ULK-1 complex moves to the membrane compartment of

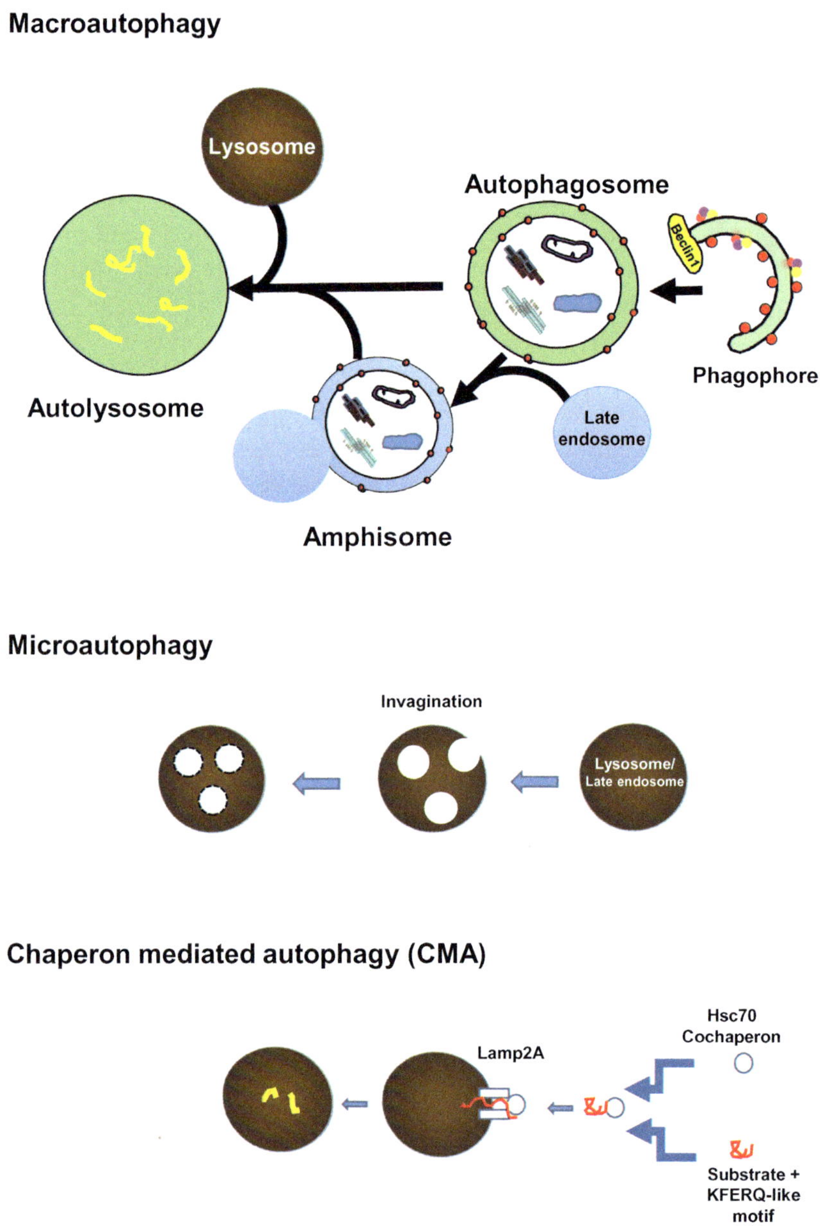

Fig. 3 Scheme of three types of autophagy; Macroautophagy, microautophagy, and the chaperon-mediated autophagy pathway. Macroautophagy: A part of the cytoplasm including the organelle is surrounded by an isolated membrane, the phagophore, which then forms an autophagosome. By the fusion of the outer membrane with a lysosome, the internal materials are broken down in the autolysosome. Microautophagy: A small amount of cytoplasmic substance is caught in the lysosome/vacuole through the random process of membrane erosion. Chaperon-mediated autophagy (CMA): Proteins with an KFERQ-like motif are identified by cytosolic Hsc 70 and co-chaperons. Then, the substrate proteins are translocated to the lumen of the lysosomes after docking with the lysosome. (Revised from Fig. 3, Hamano T, Enomoto S, Shirafuji N, et al. Autophagy and tau protein. *Int J Mol Sci*. 2021;22(14):7475.)

the autophagosome assembly along with the class III phosphatidylinositol 3-kinase (PI3K-III) complex (Fig. 4).[27]

3.3 Nucleation of phagophores

The lipid kinase PI3K-III (Vps34), its regulatory subunit P150, and Beclin1 (Atg6) are key components of the PI3K-III complex and are essential for phagophore formation and maturation into autophagosomes. Other important coregulators of the PI3K-III complex consist of Atg14L, AMBRA-1 (autophagy and Beclin 1 regulator 1), and UVRAG (UV resistance-associated gene). The activated PI3K-III complex locally generates phosphatidylinositol-3-phosphate (PI3P) at the nucleation site of the phagophore (Fig. 4).[3]

3.4 Elongation and closure

The elongation of a new organelle requires the continuous incorporation of membranous material into the expanding body. VMP1 (vacuole membrane protein 1) binds the PI3K-III complex to Beclin1 and anchors it to the surface of the expanding phagophore. Atg9 moves between membrane compartments in the cell (mitochondria, Golgi apparatus, and plasma membrane) and transfers fragments of phospholipids for incorporation into the expanded structure. Autophagosome formation requires the ubiquitin-like protein LC3 and the E3 ligase Atg12. Atg7 upregulates Atg12 and Atg10 is responsible for binding Atg12 to Atg5. The Atg-12-Atg5 heterodimer combines with Atg16L1 to form the core of the entire Atg16L1 complex. LC3-I is produced by cleavage of LC3 by the cysteine protease Atg4B. LC3-I is translocated to the inner and outer membranes of the expanding phagosome by Atg7. LC3-I binds to phosphatidylethanolamine (PE) and converts to LC3-II, which is anchored to the membrane of the autophagosome by the local action of PI3K-III and Atg16L1 complexes (Fig. 4).[3]

Mitochondria are directly taken up by mitochondrial receptors without polyubiquitination. In aggrephagy, polyubiquitinated protein aggregates (Aβ and tau oligomers) are taken up by phagophores via SLRs (sequestosome 1/p62-like receptors) such as P62 and NBR1 (neighbor of BRCA1 gene 1).

Autophagosomes form at random locations in the cytoplasm and are delivered to microtubule organizing centers where lysosomes are enriched along the microtubules.[28] The transport of autophagosomes toward the plus end of microtubules involves the interaction of FYCO1 (FYVE and coiled-coil domain containing 1), with LC3-II on the outer membrane of autophagosomes.[25]

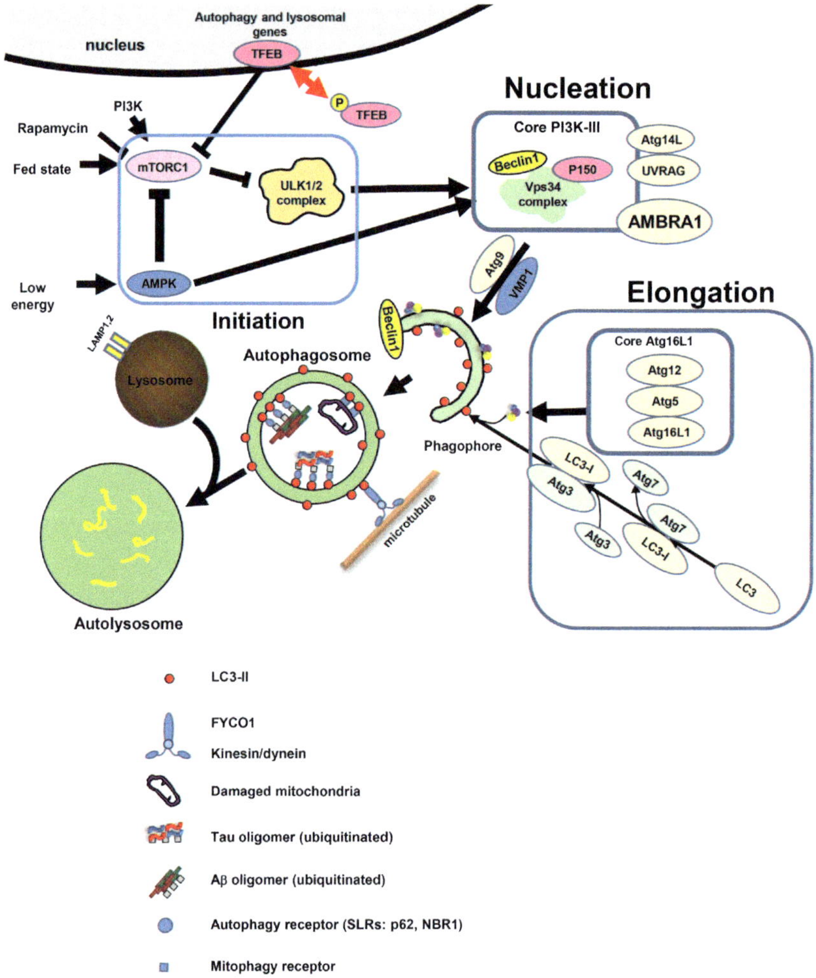

Fig. 4 The sequence of autophagy in mammals. (A) Initiation. In response to nutrient deprivation, AMPK inactivates mTORC1, releasing its inhibition on ULK, and activates the initiation complex. Downregulation of mTORC1 leads to the translocation of TFEB to the nucleus. Then, TFEB is upregulated, and autophagy- and lysosome-related genes are translated. ULK phosphorylates AMBRA-1, which promotes the interaction of Beclin1 with p150 and PI3K-III (Vps34) in PI3K-III complexes. UVRAG and Atg14L bind differently to form separate PI3K-III-complex populations in cells. AMBRA-1 interacts with PI3K-III-complexes. (B) Nucleation. ULK- and PI3K-III-complexes are translocated to early phagophore sites later with the help of Atg9 and VMP1, and local production of PI3P enlists additional effectors and the first segregating membrane is formed.

(Continued)

Fig. 4,cont'd (C) Elongation. Immediately after the activation of autophagy, Atg7 is up-regulated, promoting Atg5-Atg12 dimer formation, which connects to Atg16L1 to form the Atg16L1 elongation complex. Atg7 also directs the cleavage and lipidation of LC3 to LC3-I and then to LC3-II, which binds to the inside and outside of the formed phagosomal membrane. Mitochondria can be directly bound by mitochondrial receptors without poly-ubiquitination. In aggrephagy, poly-ubiquitinated protein aggregates (Aβ oligomer and tau oligomer) are engulfed by the phagophore via SLRs, which include P62 or NBR. (D) Maturation. Once the loaded phagophore is closed and the outer fraction of LC3-II is removed, the completed autophagosome fuses with the lysosome via a bonding process involving syntaxin members, LAMP-1, and -2, creating an autolysosome. Autophagosomes are produced in the cytoplasm at random sites. The autophagosomes are delivered to lysosomes at the microtubule organizing center. Transport of autophagosomes toward the plus end of microtubules involves the interaction of the LIR motif of FYCO1 with LC3-II on the outer membrane of the autophagosome. Abbreviations: PI3K-III, class III phosphatidylinositol 3-kinase; LC3-II, microtubule associated protein light chain 3-II; LAMP-1, LAMP-2, lysosome-associated membrane protein 1, 2; SLRs, sequestosome 1/p62-like receptors; FYCO1, FYVE and coiled-coil domain containing 1; p62, p62/sequestosome-1; NBRI-1, neighbor of BRCA1 gene 1; TFEB, transcription factor EB. (From Fig. 4, Hamano T, Enomoto S, Shirafuji N, et al. Autophagy and tau protein. *Int J Mol Sci.* 2021;22(14):7475.)

3.5 Fusion of lysosomes

When a fully developed autophagosome fuses with a lysosome, the Atg16L1 complex separates from the vesicle and the cytoplasmic fraction of LC3-II is removed by Atg4B. Fusion of autophagosomes with lysosomes requires the SNARE (sol-*N*-ethylmaleimide-sensitive factor activity part receptor) complex, which requires a complex involving multiple syntaxes and Atg14L1.[29] This complex prepares the autophagosome for fusion with the lysosome. Acidification and further maturation into mature autolysosomes take place with the help of the Rab family, small GTPases, and LAMP1 and 2 (lysosome-associated membrane proteins 1 and 2). The inner membrane components of autolysosomes are degraded by lysosomal hydrolases. A series of lysosomal proteases (cathepsins B, D, and L) is required to degrade mammalian cells. When macromolecules are degraded in the lysosome, monomeric molecules such as amino acids and lipids are returned to the cytoplasm and reused as building blocks.

4. Disturbance of the autophagy-lysosome system in AD and related disorders

4.1 Disruption of the autophagy-lysosome system in AD

In 1967, Suzuki and Terry found that swollen and dystrophic neurites surround the SPs in the AD brain.[30] In the dystrophic neurites, they found circular or oval-layered material of 0.5 μm in diameter. However, their identity was not clear at the time. In 2005, Nixon et al. revealed that they were autophagic vacuoles (Avs).[31] Avs are composed of autophagosomes, amphisomes, and autolysosomes (Fig. 3).[3]

Accumulation of lysosomes and lysosomal hydrolytic enzymes, in addition to defects in the integrity of lysosomal membranes, has been described in AD patients.[32] Increased levels of cathepsin D and LAMP1, lysosome components, were also noted in the brains of CBD and PSP patients.[33] In the altered axons of the AD brain, the accumulation of autophagosomes, Avs, multivesicular bodies (MVBs), multilayers, and cathepsin-containing autolysosomes is evident.[27,31,34]

Macroautophagy is constitutionally active and efficient in young neurons. However, the function of macroautophagy is inhibited in the AD brain. The induction of autophagy has been observed under conditions of cellular stress and neurodegeneration,[35] including AD.

Macroautophagy is induced and disturbed in AD brains, and mice overexpressing mutant human presenilin 1 (PS1) and amyloid precursor protein (APP) exhibit the induction and impairment of macroautophagy and accumulation of Avs holding Aβ.[31] In AD neurons and PS1-APP mice, autophagosomes proliferate in dendrites at an early stage, before Aβ is deposited. This suggests that the induction of macroautophagy occurs in an early phase of AD and that it is not a consequence of Aβ deposition. Boland et al. reported that when mTOR kinase activity is suppressed by rapamycin or nutrition shortage, LC3-II-positive autophagosomes, but not cathepsin D-positive autophagosomes were found in cortical primary neurons. Thus, the clearance of autophagosomes is efficient in neurons. Conversely, disruption of macroautophagy at a late stage by lysosomal protease inhibition and microtubule disruption by vinblastine treatment resulted in the accumulation of Avs in primary cortical neurons, similar to that observed in AD brains and dystrophic neuronal projections of AD model mice.

This led the authors to state that autophagic pathology in AD brains is probably due to the induction of autophagy.[36] PS1 is required for the rotation of protein substrates in lysosomes for autophagy and endocytosis. Loss of PS1 impairs the acidification of autolysosomes and inhibits the activation of cathepsins (cysteine cathepsins are proteases responsible for proteolysis in lysosomes), resulting in the dysfunction of autophagy (macroautophagy).[37] Thus, mutations in PS1, a major cause of familial AD (FAD), are directly related to the mechanisms that promote the pathogenesis of AD.[38] In AD, defects in retrograde transportation of autophagosomes from axons to cell body, and defects in the forward transport of lysosomes to the neuronal axonal dorsal end were also demonstrated.[39]

4.2 Sporadic inclusion of body myositis and tau and Aβ

Progressive muscle disorder, sporadic inclusion body myositis (s-IBM), is most commonly observed in the elderly in Europe and the United States. Patients with s-IBM are markedly disabled and no effective treatment has been established. S-IBM pathology is characterized by large amounts of ubiquitin-positive aggregates that accumulate with Avs, which contain phosphorylated tau and Aβ with a β-sheet structure. Endoplasmic reticulum stress (ERS) and 26S proteasome inhibition are also associated with s-IBM and are thought to exacerbate the deposition of misfolded proteins. Nogalska et al. reported that cathepsins D and B, enzymes of the lysosome, are inactivated in s-IBM.The authors also found that the presence of LC3-II and phosphorylation of p70S6 kinase by mTORC1 were suppressed, suggesting that autophagosomes are maturing.Thus, proteolysis in lysosomes in muscle may be inactivated in s-IBM, which may result in the deposition of misfolded proteins, independent of increased autophagosome maturation.[40] Abnormalities in mitophagy have also been reported in s-IBM, suggesting that dysfunction of mitophagy in s-IBM plays an important role.[41]

4.3 Tau and Aβ can induce autophagy dysfunction

Previous studies revealed that phosphorylated tau and Aβ disturb autophagy and mitophagy, which are major pathologies in AD.[42] Age-dependent increases in phosphorylated tau and Aβ reduce autophagy and mitophagy proteins. Aberrant interactions between phosphorylated tau and Drp-1 and Aβ and PINK1 (PTEN-induced kinase 1)/parkin induce the inability to remove defective mitochondria and other debris from neuronal cells.[43]

Fragmented tau impairs the function of CMA and lysosomes and can cause oligomerization and aggregation of tau.[44] Feng et al. also found that tau accumulation inhibits autophagosome-lysosome fusion by inhibiting the expression of IST1 (increased salt tolerance 1), which in turn inhibits ESCRT-III (an endosomal sorting complex required for transport-III) formation, revealing a vicious circle of tau deposition and autophagy impairment in the chronic progression of AD.[45]

FKBP4/FKBP52 is a member of the immunophilin protein family and functions in immunomodulation and basic cellular processes such as protein folding and transport. The encoded protein is a cis-trans prolyl isomerase, which binds to the immunosuppressive drugs FK506 and rapamycin.

FKBP52 has been reported to localize to the lysosomal system of healthy human neurons, suggesting that it is involved in lysosomal function.[46] Cleavage of the C-terminus of tau is essential for tau aggregation[3,21–23] (Figs. 1 and 2). This caspase-3-cleaved tau (D421) colocalizes with FKB52 in the autophagy-endolysosomal system of AD neurons. The amount of FKBP52 is significantly lower, and the reduction in FKBP52 correlates with a high number of NFTs in AD brain. When autophagy was inhibited in tau-induced neuroblastoma, FKBP52 was released extracellularly. In AD brain, FKBP52 is abnormally released from NFT-negative neurons, suggesting that this is related to the extent of early tau-D421 accumulation in neurons.[47] Thus, the cleavage of tau by caspases may inhibit autophagy through the release of FKBP52 and vice versa.

5. Tau degradation pathway

In the AD brain, the activity of the 20S proteasome (core particle), especially its trypsin-like activity, is inhibited.[48] When tau is ubiquitinated by CHIP, it is degraded by the proteasome.[49] The proteasome is capable of degrading endogenous tau, whether it is full length or truncated. However, PHF-tau is not digested by the proteasome, and PHF-tau interferes with the proteasome.

The relationship between tau degradation and the autophagy-lysosome system has long been unclear. Chloroquine, a malaria drug, perturbs autophagy by blocking the fusion of lysosomes and autophagosomes, rather than inhibiting lysosomal acidity or digestive function.[50] In a study of chloroquine myopathy, tau accumulated in the lysosomes of rat muscle. This implies that tau is digested by the lysosome-autophagy system.[51] We used tetracycline-off-induced M1C cells harboring wild-type tau (4R0N)[5,6,23,52–55] and treated them with the lysosomal inhibitor NH_4Cl and the autophagy inhibitors chloroquine and 3-methyladenine (3MA). Chloroquine increased the accumulation of Avs by light microscopy (Fig. 5A) and electron microscopy.[53] All chloroquine (Fig. 5 B and C), NH_4Cl, and 3MA treatments increased the accumulation of tau in the cells.[53] The following year, Wang et al. demonstrated that the degradation of tau was due to macroautophagy.[44] However, they found that the N-terminally truncated tau species were preferentially trapped by CMAs and that C-terminal truncation occurs; C-terminally truncated tau promotes tau aggregation.[6] Subsequently, it has been reported that rapamycin (Fig. 4), which induces autophagy, also reduces aggregated tau.[3,56] Therefore, the autophagy-lysosome system degrades not

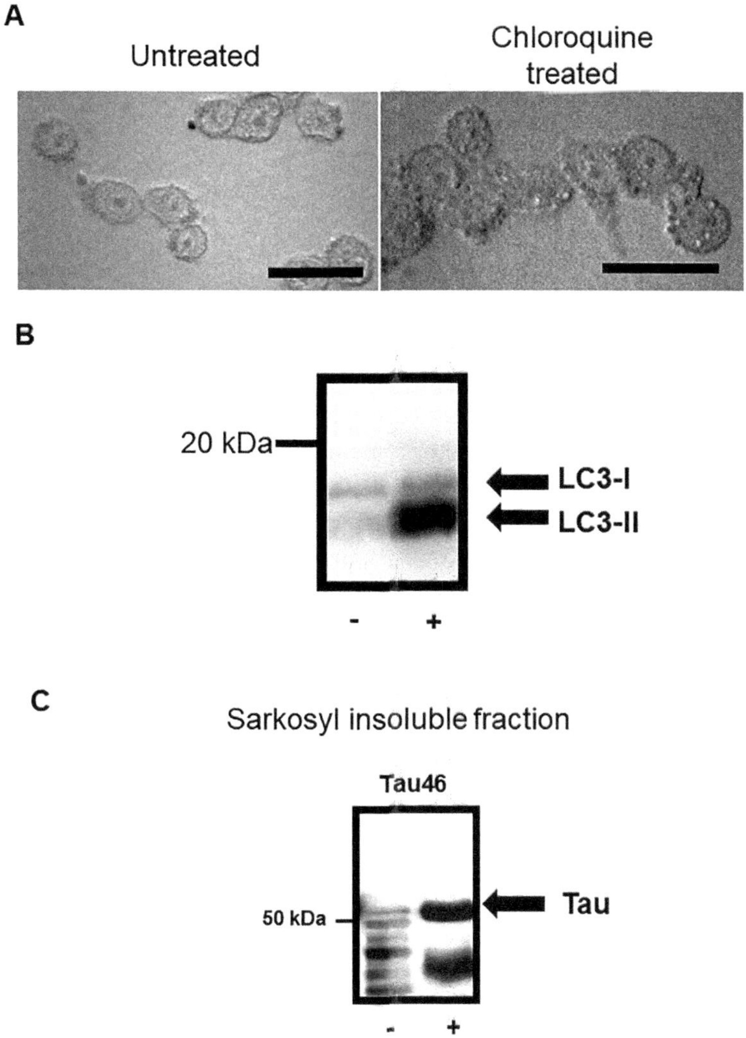

Fig. 5 Autophagic vacuole formation and tau accumulation by autophagy inhibitor chloroquine (CQ) treatment. (A) A morphological study demonstrated autophagic vacuole formation after 5 μM CQ treatment. Bar: 50 μm. (B) CQ treatment markedly increased LC-3II, suggesting that autophagosomal clearance was disturbed by CQ treatment. (C) Tau accumulation in the sarkosyl insoluble fraction detected by Tau46 after CQ treatment (−: untreated cell, +: 5 μM chloroquine treated cells).

only soluble tau but also insoluble aggregated tau. C-terminally truncated tau is favorably digested by autophagy. However, full-length tau is preferentially degraded by the proteasome.[57] Increased C-terminal cleavage of tau[6,20–23] causes tau aggregation and dysfunction of autophagy.[47] Recently, it has been proposed that the majority of tau is degraded by CMA, and when acetylated, tau is specifically degraded through macroautophagy and endocytic microautophagy.[58]

UPS, CMA, and endosome microautophagy remove neurotoxic soluble tau; however, macroautophagy breaks down intracellular insoluble tau in neurons. In addition, autophagy-dependent endolysosome degradation of ubiquitinated tau with specific phosphorylation sites has been reported.

Endolysosomal degradation of autophagy-independent tau was also recently proposed in primary cultured neurons in the hippocampus and N2a cells.[59] Of note, tau was observed in Hrs (hepatocyte growth factor-regulated tyrosine kinase substrate)/EEA1 (early endosome antigen1)/Rab5-positive early endosomes, MVBs, and LAMP1-positive vesicles. This suggests tau is transported throughout the endolysosome pathway. Tau interaction with the ESCRT protein Hrs is possibly via ubiquitination because tau interaction with Hrs increased after deubiquitylation enzyme inhibitor treatment. Tau sorting into the ESCRT pathway depended on the Rab35, a low-molecular-weight GTPase. Rab35-dependent tau degradation was especially dependent on tau phosphorylated sites, including pSer396/404 and pSer262, but not pSer202. This implies preferential selection of phosphorylated tau at the specific sites by multiple autophagy endolysosomal pathways.[34]

As mentioned above, activation of caspase-3 increases C-terminal cleavage of tau,[6,20–23] causing tau aggregation and dysfunction of autophagy.[47] Although further studies are necessary to provide direct evidence, activation of caspase-3 may be responsible for neurodegeneration, including AD.

6. Mitophagy and tau

Defective mitochondrial accumulation is a hallmark of normal aging and age-related neurodegenerative diseases, including AD. Mitophagy is inhibited in iPS cell-derived neurons from AD patients and in animal models of AD. In the *Caenorhabditis elegans* (*C. elegans*) model, stimulation of mitophagy restores cognitive function in a PINK-1, PDR-1 (Parkinson's disease-related 1;parkin), and DCT-1 (DAF-16/FOXO-controlled germline-tumorizing-1)-dependent pathway. Mitophagy activation through NAD^+ supplementation restored cognitive function through a DCT-1 (DAF-16/FOXO-controlled germline-tumorizing-1)-dependent pathway. Mitophagy reduces aggregated Aβ42 and Aβ40 and suppressed neuronal inflammation by allowing microglia to phagocytose Aβ in SPs. Mitophagy activation also suppressed tau phosphorylation in human neurons and restored cognitive function in transgenic *C. elegans* and transgenic mouse models.[60] Cummins et al. also reported that human wild-type tau (hTau) and mutant tau (hP301L) observed in human FTDP-17 inhibit mitophagy in neuroblastoma cell lines by reducing the translocation of parkin to mitochondria. Tau expression in neurons itself inhibits mitophagy and the expression of hP301L suppressed mitophagy in *C. elegans*.[61]

7. Diabetes, tau, and autophagy

Epidemiological studies confirmed that the prevalence of AD is doubled in patients with diabetes.[62] Impaired insulin signaling is associated with accelerated deposition of NFTs and SPs in AD brains.[63] Diabetic animal models also exhibit pathological changes in AD, including memory impairment and phosphorylated tau.[64] Diabetic rats treated with streptozotocin performed worse than control animals in several cognitive function tests, including the Morris water maze.

Furthermore, diabetes-induced AD-like behavioral and pathological findings (including deposition of Aβ and phosphorylated tau) were associated with reduced autophagy in neurons. The impairment of cellular autophagy was also reproduced in neuronal cell cultures under high glucose conditions.[65] Chen et al. reported that metformin ameliorated cognitive decline in a rodent model of diabetes, suppressed phosphorylated tau hyperactivity, and restored autophagy impairment in diabetic patients. In addition, both the cognitive decline and the increase in phosphorylated tau in the diabetic model were caused by the inhibition of autophagy. In HT22 cells cultured under high glucose conditions, metformin-activated autophagy via AMPK (Fig. 4).[3,66]

8. Propagation of tau by the disruption of autophagy

8.1 Pathways of pathological tau protein secretion

Microtubule-associated tau protein resides in the cytoplasm, but is also observed in the extracellular space under physiological conditions. Misfolded tau can move between cells through four mechanisms.

8.1.1 Direct secretion of tau through the cell membrane

For the direct secretion of tau via the cell membrane, the clustering of tau at the cell membrane and the interaction of cholesterol- and sphingomyelin-rich membrane microdomains with specific lipids are important. After penetrating the cell membrane, it is released from the cell membrane by heparin sulfate proteoglycans on the cell surface.[67]

8.1.2 Tau is secreted into ectosomes

Ectosomes, which are larger than exosomes, are produced by sprouting of the plasma membrane to the outside, rather from MVBs in the cells.[68] Tau was observed in ectosomes obtained from neuroblastoma cells, primary cultured neurons in the cortex, interstitial fluid (ISF) of the mouse brain,[68] and cerebrospinal fluid of AD patients and normal controls.[68,69] It has been suggested that the secretion of tau by ectosomes is a normal condition, while the secretion of tau by exosomes is predominantly observed in pathological states.[68,69]

8.1.3 Secretion of tau by exosomes and organelle hitchhiking

Tau can be taken into exosomes by inwards sprouting of the late endosome membrane leading to the development of intraluminal vesicles in MVBs that can be secreted by docking of the MVB membrane to the plasma membrane. UPS III-like organelle hitch-hiking pathways are involved in the cytosolic section of tau and other misfolded proteins, including secretory endo-lysosomes related to the autophagy-lysosomal pathway. In the MAPS (misfolding-associated protein secretion) pathway, cytosolic misfolded proteins are incorporated into the endoplasmic reticulum by chaperone-mediated capture, followed by fusion of endolysosomal vesicles with the plasma membrane and secretion, thus releasing vesicle-free tau into the extracellular space.

8.1.4 Tau seeds move between cells via tunneling nanotubes that directly connect the cytoplasm of two adjacent cells

Regardless of the secretory pathway, aggregated tau eventually reaches the cytoplasm of the receptor cell and causes template misfolding. The misfolded protein acts as a conformational template, turning a normal protein into a pathogen. At present, it is unclear which of the above mechanisms participate in tau synaptic release and whether the same mechanism is responsible for physiological and pathological synaptic release of tau.[70]

8.2 Propagation of tau by the disruption of autophagy

The propagation of tau into neurons is considered to be important in the progression of tau-mediated neurodegeneration. Tau is acetylated by the lysine acetyltransferase p300/CBP. Tau degradation and toxicity are also regulated by p300/CBP.[71] p300/CBP is aberrantly expressed in patients with tauopathy, inhibiting the autophagy-lysosomal pathway and leading to tau hypersecretion.[72]

In a mouse model of tauopathy and in cultures of human and rodent neocortex, the new p300 inhibitor SMDC37892 increased autophagic flux and reduced tau secretion. In contrast, activated p300/CBP inhibited autophagy and promoted neuronal tau secretion. Tang et al. also reported that activated mTORC promoted extracellular tau secretion. This suggests that the release of tau is mediated by mTORC through a route that does not involve exosomes.[73] It has been reported that CMA inhibition promotes cell-to-cell propagation.[58] Contrary to this, the induction of autophagy increases the secretion of both total and phosphorylated tau, and this phenomenon was inhibited by Beclin1 knockdown or autophagy inhibitors in a hTau-overexpressing human neuroblastoma cell line.[74] Furthermore, six isoforms of tau are secreted in an autophagy-dependent manner (Fig. 1).[3,74] Thus, this issue remains controversial.

9. Potential of autophagy modulators as a treatment for AD

Autophagy regulation may be a therapeutic focus of AD or a method to slow its progression. Autophagy-inducing drugs are usually divided into two major groups: mTORC-dependent and independent. For example, rapamycin[56,75,76] and its analogs,[76] Curcumin[77] and methylene blue,[78] were proposed as mTOCR1-dependent autophagy modulators. Metformin,[66] trehalose,[79] lithium,[80] clonidine,[75] curcumin,[81] nicotinamide,[82] and rifampicin[83] were proposed as mTORC1-independent autophagy modulators.[3,28] See Chapters 14 and 16 for more information on the autophagy pathway and autophagy modulators.

10. Conclusion

The ubiquitin proteasome and the autophagy-lysosome system have important roles in the degradation of tau. In particular, polymerized tau is digested by the autophagic lysosomal system. Recently, it was confirmed that tau is mainly degraded by CMA. However, acetylated tau is degraded by macroautophagy and microautophagy. Furthermore, it is clear that tau accumulation impairs autophagy and mitophagy. This vicious cycle may lead to the further progression of AD. From this point of view, the regulation of autophagy may be effective in the treatment of AD. However, Avs accumulate in the axons of AD. In other words, although autophagy is induced, the clearance of autophagy is not achieved. For these reasons, even if autophagy is induced, treatment will not be effective if clearance is not successful. Therefore, it is important to treat AD according to the stage of the disorder and to the characteristics of each patient.

Acknowledgments

A part of this study was supported by Japan Society for the Promotion of Science KAKENHI grant (No 19K07909) and Research grant from University of Fukui (LSI20306). We are grateful to Mrs. Junko Nakane, Emiko Kitagawa, Ryota Yamamoto, and Miku Imai, for their excellent help in the present study.

References

1. Kadavath H, Hofele RV, Biernat J, et al. Tau stabilizes microtubules by binding at the interface between tubulin heterodimers. *Proc Natl Acad Sci U S A*. 2015;112:7501–7506.
2. Dixit R, Ross JL, Goldman YE, Holzbaur EL. Differential regulation of dynein and kinesin motor proteins by tau. *Science*. 2008;319:1086–1089.
3. Hamano T, Enomoto S, Shirafuji N, et al. Autophagy and tau protein. *Int J Mol Sci*. 2021;22:7475.
4. Patterson KR, Remmers C, Fu Y, et al. Characterization of prefibrillar tau oligomers in vitro and in Alzheimer disease. *J Biol Chem*. 2011;286:23063–23076.
5. Hamano T, Shirafuji N, Makino C, et al. Pioglitazone prevents tau oligomerization. *Biochem Biophys Res Commun*. 2016;478:1035–1042.
6. Shirafuji N, Hamano T, Yen SH, et al. Homocysteine increases tau phosphorylation, truncation and oligomerization. *Int J Mol Sci*. 2018;19:891.
7. Gamblin TC, Chen F, Zambrano A, et al. Caspase cleavage of tau: linking amyloid and neurofibrillary tangles in Alzheimer's disease. *Proc Natl Acad Sci U S A*. 2003;100:10032–10037.
8. Chang CW, Shao E, Mucke L, et al. Tau: enabler of diverse brain disorders and target of rapidly evolving therapeutic strategies. *Science*. 2021;371:6532.
9. Ji C, Tang M, Zeidler C, et al. BAG3 and SYNPO (synaptopodin) facilitate phospho-MAPT/tau degradation via autophagy in neuronal processes. *Autophagy*. 2019;15:1199–1213.
10. Šimić G, Babić Leko M, Wray S, et al. Tau protein hyperphosphorylation and aggregation in alzheimer's disease and other tauopathies, and possible neuroprotective strategies. *Biomolecules*. 2016;6:6.
11. Neddens J, Temmel M, Flunkert S, et al. Phosphorylation of different tau sites during progression of Alzheimer's disease. *Acta Neuropathol Commun*. 2018;6:52.
12. Petrucelli L, Dickson D, Kehoe K, et al. CHIP and Hsp70 regulate tau ubiquitination, degradation and aggregation. *Hum Mol Genet*. 2004;13:703–714.
13. Cohen TJ, Guo JL, Hurtado DE, et al. The acetylation of tau inhibits its function and promotes pathological tau aggregation. *Nat Commun*. 2011;2:252.
14. Cook C, Carlomagno Y, Gendron TF, et al. Acetylation of the KXGS motifs in tau is a critical determinant in modulation of tau aggregation and clearance. *Hum Mol Genet*. 2014;23:104–116.
15. Cook C, Stankowski JN, Carlomagno Y, et al. Acetylation: a new key to unlock tau's role in neurodegeneration. *Alzheimers Res Ther*. 2014;6:29.

16. Horiguchi T, Uryu K, Giasson BI, et al. Nitration of tau protein is linked to neurodegeneration in tauopathies. *Am J Pathol.* 2003;163:1021–1031.
17. Reynolds MR, Reyes JF, Fu Y, et al. Tau nitration occurs at tyrosine 29 in the fibrillar lesions of Alzheimer's disease and other tauopathies. *J Neurosci.* 2006;26:10636–10645.
18. Liu F, Iqbal K, Grundke-Iqbal I, Hart GW, Gong CX. O-GlcNAcylation regulates phosphorylation of tau: a mechanism involved in Alzheimer's disease. *Proc Natl Acad Sci U S A.* 2004;101:10804–10809.
19. Kovacech B, Novak M. Tau truncation is a productive posttranslational modification of neurofibrillary degeneration in Alzheimer's disease. *Curr Alzheimer Res.* 2010;7:708–716.
20. Gu J, Xu W, Jin N, et al. Truncation of tau selectively facilitates its pathological activities. *J Biol Chem.* 2020;295:13812–13828.
21. de Calignon A, Fox LM, Pitstick R, et al. Caspase activation precedes and leads to tangles. *Nature.* 2010;464:1201–1204.
22. Metcalfe MJ, Huang Q, Figueiredo-Pereira ME. Coordination between proteasome impairment and caspase activation leading to TAU pathology: neuroprotection by cAMP. *Cell Death Dis.* 2012;3, e326.
23. Hamano T, Gendron TF, Ko LW, Yen SH. Concentration-dependent effects of proteasomal inhibition on tau processing in a cellular model of tauopathy. *Int J Clin Exp Pathol.* 2009;2:561–573.
24. Mizushima N, Komatsu M. Autophagy: renovation of cells and tissues. *Cell.* 2011;147:728–741.
25. Birgisdottir ÅB, Lamark T, Johansen T. The LIR motif—crucial for selective autophagy. *J Cell Sci.* 2013;126:3237–3247.
26. Ganley IG, Lam du H, Wang J, et al. ULK1.ATG13. FIP200 complex mediates mTOR signaling and is essential for autophagy. *J Biol Chem.* 2009;284:12297–12305.
27. Menzies FM, Fleming A, Caricasole A, et al. Autophagy and neurodegeneration: pathogenic mechanisms and therapeutic opportunities. *Neuron.* 2017;93:1015–1034.
28. Hamano T, Hayashi K, Shirafuji N, Nakamoto Y. The implications of autophagy in Alzheimer's disease. *Curr Alzheimer Res.* 2018;15:1283–1296.
29. Itakura E, Kishi-Itakura C, Mizushima N. The hairpin-type tail-anchored SNARE syntaxin 17 targets to autophagosomes for fusion with endosomes/lysosomes. *Cell.* 2012;151:1256–1269.
30. Suzuki K, Terry RD. Fine structural localization of acid phosphatase in senile plaques in Alzheimer's presenile dementia. *Acta Neuropathol.* 1967;8:276–284.
31. Nixon RA, Wegiel J, Kumar A, et al. Extensive involvement of autophagy in Alzheimer disease: an immuno-electron microscopy study. *J Neuropathol Exp Neurol.* 2005;64:113–122.
32. Perez SE, He B, Nadeem M, et al. Hippocampal endosomal, lysosomal, and autophagic dysregulation in mild cognitive impairment: correlation with Aβ and tau pathology. *J Neuropathol Exp Neurol.* 2015;74:345–358.
33. Piras A, Collin L, Grüninger F, et al. Autophagic and lysosomal defects in human tauopathies: analysis of post-mortem brain from patients with familial Alzheimer disease, corticobasal degeneration and progressive supranuclear palsy. *Acta Neuropathol Commun.* 2016;4:22.
34. Jiang S, Bhaskar K. Degradation and transmission of tau by autophagic-endolysosomal networks and potential therapeutic targets for tauopathy. *Front Mol Neurosci.* 2020;13, 586731.
35. Nixon RA. Autophagy, amyloidogenesis and Alzheimer disease. *J Cell Sci.* 2007;120:4081–4091.
36. Boland B, Kumar A, Lee S, et al. Autophagy induction and autophagosome clearance in neurons: relationship to autophagic pathology in Alzheimer's disease. *J Neurosci.* 2008;28:6926–6937.
37. Yu WH, Cuervo AM, Kumar A, et al. Macroautophagy—a novel Beta-amyloid peptide-generating pathway activated in Alzheimer's disease. *J Cell Biol.* 2005;171:87–98.

38. Lee JH, Yu WH, Kumar A, et al. Lysosomal proteolysis and autophagy require presenilin 1 and are disrupted by Alzheimer-related PS1 mutations. *Cell*. 2010;141:1146–1158.
39. Ferrer I, Aguiló García M, et al. Involvement of oligodendrocytes in tau seeding and spreading in tauopathies. *Front Aging Neurosci*. 2019;11:112.
40. Nogalska A, D'Agostino C, Terracciano C, et al. Impaired autophagy in sporadic inclusion-body myositis and in endoplasmic reticulum stress-provoked cultured human muscle fibers. *Am J Pathol*. 2010;177:1377–1387.
41. Askanas V, Engel WK, Nogalska A. Sporadic inclusion-body myositis: a degenerative muscle disease associated with aging, impaired muscle protein homeostasis and abnormal mitophagy. *Biochim Biophys Acta*. 2015;1852:633–643.
42. Reddy PH, Oliver DM. Amyloid beta and phosphorylated tau-induced defective autophagy and mitophagy in Alzheimer's disease. *Cell*. 2019;8:488.
43. Kuang H, Tan CY, Tian HZ, et al. Exploring the bi-directional relationship between autophagy and Alzheimer's disease. *CNS Neurosci Ther*. 2020;26:155–166.
44. Wang Y, Martinez-Vicente M, Krüger U, et al. Tau fragmentation, aggregation and clearance: the dual role of lysosomal processing. *Hum Mol Genet*. 2009;18:4153–4170.
45. Feng Q, Luo Y, Zhang XN, et al. MAPT/Tau accumulation represses autophagy flux by disrupting IST1-regulated ESCRT-III complex formation: a vicious cycle in Alzheimer neurodegeneration. *Autophagy*. 2020;16:641–658.
46. Chambraud B, Daguinot C, Guillemeau K, et al. Decrease of neuronal FKBP4/FKBP52 modulates perinuclear lysosomal positioning and MAPT/Tau behavior during MAPT/Tau-induced proteotoxic stress. *Autophagy*. 2021;1–20. https://doi.org/10.1080/15548627.2021.1875611.
47. Meduri G, Guillemeau K, Dounane O, et al. Caspase-cleaved Tau-D (421) is colocalized with the immunophilin FKBP52 in the autophagy-endolysosomal system of Alzheimer's disease neurons. *Neurobiol Aging*. 2016;46:124–137.
48. Keller JN, Hanni KB, Markesbery WR. Impaired proteasome function in Alzheimer's disease. *J Neurochem*. 2000;75:436–439.
49. Shimura H, Schwartz D, Gygi SP, Kosik KS. CHIP-Hsc70 complex ubiquitinates phosphorylated tau and enhances cell survival. *J Biol Chem*. 2004;279:4869–4876.
50. Mauthe M, Orhon I, Rocchi C, et al. Chloroquine inhibits autophagic flux by decreasing autophagosome-lysosome fusion. *Autophagy*. 2018;14:1435–1455.
51. Oyama F, Murakami N, Ihara Y. Chloroquine myopathy suggests that tau is degraded in lysosomes: implication for the formation of paired helical filaments in Alzheimer's disease. *Neurosci Res*. 1998;31:1–8.
52. Ko LW, Rush T, Sahara N, et al. Assembly of filamentous tau aggregates in human neuronal cells. *J Alzheimers Dis*. 2004;6:605–622.
53. Hamano T, Gendron TF, Causevic E, et al. Autophagic-lysosomal perturbation enhances tau aggregation in transfectants with induced wild-type tau expression. *Eur J Neurosci*. 2008;27:1119–1130.
54. Hamano T, Yen SH, Gendron T, et al. Pitavastatin decreases tau levels via the inactivation of Rho/ROCK. *Neurobiol Aging*. 2012;33:2306–2320.
55. Hamano T, Shirafuji N, Yen SH, et al. Rho-kinase ROCK inhibitors reduce oligomeric tau protein. *Neurobiol Aging*. 2020;89:41–54.
56. Ozcelik S, Fraser G, Castets P, et al. Rapamycin attenuates the progression of tau pathology in P301S tau transgenic mice. *PLoS One*. 2013;8, e62459.
57. Dolan PJ, Johnson GV. A caspase cleaved form of tau is preferentially degraded through the autophagy pathway. *J Biol Chem*. 2010;285:21978–21987.
58. Caballero B, Bourdenx M, Luengo E, et al. Acetylated tau inhibits chaperone-mediated autophagy and promotes tau pathology propagation in mice. *Nat Commun*. 2021;12:2238.
59. Vaz-Silva J, Gomes P, Jin Q, et al. Endolysosomal degradation of Tau and its role in glucocorticoid-driven hippocampal malfunction. *EMBO J*. 2018;37, e99084.

60. Fang EF, Hou Y, Palikaras K, et al. Mitophagy inhibits amyloid-β and tau pathology and reverses cognitive deficits in models of Alzheimer's disease. *Nat Neurosci.* 2019;22:401–412.
61. Cummins N, Tweedie A, Zuryn S, et al. Disease-associated tau impairs mitophagy by inhibiting Parkin translocation to mitochondria. *EMBO J.* 2019;38:e99360.6.
62. Ott A, Stolk RP, van Harskamp F, et al. Diabetes mellitus and the risk of dementia: the Rotterdam Study. *Neurology.* 1999;53:1937–1942.
63. Moloney AM, Griffin RJ, Timmons S, et al. Defects in IGF-1 receptor, insulin receptor and IRS-1/2 in Alzheimer's disease indicate possible resistance to IGF-1 and insulin signalling. *Neurobiol Aging.* 2010;31:224–243.
64. Chatterjee S, Ambegaokar SS, Jackson GR, Mudher A. Insulin-mediated changes in tau hyperphosphorylation and autophagy in a drosophila model of tauopathy and neuroblastoma cells. *Front Neurosci.* 2019;13:801.
65. Miao Y, Guo D, Li W, Zhong Y. Diabetes promotes development of Alzheimer's disease through suppression of autophagy. *J Alzheimers Dis.* 2019;69:289–296.
66. Chen JL, Luo C, Pu D, et al. Metformin attenuates diabetes-induced tau hyperphosphorylation in vitro and in vivo by enhancing autophagic clearance. *Exp Neurol.* 2019;311:44–56.
67. Merezhko M, Brunello CA, Yan X, et al. Secretion of tau via an unconventional non-vesicular mechanism. *Cell Rep.* 2018;25:2027–2035.e4.
68. Dujardin S, Bégard S, Caillierez R, et al. Ectosomes: a new mechanism for non-exosomal secretion of tau protein. *PLoS One.* 2014;9, e100760.
69. Spitzer P, et al. Microvesicles from cerebrospinal fluid of patients with Alzheimer's disease display reduced concentrations of tau and APP protein. *Sci Rep.* 2019;9, 7089.
70. Brunello CA, Merezhko M, Uronen RL, Huttunen HJ. Mechanisms of secretion and spreading of pathological tau protein. *Cell Mol Life Sci.* 2020;77:1721–1744.
71. Min S-W, Chen X, Tracy TE, et al. Critical role of acetylation in tau-mediated neurodegeneration and cognitive deficits. *Nat Med.* 2015;21:1154–1162.
72. Chen X, Li Y, Wang C, et al. Promoting tau secretion and propagation by hyperactive p300/CBP via autophagy-lysosomal pathway in tauopathy. *Mol Neurodegener.* 2020;15:2.
73. Tang Z, Ioja E, Bereczki E, et al. mTor mediates tau localization and secretion: implication for Alzheimer's disease. *Biochim Biophys Acta.* 2015;1853:1646–1657.
74. Kang S, Son SM, Baik SH, et al. Autophagy-mediated secretory pathway is responsible for both normal and pathological tau in neurons. *J Alzheimers Dis.* 2019;70:667–680.
75. Lopez A, Lee SE, Wojta K, et al. A152T tau allele causes neurodegeneration that can be ameliorated in a zebrafish model by autophagy induction. *Brain.* 2017;140:1128–1146.
76. Jiang T, Yu JT, Zhu XC, et al. Temsirolimus attenuates tauopathy in vitro and in vivo by targeting tau hyperphosphorylation and autophagic clearance. *Neuropharmacology.* 2014;85:121–130.
77. Wang C, Zhang X, Teng Z, et al. Downregulation of PI3K/Akt/mTOR signaling pathway in curcumin-induced autophagy in APP/PS1 double transgenic mice. *Eur J Pharmacol.* 2014;740:312–320.
78. Congdon EE, Wu JW, Myeku N, et al. Methylthioninium chloride (methylene blue) induces autophagy and attenuates tauopathy in vitro and in vivo. *Autophagy.* 2012;8:609–622.
79. Shimada K, Motoi Y, Ishiguro K, et al. Long-term oral lithium treatment attenuates motor disturbance in tauopathy model mice: implications of autophagy promotion. *Neurobiol Dis.* 2012;46:101–108.
80. Krüger U, Wang Y, Kumar S, Mandelkow EM. Autophagic degradation of tau in primary neurons and its enhancement by trehalose. *Neurobiol Aging.* 2012;33:2291–2305.

81. Ma QL, Zuo X, Yang F, et al. Curcumin suppresses soluble tau dimers and corrects molecular chaperone, synaptic, and behavioral deficit in aged human tau transgenic mice. *J Biol Chem.* 2013;288:40564065.
82. Liu D, Pitta M, Jiang H, et al. Nicotinamide forestalls pathology and cognitive decline in Alzheimer mice: evidence for improved neuronal bioenergetics and autophagy procession. *Neurobiol Aging.* 2013;34:1564–1580.
83. Umeda T, Ono K, Sakai A, et al. Rifampicin is a candidate preventive medicine against amyloid-β and tau oligomers. *Brain.* 2016;139:1568–1586.

CHAPTER 8

BAG3 promotes tau clearance by regulating autophagy and other vacuolar-dependent degradative processes

Changyi Ji, Heng Lin, and Gail V.W. Johnson
Department of Anesthesiology and Perioperative Medicine, University of Rochester Medical Center, Rochester, NY, United States

1. Introduction

Autophagy is an essential degradation process for a cell to remove unnecessary and damaged components, and therefore it is key for maintaining cellular homeostasis.[1] Initially, it was described as a defensive mechanism in starvation conditions[2–4]; however, it is now evident that this process is crucial for the orderly degradation and recycling of cytosolic constitutes.[1,5] In this process, long-lived proteins, lipids, carbohydrates, or damaged cellular components are targeted to lysosomes for degradation.[1] The breakdown can then serve as building blocks for cells to make new cellular components.[1] Therefore, active autophagy is a key process to rejuvenate a cell. Dysfunction of autophagy is linked to many diseases, including neurodegenerative diseases and cancer.[1]

There are four types of autophagy: macroautophagy, microautophagy, chaperone-mediated autophagy, and crinophagy.[1,6] Macroautophagy is the most thoroughly researched form of autophagy. In macroautophagy, unwanted cellular components, such as misfolded proteins and damaged mitochondria, are targeted and sequestered into a double-membraned vesicle, the autophagosome. Autophagosomes then fuse with lysosomes and form autolysosomes. The content of autolysosomes is degraded by hydrolases that are active in the acidic environment.[1] In this chapter, the term autophagy will be used throughout the text to refer to macroautophagy.

Autophagy-related (ATG) proteins orchestrate the ordered process of autophagy.[7] Initially identified in yeast and further characterized in other organisms, there are 36 ATG proteins, of which 18 belongs to the core

https://doi.org/10.1016/B978-0-323-89906-2.00011-3

machinery of autophagy.[7,8] Induction of autophagy involves dephosphorylation and activation of the Unc-51-like kinases (ULKs) (mammalian homologs of ATG1), which are part of a protein complex containing ATG13, ATG101, and FIP200.[9,10] ULK activates Beclin-1 (mammalian homologs of ATG6), which is a member of a complex with PIK3R4 (p150), ATG14L, and the class III phosphatidylinositol 3-phosphate kinase VPS34.[9,11,12] Once activated, both complexes are localized to the phagophore.[9,11,12] VPS34 phosphorylates phosphatidylinositol to generated phosphatidylinositol 3-phosphate, which serves as a docking spots for WIPI2 (WD-repeat protein interacting with phosphoinositides protein 2) and further recruits ATG16L1 to the surface of phagophore.[13] ATG16L1 interacts with ATG5 and ATG7 and forms ubiquitin-like conjugation complex.[14] This complex binds and activates ATG3, which physically attaches to ATG8 and results in lipidation of ATG8.[14] Lipidated ATG8 not only contributes to the closure of autophagosomes[15] but also enables the sequestration of specific cargos or adaptor proteins such as Sequestosome-1/p62, CALCOCO2/NDP52, and NBR1.[16,17] The mature autophagosomes then fuse with lysosomes a process that is orchestrated by various proteins including SNAREs.[18] The content inside the autolysosomes is then degraded by various hydrolases in the acidic environment, and the resulting degradation products are then released from the autolysosomes.[19]

Postmitotic cells, such as neurons, rely heavily on autophagy to maintain a healthy proteome as they cannot dilute the misfolded or aggregated proteins by cell division.[20–23] Autophagy is basally active in different subcellular compartments (e.g., axons, dendrites, and cell bodies) in neurons.[24] Loss of this activity causes neurodegeneration, significant behavioral deficits, and premature death in mouse models.[22,23] Autophagosome biogenesis is constitutively active and follows an ordered pathway.[25] In each neuronal compartment, autophagy is spatially regulated in a compartment-specific manner.[24–30]

In neurons, chaperone-assisted selective autophagy plays a significant role in maintaining proteostasis.[31] As the name implies, this form of autophagy is mediated by chaperones and cochaperones, which works in cooperation with autophagy adaptors to direct specific cargos into autophagosomes and facilitate the autophagic process.[32] For example, it has been shown that heat shock protein 70 (HSP70) and its cochaperone BCL2-associated athanogene 3 (BAG3) cooperate with the autophagy receptor SQSTM1/p62 to mediate the degradation of aggregation-prone proteins mutant huntingtin and superoxide dismutase 1.[31,33] Further, soluble tau and its phosphorylated forms in neurons are also substrates of this HSP70/BAG3-mediated autophagy.[34] Indeed, there is a growing awareness that BAG3 plays a crucial role in neuronal proteostasis.[35]

2. BAG3 protein

BAG3 belongs to the BAG family of cochaperones that were first identified as interactors of the ATPase domain of HSP70 using a yeast two-hybrid screen.[36,37] BAG3 is highly conserved in mammals and has molecular weight of 75 kDa.[38,39] A truncated BAG3 with a molecular weight of 40 kDa has also been detected in neuronal synaptosomes; however, its function has not yet been fully characterized.[40] BAG3 is predominantly located in the cytosol, with a minor fraction in nucleus under physiological conditions.[41] In addition, BAG3 expression is increased in response to stress,[41] and its subcellular localization is often altered in stress conditions. For example, BAG3 translocates to the nucleus or aggregates under acute stress or upon viral infection.[31,42] In addition, in neurons, BAG3 levels increase with age.[32]

BAG3 is a multiple-domain protein, which consists of a BAG domain at its C terminus, a tryptophan containing (WW) domain at its N terminus, two isoleucine-proline-valine (IPV) domains and a proline-rich (PXXP) domain in the central region.[37,43] BAG domain interacts with HSP70. IPV domains recruits small heat shock proteins such as HSPB6 and HSPB8.[37,43] Therefore, BAG3 serves as a hub to bring these two types of chaperones together. The ternary complex formed plays a critical role in protein quality control, in which BAG3 functions as a nucleotide exchange factor to help HSP70 refold protein clients that are delivered by HSPB or DNAJ proteins.[44] The PXXP domain of BAG3 interacts with microtubule motor dynein, which is likely to facilitate the transport of protein clients between different cellular compartments.[31] This interaction was upregulated when neurons were treated with proteasome inhibitor MG132, a condition that causes proteostatic stress (Fig. 1). One of the important functions of BAG3 is to regulate selective autophagy via specific interactions with its partner proteins.[43] Studies have shown that the WW domain of BAG3 is required for autophagy induction in glioma cells and muscle cells.[45,46] In particular, the WW domain of BAG3 interacts with a proline-rich motif (PPPY) in SYNPO2/myopodin (synaptopodin 2), an actin cytoskeleton adaptor protein, which facilitates the formation of autophagosomes during mechanical stress in muscle cells.[46]

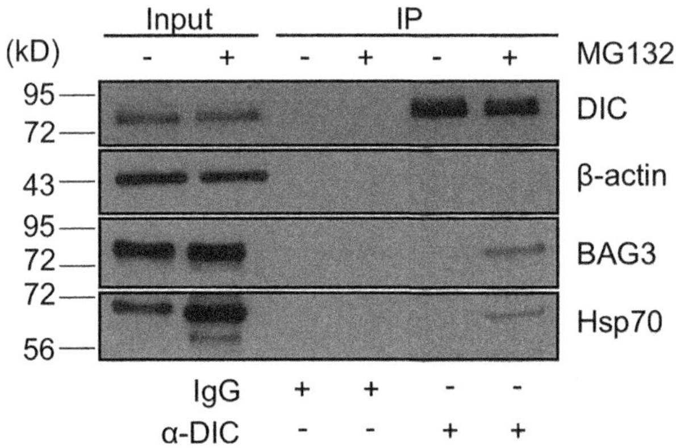

Fig. 1 BAG3 interacts with dynein intermediate chain (DIC) in mature neurons. Mature rat cortical neurons were treated with 10 μM MG132 for 24 h followed by immunoprecipitation of DIC. Both BAG3 and HSPA/Hsp70 were detected in the precipitated fraction.

2.1 BAG3 mediated tau degradation

In neurons, regulation of BAG3 expression mediates soluble tau degradation.[34] During proteostatic stress, such as proteasome inhibition, the expression of BAG3 significantly increases, which may be regulated in part by the JNK signaling pathway.[34] Upregulation of BAG3 results in decreased soluble total tau and phosphorylated tau.[34] Furthermore, overexpression of BAG3 alone is sufficient to promote soluble tau degradation in neurons.[34] BAG3 expression decreases during aging,[47,48] and in Alzheimer's disease brain the levels of BAG3 are significantly lower compared to age-matched controls.[48] Along with other autophagy-associated proteins such as the autophagy receptors NDP52, NBR1, and SQSTM1/p62, the expression of BAG3 may also be regulated by nuclear factor (erythroid-derived 2)-like 2 (Nrf2).[47] Reduction of BAG3 in Nrf2 knockout mice is closely associated with the accumulation of total tau and phosphorylated tau as these mice age.[47] Further, studies in human brain demonstrate that BAG3 is a master regulator gene of a tau "homeostasis signature", which differentiate excitatory neurons from inhibitory neurons in terms of cellular vulnerability to tau pathology.[49] In the coexpression network analysis, the BAG3 gene was identified as a hub gene among the subproteomes that modulates tau aggregation, degradation, and neurofibrillary tangle formation.[49] Excitatory neurons express relatively lower levels of BAG3 expression, which is associated with increased vulnerability to tau pathology, whereas inhibitory neurons express higher levels of BAG3 and are resistant to the accumulation of pathological tau.[34,49,50]

2.2 Mechanisms of BAG3-mediated tau degradation

Although BAG3 as a scaffold protein expressed in many different cell type, it recruit cell specific protein partners to coordinate autophagic machinery.[37,43,51] To investigate the mechanisms of BAG3-mediated tau degradation in neurons, binding partners that interact with the different domains of BAG3 were investigated for how they impacted vacuolar-dependent tau degradation.[50,52] The WW domain of BAG3 interacts with a proline-rich motif (PPPY) in its protein-binding partners.[46,53] A peptide array screen using purified BAG3 WW domain revealed 72 top hits of PPxY-containing peptides derived from the human proteome. A 12-mer peptide containing the PPSY motif of synaptopodin (SYNPO), a close homolog of SYNPO2, was identified.[46] SYNPO is a postsynaptic protein that links to the structural and functional plasticity of dendritic spines.[54–57] SYNPO functions as an organizer of the actin cytoskeleton[58] and controls calcium release from local stores at spines.[59] To verify this interaction, coimmunoprecipitation experiments were performed in Hela cells overexpressing the epitope tagged proteins, as well as in primary rat cortical neurons with endogenous levels of expression.[50] SYNPO coimmunoprecipitated with BAG3 in both experimental conditions. The WW domain is essential for BAG3 to bind SYNPO, as the tryptophan to alanine mutations in the WW domain ablated the interaction. SYNPO contains two PPxP motifs. In a similar experimental set-up, a mutated form of SYNPO with the first PPTY motif (PAAY) retains the ability to bind to BAG3, whereas inactivation of the second PPSY motif (PAAY) strongly reduces the interaction between BAG3 and SYNPO. Therefore, the PPSY motif of SYNPO is the preferential BAG3-binding site, which is consistent with the observations from the peptide array screen.[50]

Interaction between SYNPO and BAG3 raises the possibility that SYNPO may participate in cochaperone BAG3-assisted selective autophagy.[50] Supporting a role of SYNPO in autophagy, coimmunoprecipitation of SQSTM1/p62 was observed in a SYNPO affinity pull-down fraction from neurons. This interaction did not depend on the presence of BAG3.[50] Furthermore, colocalization of SYNPO with BAG3, SQSTM1/p62, as well as the chaperone HSP70 was observed in mature primary cortical neurons. Approximately 10% of SYNPO-positive puncta contain BAG3.[50] The spatial proximation of BAG3/SYNPO complex to the autophagy adaptor protein SQSTM1/p62 further suggests their involvement in neuronal autophagy. Indeed, endogenous LC3B was observed in BAG3 or SYNPO-containing puncta in neuronal processes.[50] Exogenous expression of mCherry tagged

LC3 also appears to colocalize with SQSTM1/p62, BAG3, and SYNPO-containing puncta (Fig. 2). Taken together, these data suggest SYNPO and BAG3 are physically associated with the autophagy machinery in the processes of mature cortical neurons.

The role of BAG3 and SYNPO in neuronal autophagy was further investigated by knockdown of endogenous SYNPO or BAG3.[50] Increased steady-state levels of LC3B-II were observed in both conditions. The increase did not occur in the presence of chloroquine, a reagent that inhibits the acidification and function of lysosomes.[50] This observation suggests that LC3B accumulation in the absence of BAG3 or SYNPO is not due to an enhancement of autophagy initiation, but rather a block of LC3B degradation. Decreased autophagy flux in neuronal processes was detected by a tandem-fluorescence tagged reporter to quantify autophagosomes and autolysosomes.[50] This blockage of autophagic flux was due to an inhibition of autophagosome-lysosome fusion as suggested by less colocalization between LC3B and LAMP1 positive puncta in neuronal processes.[50] BAG3 and SYNPO do not directly regulate lysosome function as knockdown of BAG3 or SYNPO did not affect the maturation of the lysosomal protease cathepsin L or the total level of LAMP1.[50] Consistent with the role of BAG3 and SYNPO in autophagosome-lysosome fusion, both proteins appear to localize to the outer membrane of autophagic vesicles as revealed by the proteinase K protection assay.[50]

Phosphorylated tau is one of the client proteins targeted by BAG3-associated autophagy.[34,49,50,52] In mature cortical neurons, suppression of BAG3 or SYNPO did not change the level of phospho-tau Thr231 or Ser396/Ser404, which are enriched in somatic and axonal compartments. However, the levels of phospho-tau Ser262 significantly increased when either BAG3 or SYNPO was knockdown.[50] Phospho-tau Ser262 appears to associate with dendritic compartments compared to other tau species.[60,61] When the expression of BAG3 or SYNPO was decreased in neurons, increased overlap between phospho-tau Ser262 and LC3B positive vesicles was observed at PSD95-positive postsynaptic densities.[50] Therefore, we conclude that BAG3- and SYNPO-mediated autophagy facilitates the removal of phosphorylated tau from postsynaptic terminals. This result also indicates that BAG3, in cooperation with SYNPO, participates in compartment-specific regulation of neuronal autophagy. Efficient removal of unwanted proteins at the distal processes and dynamic synapses is crucial to maintain a healthy neuronal proteome.

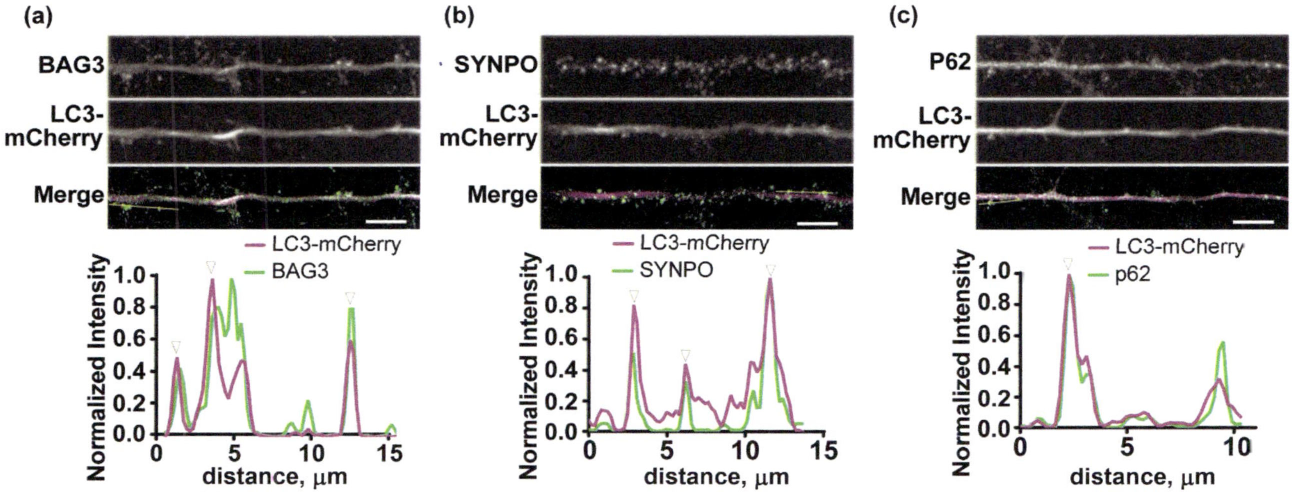

Fig. 2 Colocalization of BAG3, SYNPO, or SQSTM1/p62 with LC3-mCherry-positive puncta in neuronal processes. LC3-mCherry was expressed in neurons by lentivirus. Neurons were coimmunostained for BAG3, SYNPO, or SQSTM1/p62, respectively. Overlap of BAG3 (A), SYNPO (B) or SQSTM1/p62, (C) with LC3-mCherry puncta was observed in neuronal processes. The corresponding line scans are shown on the bottom. Arrowheads denote areas of overlap. Scale bar: 10 μm.

To further uncover potential BAG3 interactors in neurons, an unbiased proteomic screen was performed by immunoprecipitation of endogenous BAG3.[52] A total of 127 proteins were found to selectively coprecipitate with BAG3 from neurons.[52] KEGG pathway analysis revealed the top pathway represented by the BAG3-associated proteins was endocytosis.[52] Further analysis of this subgroup of proteins identified a cluster that interact with each other and closely associate with chaperones and cochaperones.[52] These proteins can be further divided into three subgroups: chaperones, cytoskeletal proteins, and proteins that modulate the activity of small GTPases.[52] Among the small GTPase regulators, TBC1 Domain Family Member 10B (TBC1D10B) was identified as a bona fide BAG3 interactor.[52] Given the role of TBC1D10B as the primary GTPase-activating protein for Rab35-an endosomal small GTPase,[62,63] the interaction between TBC1D10B and BAG3 and the involvement of BAG3 in endosome-lysosome pathway were further investigated in neurons.

BAG3 forms a tripartite complex with TBC1D10B and HSP70.[52] The interaction of BAG3 with TBC1D10B is facilitated by HSP70.[52] Through the association with TBC1D10B, BAG3 modulates Rab35 activity by preventing TBC1D10B from promoting GTP hydrolysis and therefore prolonging Rab35 in a GTP-bounded state.[52] Reduction of BAG3 levels or a BAG3 mutant deficient in HSP70 binding leads to lower Rab35 activity.[52] Since Rab35 plays a critical role in the endosomal sorting complex required for transport (ESCRT) system,[63,64] we examined the impact of BAG3 on ESCRT function. The recruitment of Hrs, an ESCRT 0 component, to the surface of the endosome is mediated by Rab35.[65,66] ESCRT 0 plays a role in directing cargo to endosomes for engulfment, and also facilitating the recruitment of ESCRT machinery to form intraluminal vesicles.[65] The mobility of Hrs and its association to Rab35 are substantially enhanced by the presence of BAG3, whereas loss of BAG3 decreases Hrs and Rab35 interaction.[52]

Phospho-tau is one cargo that is directed to *endo*-lysosomal degradation through the ESCRT system.[66,67] The modulation of the ESCRT system by BAG3 through Rab35/Hrs also plays an important role in regulating the degradation of phosphorylated tau through this pathway.[52] Reduction of BAG3 increases the accumulation of phosphorylated tau, whereas depletion of TBC1D10B decreases the level of phosphorylated tau in neurons.[52] Consistent with the notion that BAG3/TBC1D10B plays a role in regulate tau sorting into ESCRT system, BAG3 knockdown decreases colocalization between phospho-tau and CHMP2B, an ESCRT-III component; whereas, depletion of TBC1D10B or BAG3/TBC1D10B increases the localization of phospho-tau to the ESCRT system.[52]

The important role of BAG3 in modulating endolysosomal degradation of pathological tau was further confirmed in a tauopathy mice model overexpressing human tau with the P301S mutation (PS19 line).[52] Increased BAG3 expression in hippocampi of these mice significantly reduces the level of pathological tau.[52] Further, significantly more Hrs was recruited to Rab35, and significantly more phospho-tau was sorted to the *endo*-lysosomal pathway via the ESCRT system.[52] As a result, the integrity of neuronal dendrites and postsynaptic densities was supported in the PS19 mice with BAG3 overexpression compared to controls.[52] This study further underscores the role of BAG3 in maintaining tau homeostasis in neurons. It is clear that in addition to the regulation of the autophagic pathway,[34,50] BAG3 also modulates endolysosomal pathway to facilitate the removal of phosphorylated tau.[52] As the level of BAG3 decreases as a function of age or in pathological conditions such as AD,[47,48] increasing the expression level of BAG3 would be beneficial to prevent the accumulation of pathological tau species, and therefore maintain a health neuronal processes.

3. Summary

BAG3 along with its partner proteins plays a vital role in protein quality control.[32,37,43] These roles include, but not limited to, support of proper protein folding, prevention of protein aggregation, and targeting unwanted proteins for vacuolar-mediated degradation.[37] One likely mechanism that enables BAG3 to perform such versatile roles in protein quality control is the ability to recruit various protein partners through its modular domains. The identities of BAG3 protein partners appear to be cell-type dependent.[45,46,48,50,53,68,69] In neurons, BAG3 interacts with an array of proteins that are specifically adapted for maintaining proteases in the unique polar structure of the neuron (Fig. 3).[50,52] Efficient protein turnover mechanisms in distal processes, especially at dynamic synapses, are essential to support neuronal activities.[24,25,48] Given their postmitotic nature, an efficient protein degradation system is crucial to maintaining a healthy proteome in neurons. In mature cortical neurons, BAG3 cooperates with SYNPO to recruit autophagy machinery to remove phospho-tau in neuronal processes and postsynaptic densities.[50] Beyond its role as a regulator of autophagy, BAG3 also facilitates tau degradation via the endolysosomal system through an interaction with TBC1D10B to regulate the activity of endosomal GTPase

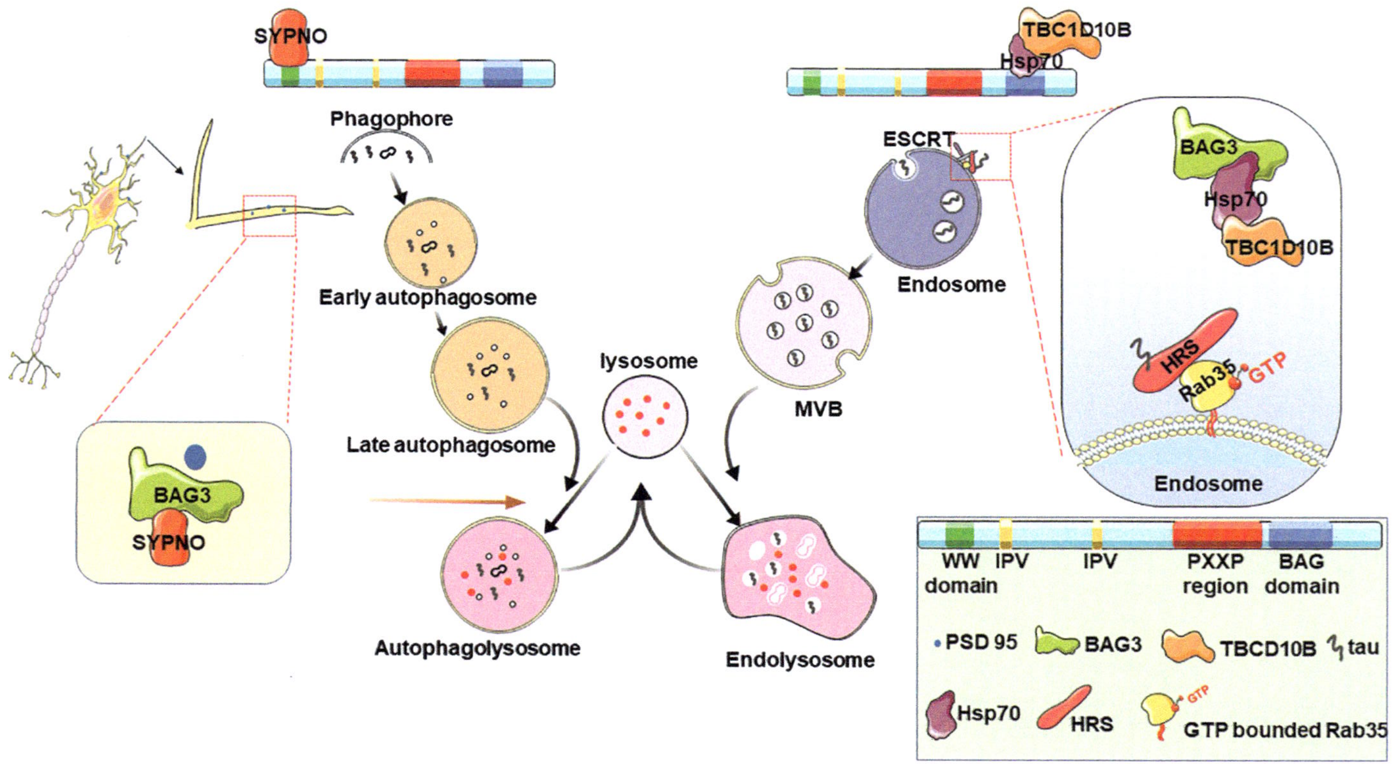

Fig. 3 BAG3-mediated vacuolar-dependent tau degradation pathways in neurons. BAG3 cooperates with SYNPO to mediated autophagosome-lysosome-dependent degradation of phospho-tau in dendrites. In addition, BAG3 interacts with TBC1D10B through Hsp70 to modulate Rab35 activity, which promote tau degradation through the *endo*-lysosome dependent pathway.

Rab35.[52] Both degradation pathways are important for the removal of potentially pathological tau species. These diverse roles of BAG3 in these different, lysosome-dependent tau clearance pathways further support the notion that this protein is a master regulator of tau homeostasis in neurons.[49] Indeed, there is a growing body of evidence that dysregulation of BAG3 is likely a contributing factor in the progression of neurodegenerative proteinopathies in general, and more specifically tauopathies such as Alzheimer's disease and related frontotemporal dementias.[48,49] Further studies are needed to be more fully delineate the role of BAG3 dysregulation in these diseases and to develop intervention strategies based on these knowledge to ameliorate the progression of these diseases.

Acknowledgments

The work from the authors' lab was supported by National Institute of Health (NIH) grants R56 AG067739, R01 NS098769, and R01 AG073121.

References

1. Mizushima N, Komatsu M. Autophagy: renovation of cells and tissues. *Cell.* 2011;147(4):728–741.
2. Mizushima N, et al. In vivo analysis of autophagy in response to nutrient starvation using transgenic mice expressing a fluorescent autophagosome marker. *Mol Biol Cell.* 2004;15(3):1101–1111.
3. Kaushik S, et al. Autophagy in hypothalamic AgRP neurons regulates food intake and energy balance. *Cell Metab.* 2011;14(2):173–183.
4. Kuma A, et al. The role of autophagy during the early neonatal starvation period. *Nature.* 2004;432(7020):1032–1036.
5. Klionsky DJ. Autophagy revisited: a conversation with Christian de Duve. *Autophagy.* 2008;4(6):740–743.
6. Csizmadia T, Juhász G, Martinez AB. Crinophagy mechanisms and its potential role in human health and disease. In: Galluzzi L, ed. *Progress in Molecular Biology and Translational Science.* Academic Press; 2020:239–255. [Chapter Eleven].
7. Mizushima N, Yoshimori T, Ohsumi Y. The role of Atg proteins in autophagosome formation. *Annu Rev Cell Dev Biol.* 2011;27:107–132.
8. Galluzzi L, et al. Molecular definitions of autophagy and related processes. *EMBO J.* 2017;36(13):1811–1836.
9. Russell RC, et al. ULK1 induces autophagy by phosphorylating Beclin-1 and activating VPS34 lipid kinase. *Nat Cell Biol.* 2013;15(7):741–750.
10. Hara T, et al. FIP200, a ULK-interacting protein, is required for autophagosome formation in mammalian cells. *J Cell Biol.* 2008;181(3):497–510.
11. Itakura E, et al. Beclin 1 forms two distinct phosphatidylinositol 3-kinase complexes with mammalian Atg14 and UVRAG. *Mol Biol Cell.* 2008;19(12):5360–5372.
12. Kang R, et al. The Beclin 1 network regulates autophagy and apoptosis. *Cell Death Differ.* 2011;18(4):571–580.
13. Proikas-Cezanne T, et al. WIPI proteins: essential PtdIns3P effectors at the nascent autophagosome. *J Cell Sci.* 2015;128(2):207–217.

14. Dooley HC, et al. WIPI2 links LC3 conjugation with PI3P, autophagosome formation, and pathogen clearance by recruiting Atg12-5-16L1. *Mol Cell.* 2014;55(2):238–252.
15. Fujita N, et al. An Atg4B mutant hampers the lipidation of LC3 paralogues and causes defects in autophagosome closure. *Mol Biol Cell.* 2008;19(11):4651–4659.
16. Kabeya Y, et al. LC3, GABARAP and GATE16 localize to autophagosomal membrane depending on form-II formation. *J Cell Sci.* 2004;117(Pt 13):2805–2812.
17. Park S, et al. Choline dehydrogenase interacts with SQSTM1/p62 to recruit LC3 and stimulate mitophagy. *Autophagy.* 2014;10(11):1906–1920.
18. Furuta N, et al. Combinational soluble N-ethylmaleimide-sensitive factor attachment protein receptor proteins VAMP8 and Vti1b mediate fusion of antimicrobial and canonical autophagosomes with lysosomes. *Mol Biol Cell.* 2010;21(6):1001–1010.
19. Yang Z, et al. Atg22 recycles amino acids to link the degradative and recycling functions of autophagy. *Mol Biol Cell.* 2006;17(12):5094–5104.
20. Ciechanover A, Kwon YT. Degradation of misfolded proteins in neurodegenerative diseases: therapeutic targets and strategies. *Exp Mol Med.* 2015;47, e147.
21. Boland B, et al. Autophagy induction and autophagosome clearance in neurons: relationship to autophagic pathology in Alzheimer's disease. *J Neurosci.* 2008;28(27):6926–6937.
22. Komatsu M, et al. Loss of autophagy in the central nervous system causes neurodegeneration in mice. *Nature.* 2006;441(7095):880–884.
23. Hara T, et al. Suppression of basal autophagy in neural cells causes neurodegenerative disease in mice. *Nature.* 2006;441(7095):885–889.
24. Maday S, Holzbaur EL. Compartment-specific regulation of autophagy in primary neurons. *J Neurosci.* 2016;36(22):5933–5945.
25. Maday S, Holzbaur EL. Autophagosome biogenesis in primary neurons follows an ordered and spatially regulated pathway. *Dev Cell.* 2014;30(1):71–85.
26. Okerlund ND, et al. Bassoon controls presynaptic autophagy through Atg5. *Neuron.* 2017;93(4):897–913.e7.
27. Soukup S-F, et al. A LRRK2-dependent endophilinA phosphoswitch is critical for macroautophagy at presynaptic terminals. *Neuron.* 2016;92(4):829–844.
28. Stavoe AK, et al. KIF1A/UNC-104 transports ATG-9 to regulate neurodevelopment and autophagy at synapses. *Dev Cell.* 2016;38(2):171–185.
29. Vanhauwaert R, et al. The SAC1 domain in synaptojanin is required for autophagosome maturation at presynaptic terminals. *EMBO J.* 2017;36(10):1392–1411.
30. Wang T, et al. Control of autophagosome axonal retrograde flux by presynaptic activity unveiled using Botulinum neurotoxin type A. *J Neurosci.* 2015;35(15):6179–6194.
31. Gamerdinger M, et al. BAG3 mediates chaperone-based aggresome-targeting and selective autophagy of misfolded proteins. *EMBO Rep.* 2011;12(2):149–156.
32. Gamerdinger M, et al. Protein quality control during aging involves recruitment of the macroautophagy pathway by BAG3. *EMBO J.* 2009;28(7):889–901.
33. Minoia M, et al. BAG3 induces the sequestration of proteasomal clients into cytoplasmic puncta: implications for a proteasome-to-autophagy switch. *Autophagy.* 2014;10(9):1603–1621.
34. Lei Z, Brizzee C, Johnson GVW. BAG3 facilitates the clearance of endogenous tau in primary neurons. *Neurobiol Aging.* 2015;36(1):241–248.
35. Lin H, et al. The role of BAG3 in health and disease: a "magic BAG of tricks". *J Cell Biochem.* 2022;123(1):4–21 [accepted for publication].
36. Takayama S, Xie Z, Reed JC. An evolutionarily conserved family of Hsp70/Hsc70 molecular chaperone regulators. *J Biol Chem.* 1999;274(2):781–786.
37. Behl C. Breaking BAG: the co-chaperone BAG3 in health and disease. *Trends Pharmacol Sci.* 2016;37(8):672–688.

38. Wada S, Hamada M, Satoh N. A genomewide analysis of genes for the heat shock protein 70 chaperone system in the ascidian *Ciona intestinalis*. *Cell Stress Chaperones.* 2006;11(1):23–33.
39. Lee J-H, et al. Bis, a Bcl-2-binding protein that synergizes with Bcl-2 in preventing cell death. *Oncogene.* 1999;18(46):6183–6190.
40. Bruno AP, et al. Identification of a synaptosome-associated form of BAG3 protein. *Cell Cycle.* 2008;7(19):3104–3105.
41. Sturner E, Behl C. The role of the multifunctional BAG3 protein in cellular protein quality control and in disease. *Front Mol Neurosci.* 2017;10:177.
42. Kyratsous CA, Silverstein SJ. The co-chaperone BAG3 regulates herpes simplex virus replication. *Proc Natl Acad Sci U S A.* 2008;105(52):20912–20917.
43. Klimek C, et al. BAG3-mediated proteostasis at a glance. *J Cell Sci.* 2017;130(17):2781 2788.
44. Rauch JN, et al. BAG3 is a modular, scaffolding protein that physically links heat shock protein 70 (Hsp70) to the small heat shock proteins. *J Mol Biol.* 2017;429(1):128–141.
45. Merabova N, et al. WW domain of BAG3 is required for the induction of autophagy in glioma cells. *J Cell Physiol.* 2015;230(4):831–841.
46. Ulbricht A, et al. Cellular mechanotransduction relies on tension-induced and chaperone-assisted autophagy. *Curr Biol.* 2013;23(5):430–435.
47. Tang M, et al. Nrf2 mediates the expression of BAG3 and autophagy cargo adaptor proteins and tau clearance in an age-dependent manner. *Neurobiol Aging.* 2017;63:128–139.
48. Zhou J, et al. Cyclin-dependent kinase 5–dependent BAG3 degradation modulates synaptic protein turnover. *Biol Psychiatry.* 2020;87(8):756–769.
49. Fu H, et al. A tau homeostasis signature is linked with the cellular and regional vulnerability of excitatory neurons to tau pathology. *Nat Neurosci.* 2019;22(1):47–56.
50. Ji C, et al. BAG3 and SYNPO (synaptopodin) facilitate phospho-MAPT/Tau degradation via autophagy in neuronal processes. *Autophagy.* 2019;15(7):1199–1213.
51. Rosati A, et al. BAG3: a multifaceted protein that regulates major cell pathways. *Cell Death Dis.* 2011;2(4):e141.
52. Lin H, et al. BAG3 regulation of Rab35 mediates the ESCRT/endolysosome pathway and tau clearance. *bioRxiv.* 2021. p. 2021.01.25.428055.
53. Iwasaki M, et al. BAG3 directly associates with guanine nucleotide exchange factor of Rap1, PDZGEF2, and regulates cell adhesion. *Biochem Biophys Res Commun.* 2010;400(3):413–418.
54. Mundel P, et al. Synaptopodin: an actin-associated protein in telencephalic dendrites and renal podocytes. *J Cell Biol.* 1997;139(1):193–204.
55. Deller T, et al. Synaptopodin-deficient mice lack a spine apparatus and show deficits in synaptic plasticity. *Proc Natl Acad Sci U S A.* 2003;100(18):10494–10499.
56. Kremerskothen J, et al. Synaptopodin, a molecule involved in the formation of the dendritic spine apparatus, is a dual actin/alpha-actinin binding protein. *J Neurochem.* 2005;92(3):597–606.
57. Vlachos A, et al. Synaptopodin regulates plasticity of dendritic spines in hippocampal neurons. *J Neurosci.* 2009;29(4):1017–1033.
58. Asanuma K, et al. Synaptopodin orchestrates actin organization and cell motility via regulation of RhoA signalling. *Nat Cell Biol.* 2006;8(5):485–491.
59. Korkotian E, Frotscher M, Segal M. Synaptopodin regulates spine plasticity: mediation by calcium stores. *J Neurosci.* 2014;34(35):11641–11651.
60. Zempel H, et al. Abeta oligomers cause localized Ca(2+) elevation, missorting of endogenous tau into dendrites, tau phosphorylation, and destruction of microtubules and spines. *J Neurosci.* 2010;30(36):11938–11950.

61. Xia D, Li C, Gotz J. Pseudophosphorylation of Tau at distinct epitopes or the presence of the P301L mutation targets the microtubule-associated protein Tau to dendritic spines. *Biochim Biophys Acta*. 2015;1852(5):913–924.
62. Hsu C, et al. Regulation of exosome secretion by Rab35 and its GTPase-activating proteins TBC1D10A–C. *J Cell Biol*. 2010;189(2):223–232.
63. Klinkert K, Echard A. Rab35 GTPase: a central regulator of Phosphoinositides and F-actin in endocytic recycling and beyond. *Traffic*. 2016;17(10):1063–1077.
64. Sheehan P, et al. Activity-dependent degradation of synaptic vesicle proteins requires Rab35 and the ESCRT pathway. *J Neurosci*. 2016;36(33):8668–8686.
65. Bache KG, et al. Hrs regulates multivesicular body formation via ESCRT recruitment to endosomes. *J Cell Biol*. 2003;162(3):435–442.
66. Vaz-Silva J, et al. Endolysosomal degradation of Tau and its role in glucocorticoid-driven hippocampal malfunction. *EMBO J*. 2018;37(20).
67. Chen JJ, et al. Compromised function of the ESCRT pathway promotes endolysosomal escape of tau seeds and propagation of tau aggregation. *J Biol Chem*. 2019;294(50):18952–18966.
68. Fontanella B, et al. The co-chaperone BAG3 interacts with the cytosolic chaperonin CCT: new hints for actin folding. *Int J Biochem Cell Biol*. 2010;42(5):641–650.
69. Iorio V, et al. BAG3 regulates formation of the SNARE complex and insulin secretion. *Cell Death Dis*. 2015;6, e1684.

CHAPTER 9

Tau propagation and autophagy

Mohammad Nasir Uddin[a,c], Shotaro Shimonaka[a,d], and Yumiko Motoi[a,b]

[a]Department of Diagnosis, Prevention, and Treatment of Dementia, Graduate School of Medicine, Juntendo University, Bunkyo-ku, Tokyo, Japan
[b]Department of Neurology, Graduate School of Medicine, Juntendo University, Bunkyo-ku, Tokyo, Japan
[c]Genome and Regenerative Medicine Center, Graduate School of Medicine, Juntendo University, Bunkyo-ku, Tokyo, Japan
[d]Research Institute for Diseases of Old Age, Graduate School of Medicine, Juntendo University, Bunkyo-ku, Tokyo, Japan

1. Introduction

Neurodegenerative disorders showing abundant intracellular Tau aggregates are collectively referred to as tauopathies that include Alzheimer's disease (AD), progressive supranuclear palsy (PSP), corticobasal degeneration (CBD), Pick's disease (PiD), and argyrophilic grain disease (AGD). Tau mainly accumulates in neurons in AD and PiD, while the pathologies of PSP and CBD are characterized by glial Tau pathologies, such as tufted astrocytes and astrocytic plaques. Tauopathies have distinct biochemical signatures; all six isoforms of Tau accumulate in AD, while PiD is characterized by a predominance of 3 repeat Tau and CBD, PSP, and AGD are tauopathies with a predominance of 4 repeat Tau.[1] Cryo-electron microscopy (cryo-EM) analyses recently revealed that each tauopathy has a different structural conformation and is called a strain.[2–4] These findings led to the strain hypothesis, which suggests that the prion-like propagation of strains accounts for neuropathological heterogeneity.[5]

Autophagy is an evolutionarily conserved cellular mechanism for the degradation and recycling of cellular components via the lysosomal pathway. In neurons, autophagy is crucial for proper axon guidance, vesicular release, the dendritic spine architecture, spine pruning, and synaptic plasticity, and the dysregulation of autophagy is associated with brain disorders, including autism spectrum disorders, and neurodegenerative diseases, such as Parkinson's disease and AD.[6,7] The clearance of the physiological and pathological forms of Tau is mediated by proteasomal and autophagic degradative

Autophagy Dysfunction in Alzheimer's Disease and Dementia
https://doi.org/10.1016/B978-0-323-89906-2.00014-9

systems. Monomeric Tau is natively unfolded, which may make it a target for 20S proteasomes, while Tau aggregates are selectively degraded by autophagy.[8] Autophagy is an ongoing clearance mechanism for larger, longer lived proteins and aggregates as well as organelles, such as mitochondria and peroxisomes. We herein describe strain-mediated Tau propagation and then discuss the role of autophagy in tauopathies.

2. Tau propagation mechanisms

Tauopathies are characterized by the progressive intraneuronal deposition of aggregated Tau proteins. In AD, Tau stereotypically extends as aggregate deposits along anatomically connected networks from the transentorhinal cortex or neocortical areas.[9] The spatiotemporal progression of Tau aggregates correlates with cognitive impairment. Along with this classical observation, an intracerebral injection of brain extracts containing aggregated Tau or recombinant Tau seeds induces tauopathy in Tau-transgenic (Tg) or nontransgenic (NTg) mice. The Tau lesions induced propagate from the injection site to other regions via a connectome-dependent manner. This phenomenon strongly supports the transfer of Tau, the cell-to-cell "prion-like" particle, from one cell to another, similar to the aggregated form of the prion protein in Creutzfeldt-Jacob disease. Along with cell-to-cell spreading, internalized Tau is considered to induce the fibrous aggregation of endogenous Tau by templated self-assembly (Fig. 1).

2.1 Experimental methods for Tau propagation

2.1.1 In vitro Tau propagation

Self-assembly of Tau

Tau protein is highly hydrophilic and natively unfolded. Under pathological conditions, Tau becomes misfolded, attains a β-structure, and assembles into paired helical filaments (PHF). In vitro aggregation is triggered by the addition of polyanionic cofactors, including heparin and RNA, and followed continuously by thioflavin T (Th T) fluorescence. The Tau aggregation assay is highly reproducible and has been used as a robust tool for screening drugs that may interfere with the pathogenesis of Tau.[10]

Seed-dependent Tau propagation

Internalized Tau is considered to induce the fibrous aggregation of endogenous Tau by templated self-assembly (Fig. 1). Recent studies using

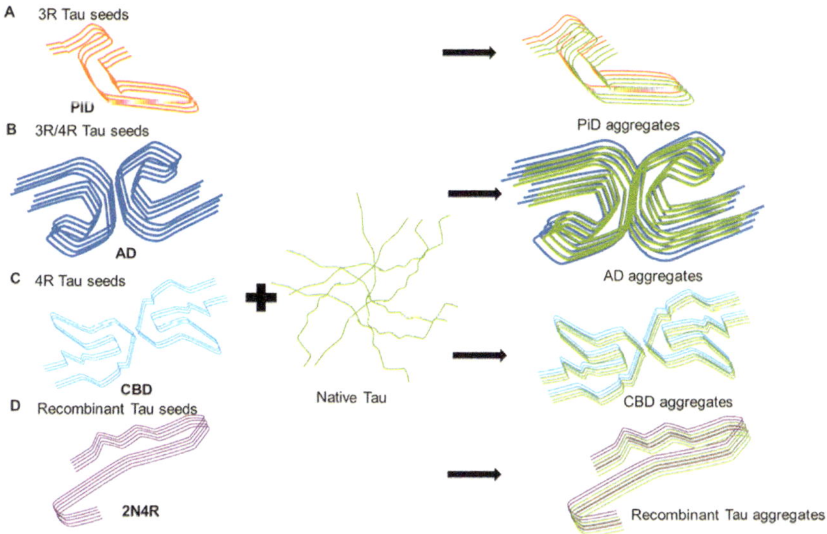

Fig. 1 Schematic model of in vitro Tau propagation based on cryo-EM structures. Pick's disease adopts a single novel fold of 3R Tau (A), which is distinct from the Tau filament fold found in AD that contains C-shaped protofilaments (B). 4R tauopathy CBD comprises an antiparallel four-layered fold that undergoes the self-assembly of naïve unfolded Tau and forms CBD Tau aggregates (C). Recombinant Tau seeds have distinct structures from other patient-derived Tau seeds (D). Fibril structures were taken from PDB files originally published in Ref. 2–4.

cryo-EM to elucidate the structure of AD Tau filaments discovered a C-shaped filament core encompassing residues 204/306–378/380 of Tau protein.[3] Abnormal forms of Tau, such as AD Tau filaments, appear to act as seeds that recruit and convert normal Tau to a pathological species in a prion-like manner. The recombinant AD Tau core exhibited potent seeding activity to recruit normal full-length Tau to filaments, thereby creating a pathogenic species with enhanced seeding capability.[11] The structures of other Tau filaments in CBD, PiD, and chronic traumatic encephalopathy have been elucidated using cryo-EM.[2,4,12,13] The self-propagating prion-like properties of these different tauopathies have been used to develop a diagnostic marker using an ultrasensitive real-time quaking-induced conversion assay.[14,15]

Seed-dependent Tau propagation using a cell-based assay

Sonicated recombinant Tau aggregates were previously shown to enter C17.2 cells and induce the fibrillization of intracellular-transfected Tau.[16]

PHF extracted from the AD brain developed intracellular aggregates that formed aggresomes.[17] Using SH-SY5Y cells, we examined the seed-dependent propagation properties of a 24-kDa fragment (Tau-CTF24) of Tau that increased with aging in tauopathy model mice and human tauopathies.[18] The findings obtained showed that Tau-CTF24 consisted of four binding domains and a C terminus (243–441). When insoluble Tau from AD, PSP, and CBD brains or aggregated recombinant Tau was introduced as seeds into SH-SY5Y cells, a larger amount of insoluble Tau formed in cells overexpressing Tau-CTF24 than in those overexpressing full-length Tau. Using SH-SY5Y cells expressing a series of 10 partial deletions (Del 1–19, covering 244–400 amino acids) in Tau-CTF24 mutants, we attempted to identify the key sequences of seed-dependent propagation and found that aggregation was significantly decreased in the Del 8 mutant lacking 353–368 amino acids.[19] In Del 8, the Asn-368 mutation suppressed Tau aggregation induced by an AD Tau seed. The effects of Asn-368 were observed in AD, but not recombinant, PSP, or CBD seeds (Fig. 2). These findings imply that the Asn-368 residue plays a role in strain-specific Tau aggregation in different tauopathies.

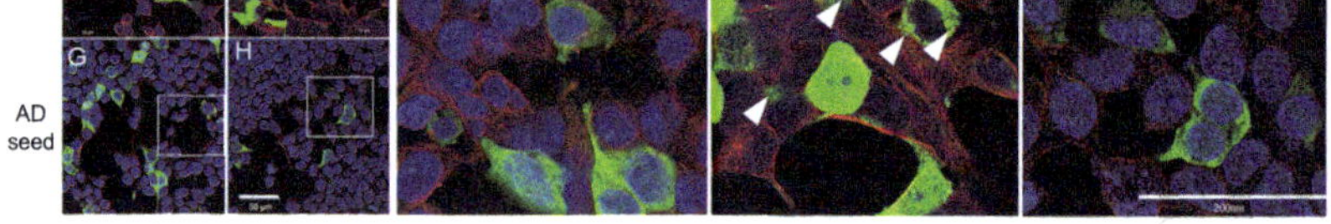

Fig. 2 Effects of Δ368 on Tau aggregation were only observed in AD seeds. HA-tagged Tau-CTF24 WT or the Δ368 mutant was expressed in SH-SY5Y cells and treated with AD brain lysates or heparin-induced recombinant full-length Tau fibrils. Aggregate-like structures were absent in the no seed panel (A, B, and C), and only diffuse staining was noted in soma. However, the introduction of recombinant seeds (D, E, and F) or AD seeds induced aggregate-like structures (white arrowheads) (G and I upper). The number of aggregates was reduced in cells expressing Tau-CTF24 with Δ368 induced by AD seeds, (H and I lower), but was maintained in wild-type recombinant Tau seeds (E and F lower). The insets in A, B, D, E, G, and H indicate areas of higher magnification as C, F, and I. Cells were stained with an anti-HA antibody (*green*), antitubulin antibody (*red*), and DAPI (*blue*).

2.1.2 In vivo Tau propagation

Tau aggregates injected into the unilateral hippocampus and the overlying cerebral cortex of Tg or NTg mouse brains induced the Tau pathology via connected networks in the contralateral hippocampus, fimbria, fornix, lateral septal nucleus, mammillary area, EC, and locus coeruleus.[20–22] In our laboratory, a sarkosyl-insoluble fraction of AD brains was injected into the unilateral hippocampus of Tg601 mice, which overexpress wild-type human Tau, and NTg mice. The Barnes maze test conducted 17–19 months after the injection showed increased primary latency to reach the target hole in the 3rd and 5th sessions in AD brain-injected mice (Fig. 3A).[1] Seventeen to nineteen months incubation of AD brain induced AT8-positive neurons and threads more strikingly in Tg601 mice (Fig. 3B and C). These findings indicate that Tau propagation contributes to functional decline and also suggest that long-term changes in nonmutated Tau mice reflect human pathological conditions.

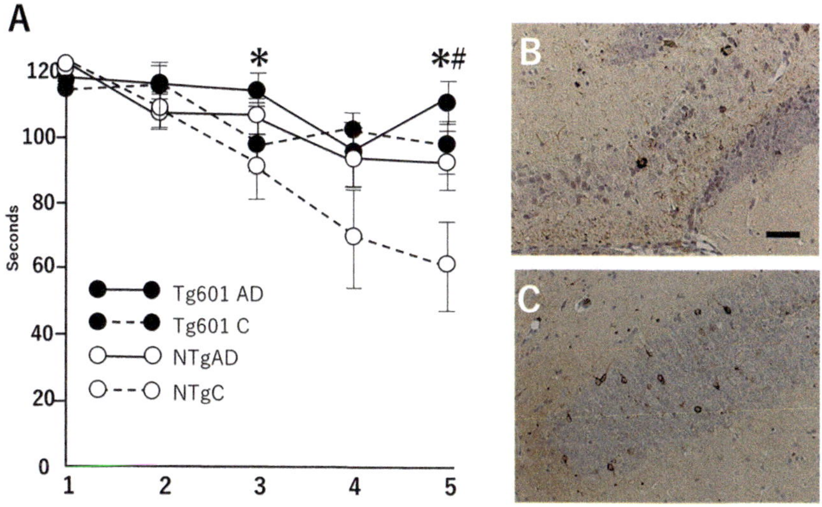

Fig. 3 AD brain injection-induced learning disability and Tau pathology. The Barnes maze test was originally developed by Carol Barnes to overcome the stress induced by swimming in the Morris water maze test. A two-way factorial ANOVA for each training session revealed the significant effect of the AD brain injection in the 3rd session (F (3, 61) = 4.1277, P = .0469) and 5th session (F (3, 61) = 5.6253, P = .021)[1] (A). An AT8 antibody, which recognizes phosphorylated serine 202 and threonine 205, showed Tau-positive threads and neurons in the hippocampus of NTg mice (B). In the dentate gyrus of the ipsilateral hippocampus, the number of AT8-positive cells was shown for Tg601 mice (C). Bar = 100 mm.

The Tau pathology observed in mice after a tauopathy brain injection extended in a Tau strain-specific manner. PSP-Tau and CBD-Tau strains induced astroglial and oligodendroglial Tau inclusions, recapitulating the diversity of neuropathology in human tauopathies.[22] In our laboratory, a PSP brain homogenate injection into the unilateral hippocampus of NTg mice after 22 months incubation induced astrocytic plaque-like structures in the ipsilateral retrosplenial granular cortex (Fig. 4). An AGD brain homogenate injection also induced grain-like structures.[23] These strain-specific propagation properties may be related to the different cryo-EM structures (Fig. 1).[2–4,24]

2.2 Potential routes for the intercellular spread of Tau aggregates

The intercellular spread of Tau aggregates occurs in a hierarchical pattern; they initially appear in the entorhinal cortex (EC) and then spread to the hippocampus and surrounding areas. The mechanisms and agents underlying the progression of the Tau pathology remain controversial. One of the most interesting and poorly examined questions in this field is whether a pathology and/or dysfunction in the EC initiates the anatomical progression of disease or if a pathology and/or dysfunction in the extrahippocampal areas develop independently and are independent of events occurring in

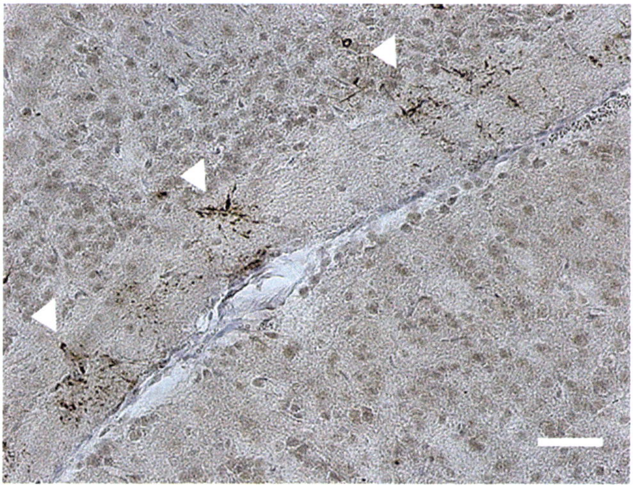

Fig. 4 Glial-like Tau pathology in the ipsilateral retrosplenial cortex after a PSP brain injection. A PSP brain homogenate was injected into the unilateral hippocampus of 2-month-old NTg mice. The retrosplenial cortex of mice showed AT8-positive astrocytic plaque-like structures (white arrow heads) at 19 months.

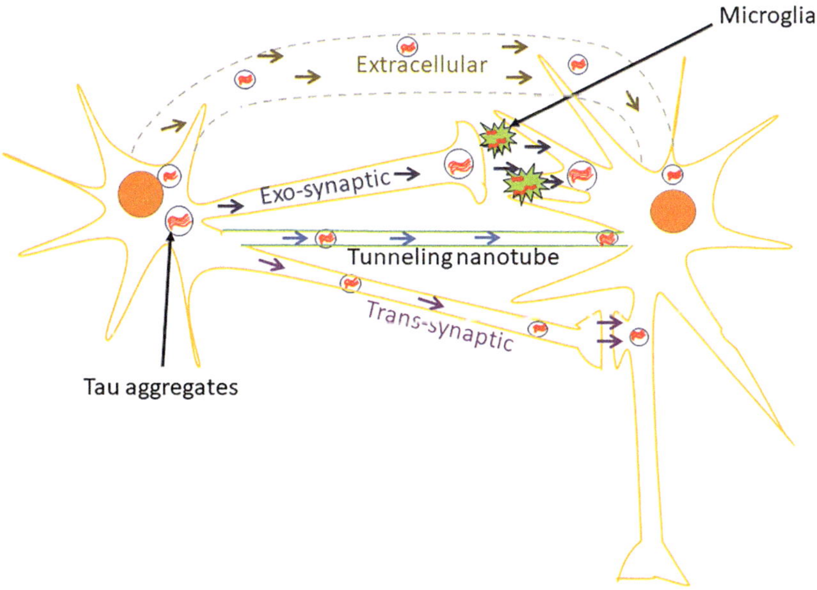

Fig. 5 Schematic representation of routes for the intercellular spread of Tau aggregates. Aggregated Tau is transmitted or spreads from neuron to neuron trans-synaptically, exo-synaptically, extracellularly, and via tunneling nanotubes.

the EC. To the best of our knowledge, 4 routes have been proposed for the intercellular spread of Tau aggregates (Fig. 5).

2.2.1 Trans-synaptic Tau propagation

A number of circumstantial observations support the trans-synaptic spread hypothesis for AD in terms of both pathology development and functional outcomes. A tauopathy mouse model study demonstrated the propagation of pathology from the EC and reported a trans-synaptic mechanism of spread along anatomically connected networks between connected and vulnerable neurons.[25] Hyman and colleagues showed that PBS-soluble phosphorylated high-molecular-weight Tau, though very low in abundance, was taken up, axonally transported, and passed onto synaptically connected neurons.[26] Using microfluidic devices, exosomes have been shown to mediate the trans-neuronal transfer of Tau depending on synaptic connectivity. The spread of Tau is achieved by the direct transmission of exosomes between neurons. In organotypic hippocampal slices, Tau-containing exosomes in conditioned medium are taken up by neurons and microglia, not astrocytes. In N2a cells, Tau assemblies were shown to be released via exosomes.[27]

2.2.2 Exo-synaptic Tau propagation

Ikezu and co-workers reported another nonsynaptic route for the spread of Tau aggregates because Tau accumulation in layer II of the EC region does not directly connect to the CA1 region of the hippocampus, in which Tau accumulation is commonly detected as the pathology progresses.[28] In two independent models of tauopathy, the DG and EC regions of PS19 mice and the DG of the rapid AAV model, microglial-mediated Tau propagation may explain the evolution of Braak staging, which shows the lateral propagation of the Tau pathology in neurons that are not connected by synapses, such as those from the transentorhinal cortex to the higher-order sensory association areas of the occipitotemporal gyrus in Braak III stage AD. The lateral (nonsynaptic) propagation of Tau has also been reported in rTgECTau mice, which reversibly express the human variant Tau P301L primarily in layer II of the EC.[28,29]

2.2.3 Extracellular Tau propagation

The extracellular localization of Tau implies novel physiological functions that are altered during the neurodegeneration process.[30] Naked Tau assemblies may be secreted into the extracellular environment by exocytosis and taken up by naïve cells via endocytosis. Evidence for the secretion of Tau species within extracellular vesicles, e.g., exosomes (~30–100 nm), which originate from the fusion of multivesicular bodies (MVB) with the plasma membrane, and ectosomes (~100–1000 nm), which form directly from the plasma membrane, has been reported; however, the contribution of these systems to the spread of disease has not yet been confirmed or defined.[31] Tau is also present in the cerebrospinal fluid (CSF) and interstitial fluid as well as in the conditioned medium of neuronal cultures. Extracellular Tau species may be toxic to neurons[32] and, more importantly, may serve as pathogenic "seeds" to be taken up by surrounding neurons, thereby enabling prion-like propagation.[33]

2.2.4 Tau propagation by tunneling nanotubes (TNTs)

TNTs are filamentous actin-containing membranous structures that bridge and connect cells. Previous studies reported the presence of misfolded protein assemblies associated with disease within TNTs directly connecting the cytoplasm of distant cells, such as from immune to neuronal cells.[34] In addition to allowing the long-range intercellular transfer of cytoplasmic macromolecules, plasma membrane components, vesicles, and organelles, TNTs transfer pathogens, such as HIV, infectious prion particles, and huntingtin-Exon 1 aggregates, from affected donor to naive recipient cells.

3. Autophagic impairment promotes Tau aggregation

Autophagy plays a vital role in regulating cellular homeostasis and disease states. Therefore, dysfunctional autophagy results in the accumulation of pathogenic proteins inside cells.[35] However, autophagy in neurons is more than a simple housekeeper, it has several neuronal-specific functions, such as axonal guidance, synaptic transmission, proper neuronal connectivity, and neural stem cell development.[6,36] Impairments in these degradative systems leads to the intracellular accumulation of Tau, neuronal deficits, and the formation of aggregates.[37] Three types of autophagy have been described in mammals according to the delivery mechanisms of cargo to lysosomes (Fig. 6). In macroautophagy, cytosolic cargo is initially trapped inside double-membrane vesicles (autophagosomes), which then fuse with lysosomes to completely degrade the cargo.[38] In chaperone-mediated autophagy (CMA), proteins are identified one by one by a cytosolic chaperone that transports them to the lysosomal surface for translocation across this membrane.[39] In microautophagy and endosomal microautophagy, cargo is internalized through invaginations in lysosomal and endosomal membranes.[37]

3.1 Macroautophagic dysfunction in tauopathies

Macroautophagy, the most characterized form of autophagy, is classified into two main types. Basal macroautophagy is a quality control mechanism that prevents metabolic and oxidative stress by degrading aggregated or aggregate-prone proteins. Starvation-induced autophagy occurs in response to nutrient deprivation and recycles macromolecules to provide substrates for energy metabolism.[38] Autophagy is a highly regulated process that involves the sequential activation of protein complexes encoded by autophagic genes (ATG). Briefly, in mammalian cells, autophagy is initiated by the activation of the class III phosphatidylinositol 3-kinase (PI3K) complex by the unc-51-like autophagy-activating kinases 1/2 (ULK 1/2) complex, leading to the primary nucleation of the membrane and recruitment. Membrane elongation then occurs under the direction of two ubiquitin-like conjugation systems, ATG12–ATG5 and MAP1LC3B/LC3 (microtubule-associated protein 1 light chain 3). ATG12 binds to ATG5 and ATG16L1 in order to facilitate the recruitment and lipidation of the cytosolic form of LC3 (LC3-I) with phosphatidylethanolamine to become the phagophore-specific and autophagosome membrane-associated form LC3-II.[39] Cytoplasmic substrates are recognized by autophagic receptors, such as SQSTM1/p62 (sequestosome 1) or NBR1 (NBR1 autophagy cargo receptor), which also recognize the LC3-II protein.[40] The autophagosome

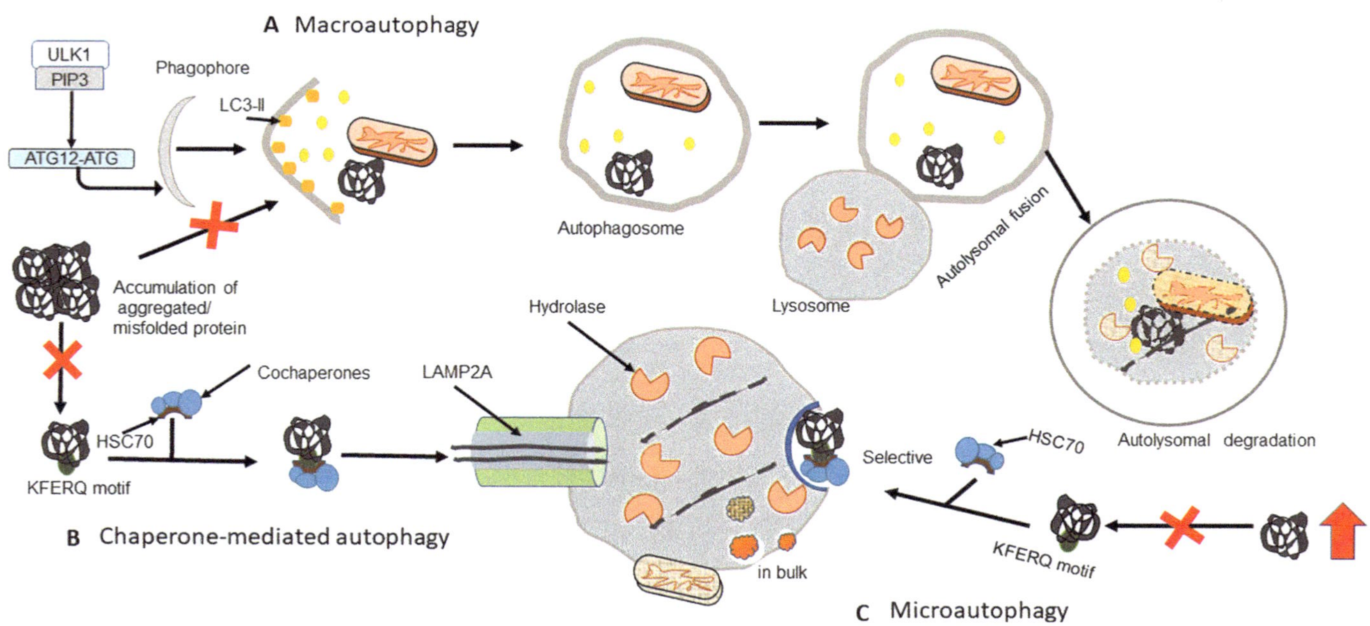

Fig. 6 Three main types of autophagy impaired in AD. In macroautophagy, cytosolic cargo is initially trapped inside double-membrane vesicles (autophagosomes) that then fuse with lysosomes for the complete degradation of cargo (A). In chaperone-mediated autophagy (CMA), proteins are identified by a cytosolic chaperone that transports them to the lysosomal surface for translocation across this membrane (B) In microautophagy and endosomal microautophagy, cargo is internalized through invaginations in lysosomal and endosomal membranes (C).

fuses its outer membrane with a lysosome, forming a single membrane autolysosome, the contents of which are degraded by acidic lysosomal hydrolases, thereby restoring the lysosome (Fig. 6A). When the mechanisms underlying autophagy are disrupted, cells become more prone to the accumulation of defects, which reduces cell viability. If autophagic pathways are not rescued, the accumulation of cellular debris may occur and accelerate physiological tauopathies and related neurodegenerative diseases.[37]

A prominent feature of AD is the massive accumulation of lysosomal vesicular structures in degenerating neurons, and pathological studies have suggested the impairment of macroautophagy in neurodegenerative disorders, including AD and other tauopathies.[7] Atg7-deficient mice were found to be defective for the initiation of macroautophagy and displayed progressive neurodegeneration with prominent inclusions positive for ubiquitin, p62, phosphorylated Tau, and GSK3β.[41] A previous study reported that miR-132/212 targeted Tau mRNA to regulate Tau expression, the downregulation of miR-132/212 expression in the brains of AD patients led to the aggregation of Tau, and the regulation of Tau aggregation by miR-132/212 was associated with dysfunctional autophagy.[42] A transcriptomic analysis of APOE3/4-targeted replacement mice identified the upregulation of genes associated with vesicle function, including nine vacuolar H^+ATPases and Rab GTPases, which are vital for early and late endosome (LE) functions. The dysregulation of autophagy within the EC may facilitate the initial aggregation of Tau, which ultimately results in the pathological spread of Tau to other vulnerable brain regions.[43] AD neurons may be relevant to the progression of tauopathy by hindering correct lysosomal positioning and facilitating Tau release in the extracellular space during short-lived and acute Tau proteotoxic stress, and also by triggering the accumulation of insoluble Tau under a long-term Tau pathology.[44]

3.2 CMA dysfunction in tauopathies

Initial investigations on CMA indicated that the degradation of intracellular components by lysosomes was selective; however, this concept is now recognized as one type of autophagy. Lysosomes not only degrade the cellular cytosol in a nonspecific manner but also discriminate between what to target for degradation with the involvement of a degradation tag, a chaperone, and this sophisticated mechanism induces selected proteins to cross the lysosomal membrane through a dedicated translocation complex.[45] The selectivity of CMA is conferred by the ability of the heat shock cognate protein of 70 kDa (hsc70) to recognize a pentapeptide motif (biochemically related to the pentapeptide KFERQ) in CMA substrate proteins and deliver them to the lysosomal surface for binding to lysosome-associated membrane protein type 2A (LAMP2A).[46,47] Following the binding of substrates, LAMP2A multimerizes into a translocation complex, allowing substrate proteins access to the lysosomal lumen. Substrate entry requires a form of hsc70 residing inside lysosomes that completes substrate translocation (Fig. 6B).[48] Tau contains two targeting motif homologs (336QVEVK340 and 347KDRVQ351) in microtubule-binding domains, which are involved in the direct delivery of cytosolic Tau by chaperones to lysosomal membranes. CMA contributes to the degradation of pathogenic proteins involved in neurodegenerative disorders, such as a-synuclein (a-syn) and Tau.[49] The experimental blockage of CMA resulted in the rerouting of oligomers of acetylated Tau toward LE/MVB by endosomal microautophagy, and this mechanism may lead to the extracellular release and cell-to-cell propagation of Tau.[50] The loss of CMA in a mouse model of AD exerted synergistic negative effects on the proteome at risk of aggregation, thereby increasing neuronal disease vulnerability and accelerating disease progression.[45] An inducible mouse neuroblastoma N2a cell model of tauopathy expressing the repeat domain of Tau with a FTDP-17 mutation ($Tau_{RD}\Delta K280$) and its F1 fragment interacting with the cytosolic hsc70 chaperone was examined, and $Tau_{RD}\Delta K280$ mutants facilitated its binding to lysosomes in a manner that depended on the CMA receptor LAMP-2A.[51] Besides Tau, a connection between CMA and an AD-related protein, the regulator of calcineurin 1 (RCAN1) has also been established. CMA degrades RCAN1, the high expression of which in AD brains has been linked to neuronal demise.[52]

3.3 Microautophagic dysfunction in tauopathies

Microautophagy is a type of autophagy that is characterized by direct enclosure with the vacuolar/lysosomal membrane, which completes the isolation and uptake of cell components in the vacuole (Fig. 6C). Microautophagy involves the invagination of the endosomal membrane by a series of Endosomal Sorting Complex Required for Transport (ESCRT) complexes to generate intraluminal vesicles, which capture cytosolic- and membrane-bound molecules.[53] Microautophagy may be categorized into three types depending on the morphology of the vacuolar/lysosomal membrane deformation: (i) with lysosomal/vacuolar membrane protrusions, (ii) invaginations, and recently discovered (iii) LE invaginations.[54] Selective organelle degradation by microautophagy occurs via interactions between organelle proteins and surface proteins on vacuoles, lysosomes, or LE.[51] Selective and nonselective microautophagy was differentiated based on the selectivity of the engulfment of cargo during these processes; cargo may consist of nonselective soluble components in the cytosol.[55] However, particular organelles and cellular components may also be sequestered in selective microautophagy, such as micropexophagy (peroxisomes),[56] piecemeal microautophagy of the nucleus (part of the nucleus),[57] and micro-autophagy (mitochondria).[58] A previous study reported increased levels of mitophagy markers, including COX IV, TOMM20, and the ratio of mtDNA to genomic DNA indexed as mt-Atp6/Rpl13, in AD brains and in human wild-type Tau-overexpressing Tg mice.[59] A direct toxic effect of pathogenic aggregate proteins on Hsc70, which is responsible for cargo recognition in CMA and endosomal microautophagy, has not yet been demonstrated. However, a model of chaperone depletion has been suggested in which the aggregate-mediated sequestration of Hsc70 reduces its availability for other cellular functions, such as endocytosis and protein folding. Moreover, immunohistochemistry of the EC and cingulate cortex of APOE4/4-targeted replacement mice revealed an increase in early endosomes, which has also been observed in AD.[43] Furthermore, the knockdown of several components of the ESCRT machinery, including charged multivesicular body protein 6 (CHMP6) or CHMP2A in combination with CHMP2B (the gene of which is linked to familial FTD), promoted the propagation of Tau aggregation.[60] In a human cell system for seeding-dependent tau aggregation, the molecular chaperones heat-shock cognate 71-kDa protein (HSC71)/heat-shock protein 70 (HSP70), HSP90, and J-domain co-chaperones were sequestered by tau aggregates. A single-cell analysis of protein-folding and clathrin-mediated endocytosis revealed that chaperone-dependent cellular activities were significantly impaired by tau aggregation and may be reversed by a treatment with small-molecule regulators of heat-shock transcription factor 1 proteostasis, which induce the expression of cytosolic chaperones.[61]

4. Pharmacological agents modulating Tau propagation

4.1 Autophagic modulators attenuated Tau propagation

The autophagy-lysosomal pathway (ALP) is recognized as the main mechanism responsible for the clearance of protein aggregates that cannot be proteolyzed by proteasomes. Accumulating evidence has demonstrated the pharmacological enhancement of autophagy activity in patient-derived neurons and in vivo reductions in oligomeric and aggregated Tau, the mitigation of Tau neuronal transmission, and reduced cell loss, which support the potential of autophagy modulators as therapeutic agents (Table 1).[73] The stimulation of autophagy by trehalose in the Tg Tau-P301S mouse model was shown to significantly reduce Tau inclusions in the brain and promote neuronal survival in the cerebral cortex and brainstem. However, trehalose failed to activate autophagy in the spinal cord and did not affect motor impairments in Tau-P301S mice.[62] Studies on mouse models treated with rapamycin, an inhibitor of mammalian target of rapamycin (mTOR), revealed a reduction in pathogenic Tau levels and improved cognition by the upregulation of autophagy.[63] The pharmacological upregulation of autophagy was recently shown to rescue the Tau phenotype in tauopathy patient iPSC-derived neurons. Three ATP-competitive mTOR kinase inhibitors, OSI-027, AZD2014, and AZD8055, exert stronger effects on Tau clearance than rapamycin without affecting viability. These compound treatments downregulated P-Tau and insoluble Tau, which reduced Tau-mediated neuronal stress vulnerability. The most notable finding in this study was the discovery that a single-dose 24-h treatment reduced Tau for 12–16 days without the loss of cell viability or integrity.[64] However, the induction of autophagy in tauopathies does not necessarily have a straightforward beneficial outcome. Research has shown that the induction of new autophagosome formation is not necessarily impaired in disease states, the lysosomal-dependent proteolytic system is the main cause of disruption, which leads to the "backed up" accumulation of autophagosomes carrying misfolded proteins.[74] In many cases, the pharmacological enhancement of autophagosome formation also activates transcription factor EB (TFEB), which simultaneously coordinates lysosomal biogenesis as well as the genes required for autophagosome formation, fulfilling the criteria for the upregulation of ALP.[75] These findings also suggest the potential of TFEB as a therapeutic target. On the other hand, pharmacological treatments that improve the catalytic performance of lysosomal enzymes and reduce the load of autolysosomes may also rescue lysosomal functions. A previous study discovered that pathological Tau reduced the effects of fisetin (3,7,3′,4′-tetrahydoxyflavone), an organic flavonoid present in many types of fruits and vegetables. The treatment of mouse primary cortical neurons and rat primary cortical neurons with fisetin reduced S262 and S396/S404 hyperphosphorylated Tau and decreased sarkosyl-insoluble Tau in 293T cells by enhancing autophagy through increases in the transcriptional activity of TFEB and Nrf2.[65] Pimozide, an antipsychotic, that is used to manage

Table 1 Pharmacological agents promoting the clearance of Tau propagation by activating autophagy.

Pharmacological agents	In vitro/in vivo models	Molecular mechanisms	Influence of Tau clearance	Additional outcome
Trehalose[62]	P301S tau Tg mice, C57BL/6 mice	m-TOR independent	↓ Sarkosyl insoluble tau	↑ Neural activity
Rapamycin[63]	7PA2, 3X Tg-AD mice	Inhibited m-TOR	↓ Tau pathology	↓ Aβ pathology
OSI-027, AZD2014, and AZD8055[64]	Patient iPSC-derived neuronal cells	Inhibited m-TOR	↓ Phosphorylated and insoluble tau	↓ Neuronal stress vulnerability
Fisetin (3,7,3′,4′-tetrahydroxyflavone)[65]	Mouse cortical neuronal cells (T4)	↓ m-TOR, ↑ TFEB, Nrf2	↓ Tau phosphorylation, insoluble tau	
Pimozide[66]	TauC3 mice	AMPK-mediated, non-m-TOR pathway	↓ Phosphorylation, Tau aggregates	↑ Memory deficits
Berberine[67]	3XTgAD mice	Class III PI3K/beclin-1 pathway	↓ Tau phosphorylation	↑ Learning and memory
Compound 22[68] (derivative of nitazoxanide (NTZ))	SH-SY5Y cells, BV2 cells, APP/PS1 mice	Inhibited m-TOR	↓ Tau phosphorylation	↑ Aβ clearance
Cerium oxide nanoparticles ($nCeO_2$)[69]	*Drosophila* AD model	↑ ATG1 and ATG 18	↓ hTau gene expression	
Bis(ethylmaltolato) oxidovanadium (IV) (BEOV)[70]	APPSwe/PS1dE9 mice, N2asw cells	Tyrosine phosphatase-1B (PTP1B) and Akt/GSK3β pathway	↓ Tau phosphorylation	↓ Aβ production
Metformin[71]	APP/PS1 Tg mice	↑ m-TOR-dependent microglial autophagy	↓ Tau pathology	↓ Aβ pathology
Lithium[72]	SH-SY5Y cells	m-TOR independent	↓ Sarkosyl insoluble tau	

debilitating motor and phonic tics in patients with Tourette's disorder, also increased autophagic flux through the activation of the AMPK-ULK1 axis, but not mTOR, and reduced the levels of abnormally phosphorylated Tau aggregates in neuronal cells.[66] In the triple-Tg mouse model of AD (3×Tg AD), berberine, a natural isoquinoline alkaloid isolated from *Rhizoma coptidis*, promoted the autophagic clearance of Tau by enhancing the activity of autophagy via the class III PI3K/beclin-1 pathway.[67] Moreover, through high-throughput compound screening, a novel p300/CBP inhibitor was identified that promotes autophagic flux, decreases Tau secretion, and reduced Tau propagation in fibril-induced Tau spreading models both in vitro and in vivo.[76] An inducible tauopathy cell model with a repeat domain with the K280 mutation cleared Tau aggregation through the activation of autophagy, and not by the ubiquitin proteasomal system.[77] Nitazoxanide is an FDA-approved antiparasite drug and its derivative, compound **22** inhibited the phosphorylation of p70S6K, which was the direct substrate of mTOR, and its phosphorylation level may reflect mTOR-dependent autophagy in vivo as well as in vitro.[68] Cerium oxide nanoparticles ($nCeO_2$) act as antioxidants by mimicking the activities of catalase and superoxide dismutase and also as scavengers of free radicals. A one-month synthesized $nCeO_2$ diet significantly improved climbing activity in a pan-neuronal fly model of AD. nCeO2 also induced a significant decrease in hTau gene expression and increased the mRNA expression of the key autophagy genes ATG1 and ATG18.[69] Bis(ethylmaltolato) oxidovanadium (IV) (BEOV) was shown to significantly reduce the generation of Aβ by increasing the expression of peroxisome proliferator-activated receptor-gamma and insulin-degrading enzyme and by decreasing β-secretase 1 in the hippocampus and cortex of AD mice. BEOV also reduced Tau hyperphosphorylation by inhibiting protein tyrosine phosphatase-1B and regulating the pathway of insulin receptor/insulin receptor substrate-1/protein kinase B/glycogen synthase kinase 3 beta.[70] The administration of metformin ameliorated microglial autophagy impairment, enhanced the activation of microglia around Aβ plaques, promoted the phagocytosis of Tau aggregates in dystrophic neurites surrounding Aβ plaques, and attenuated the Aβ load.[71] Noninvasive autophagic modulators, such as near infrared light and ultrasound, have been suggested to reduce or alleviate the Tau burden in tauopathy mouse models.[78,79] However, the lack of highly specific and safe autophagy modulators is still a major obstacle to the development of autophagy modulators for clinical applications. Among the hundreds of autophagy modulator candidates being reported, only a few may ultimately progress to clinical trials. Therefore, the development of novel autophagy modulators with high specificity and low toxicity is urgently needed.

4.2 Lithium-induced autophagic promotion of aggregated Tau clearance

Lithium, an FDA-approved mood stabilizer that has been safely used to treat bipolar disorders for more than 50 years, is a validated autophagy inducer with potent efficacy.[80–82] Autophagic activity has been associated with the beneficial effects of lithium in several disease models. Lithium was shown to reduce neuronal damage and promotes neurological recovery by inducing autophagy in a rat model of spinal cord injury.[83] Lithium-triggered autophagy may be an attractive strategy for developing novel pharmacological therapies in Parkinson's disease.[84] Sarkar et al. demonstrated that lithium-induced autophagy, independently of mTOR, by inhibiting inositol monophosphatase (IMPase), resulting in the clearance of huntingtin and α-syn in nonneuronal and neural precursor cells.[85,86] The generation of conformationally distinct Tau aggregates (or strains) has been reported.[87] We recently showed that a lithium treatment reduced AD seed-dependent Tau aggregation in SH-SY5Y cells by promoting autophagy. The lithium-induced promotion of autophagy was weaker in the case of CBD seeds than in AD or PSP seeds.[72] A transmission electron microscopy analysis showed a markedly higher number of autophagic vacuoles in AD- and PSP-seeded cells than in SH-SY5Y cells seeded by CBD brain lysates (Fig. 7). In contrast to AD and PSP seed-treated cells, CBD-seeded cells have not followed the IMPase inhibition pathway. We speculate that differences in filament structures may influence enzyme-protein interactions, leading to the strain-specific lithium-induced promotion of autophagy.

5. Conclusion

Different strains of tau exhibit the distinct capacity for propagation, which may lead to the different phenotypes of tauopathies. Our findings on lithium-induced autophagy indicate the possibility of strain-specific degradation.

Author contributions

MNU and YM developed the study concept and design and wrote the manuscript. SS provided technical support.

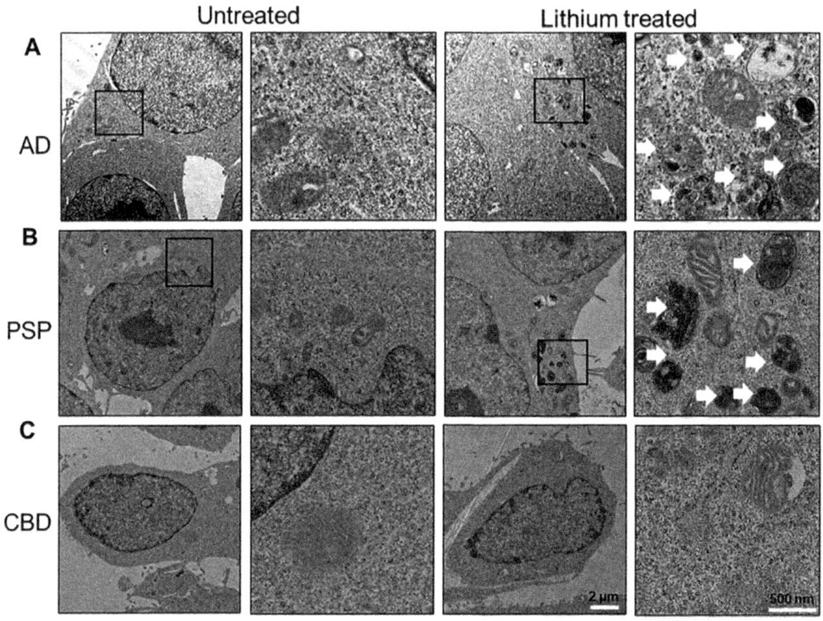

Fig. 7 Autophagic clearance of aggregated Tau in a strain-specific manner. SH-SY5Y cells overexpressing Tau-CTF24 were treated with AD, PSP, or CBD brain lysates (A, B, and C). After a 48-h incubation with 10 mM LiCl, 2.5% glutaraldehyde was added as a preservative and transmission electron microscopy was performed. The right panels in A and B represent high-magnification views of the insets. The lithium treatment increased the number of autophagic vacuoles in AD and PSP, but not in CBD. Scale bars: 500 nm in D, H, and L at a high magnification and 2 μm at a low magnification.

Conflict of interest

There is no conflict of interest.

References

1. Hayashi T, et al. Learning deficits accompanied by microglial proliferation after the long-term post-injection of Alzheimer's disease brain extract in mouse brains. *J Alzheimers Dis.* 2021;79(4):1701–1711.
2. Falcon B, et al. Structures of filaments from Pick's disease reveal a novel tau protein fold. *Nature.* 2018;561(7721):137–140.
3. Fitzpatrick AWP, et al. Cryo-EM structures of tau filaments from Alzheimer's disease. *Nature.* 2017;547(7662):185–190.
4. Zhang W, et al. Novel tau filament fold in corticobasal degeneration. *Nature.* 2020; 580(7802):283–287.
5. Falcon B, et al. Galectin-8-mediated selective autophagy protects against seeded tau aggregation. *J Biol Chem.* 2018;293(7):2438–2451.

6. Hwang JY, Yan J, Zukin RS. Autophagy and synaptic plasticity: epigenetic regulation. *Curr Opin Neurobiol*. 2019;59:207–212.
7. Nassif M, Hetz C. Autophagy impairment: a crossroad between neurodegeneration and tauopathies. *BMC Biol*. 2012;10:78.
8. Chesser AS, Pritchard SM, Johnson GV. Tau clearance mechanisms and their possible role in the pathogenesis of Alzheimer disease. *Front Neurol*. 2013;4:122.
9. Braak H, Braak E. Neuropathological stageing of Alzheimer-related changes. *Acta Neuropathol*. 1991;82(4):239–259.
10. Crespo R, Koudstaal W, Apetri A. In vitro assay for studying the aggregation of tau protein and drug screening. *J Vis Exp*. 2018;141. https://doi.org/10.3791/58570.
11. Ash PEA, et al. TIA1 potentiates tau phase separation and promotes generation of toxic oligomeric tau. *Proc Natl Acad Sci U S A*. 2021;118(9).
12. Arakhamia T, et al. Posttranslational modifications mediate the structural diversity of tauopathy strains. *Cell*. 2020;180(4):633–644 e12.
13. Falcon B, et al. Novel tau filament fold in chronic traumatic encephalopathy encloses hydrophobic molecules. *Nature*. 2019;568(7752):420–423.
14. Saijo E, et al. Ultrasensitive and selective detection of 3-repeat tau seeding activity in pick disease brain and cerebrospinal fluid. *Acta Neuropathol*. 2017;133(5):751–765.
15. Saijo E, et al. 4-repeat tau seeds and templating subtypes as brain and CSF biomarkers of frontotemporal lobar degeneration. *Acta Neuropathol*. 2020;139(1):63–77.
16. Frost B, Jacks RL, Diamond MI. Propagation of tau misfolding from the outside to the inside of a cell. *J Biol Chem*. 2009;284(19):12845–12852.
17. Santa-Maria I, et al. Paired helical filaments from Alzheimer disease brain induce intracellular accumulation of Tau protein in aggresomes. *J Biol Chem*. 2012;287(24):20522–20533.
18. Ando F, et al. Preparation of a PtPb/TiO_2/cup-stacked carbon nanotube composite for enhancement of the electrocatalytic reaction of the oxygen reduction reaction. *Chem Lett*. 2015;44(12):1741–1743.
19. Shimonaka S, et al. Asparagine residue 368 is involved in Alzheimer's disease tau strain-specific aggregation. *J Biol Chem*. 2020;295(41):13996–14014.
20. Ahmed Z, et al. A novel in vivo model of tau propagation with rapid and progressive neurofibrillary tangle pathology: the pattern of spread is determined by connectivity, not proximity. *Acta Neuropathol*. 2014;127(5):667–683.
21. Iba M, et al. Synthetic tau fibrils mediate transmission of neurofibrillary tangles in a transgenic mouse model of Alzheimer's-like tauopathy. *J Neurosci*. 2013;33(3):1024–1037.
22. Narasimhan S, et al. Pathological tau strains from human brains recapitulate the diversity of tauopathies in nontransgenic mouse brain. *J Neurosci*. 2017;37(47):11406–11423.
23. Clavaguera F, et al. Brain homogenates from human tauopathies induce tau inclusions in mouse brain. *Proc Natl Acad Sci U S A*. 2013;110(23):9535–9540.
24. Pooler AM, et al. Propagation of tau pathology in Alzheimer's disease: identification of novel therapeutic targets. *Alzheimers Res Ther*. 2013;5(5):49.
25. Liu L, et al. Trans-synaptic spread of tau pathology in vivo. *PLoS ONE*. 2012;7(2), e31302.
26. Takeda S, et al. Neuronal uptake and propagation of a rare phosphorylated high-molecular-weight tau derived from Alzheimer's disease brain. *Nat Commun*. 2015;6:8490.
27. Wang Y, et al. The release and trans-synaptic transmission of Tau via exosomes. *Mol Neurodegener*. 2017;12(1):5.
28. Asai H, et al. Depletion of microglia and inhibition of exosome synthesis halt tau propagation. *Nat Neurosci*. 2015;18(11):1584–1593.
29. Wood H. Alzheimer disease: evidence for trans-synaptic and exo-synaptic tau propagation in Alzheimer disease. *Nat Rev Neurol*. 2015;11(12):665.
30. Pooler AM, et al. Physiological release of endogenous tau is stimulated by neuronal activity. *EMBO Rep*. 2013;14(4):389–394.

31. Dujardin S, et al. Ectosomes: a new mechanism for non-exosomal secretion of tau protein. *PLoS ONE*. 2014;9(6), e100760.
32. Diaz-Hernandez M, et al. Tissue-nonspecific alkaline phosphatase promotes the neurotoxicity effect of extracellular tau. *J Biol Chem*. 2010;285(42):32539–32548.
33. Wu JW, et al. Neuronal activity enhances tau propagation and Tau pathology in vivo. *Nat Neurosci*. 2016;19(8):1085–1092.
34. Tardivel M, et al. Tunneling nanotube (TNT)-mediated neuron-to neuron transfer of pathological Tau protein assemblies. *Acta Neuropathol Commun*. 2016;4(1):117.
35. Eskelinen EL, Saftig P. Autophagy: a lysosomal degradation pathway with a central role in health and disease. *Biochim Biophys Acta*. 2009;1793(4):664–673.
36. Wu X, et al. Autophagy regulates Notch degradation and modulates stem cell development and neurogenesis. *Nat Commun*. 2016;7:10533.
37. Kaushik S, Cuervo AM. Proteostasis and aging. *Nat Med*. 2015;21(12):1406–1415.
38. Kenney DL, Benarroch EE. The autophagy-lysosomal pathway: general concepts and clinical implications. *Neurology*. 2015;85(7):634–645.
39. Xie Z, Klionsky DJ. Autophagosome formation: core machinery and adaptations. *Nat Cell Biol*. 2007;9(10):1102–1109.
40. Kirisako T, et al. The reversible modification regulates the membrane-binding state of Apg8/Aut7 essential for autophagy and the cytoplasm to vacuole targeting pathway. *J Cell Biol*. 2000;151(2):263–276.
41. Inoue K, et al. Macroautophagy deficiency mediates age-dependent neurodegeneration through a phospho-tau pathway. *Mol Neurodegener*. 2012;7:48.
42. Schubert CR, et al. BrainSeq: neurogenomics to drive novel target discovery for neuropsychiatric disorders. *Neuron*. 2015;88(6):1078–1083.
43. Nuriel T, et al. The endosomal-lysosomal pathway is dysregulated by APOE4 expression in vivo. *Front Neurosci*. 2017;11:702.
44. Chambraud B, et al. Decrease of neuronal FKBP4/FKBP52 modulates perinuclear lysosomal positioning and MAPT/Tau behavior during MAPT/Tau-induced proteotoxic stress. *Autophagy*. 2021;17(11):3491–3510.
45. Bourdenx M, Gavathiotis E, Cuervo AM. Chaperone-mediated autophagy: a gatekeeper of neuronal proteostasis. *Autophagy*. 2021;17(8):2040–2042.
46. Chiang HL, et al. A role for a 70-kilodalton heat shock protein in lysosomal degradation of intracellular proteins. *Science*. 1989;246(4928):382–385.
47. Dice JF. Peptide sequences that target cytosolic proteins for lysosomal proteolysis. *Trends Biochem Sci*. 1990;15(8):305–309.
48. Bermejo-Millo JC, et al. High-fructose consumption impairs the redox system and protein quality control in the brain of Syrian hamsters: therapeutic effects of melatonin. *Mol Neurobiol*. 2018;55(10):7973–7986.
49. Kirchner P, et al. Proteome-wide analysis of chaperone-mediated autophagy targeting motifs. *PLoS Biol*. 2019;17(5), e3000301.
50. Caballero B, et al. Acetylated tau inhibits chaperone-mediated autophagy and promotes tau pathology propagation in mice. *Nat Commun*. 2021;12(1):2238.
51. Wang Y, et al. Tau fragmentation, aggregation and clearance: the dual role of lysosomal processing. *Hum Mol Genet*. 2009;18(21):4153–4170.
52. Liu H, et al. Degradation of regulator of calcineurin 1 (RCAN1) is mediated by both chaperone-mediated autophagy and ubiquitin proteasome pathways. *FASEB J*. 2009;23(10):3383–3392.
53. Hurley JH, Emr SD. The ESCRT complexes: structure and mechanism of a membrane-trafficking network. *Annu Rev Biophys Biomol Struct*. 2006;35:277–298.
54. Oku M, Sakai Y. Three distinct types of microautophagy based on membrane dynamics and molecular machineries. *Bioessays*. 2018;40(6), e1800008.

55. Sahu R, et al. Microautophagy of cytosolic proteins by late endosomes. *Dev Cell.* 2011;20(1):131–139.
56. Mukaiyama H, et al. Modification of a ubiquitin-like protein Paz2 conducted micropexophagy through formation of a novel membrane structure. *Mol Biol Cell.* 2004;15(1):58–70.
57. Roberts P, et al. Piecemeal microautophagy of nucleus in *Saccharomyces cerevisiae. Mol Biol Cell.* 2003;14(1):129–141.
58. Lemasters JJ. Variants of mitochondrial autophagy: types 1 and 2 mitophagy and micromitophagy (type 3). *Redox Biol.* 2014;2:749–754.
59. Ando K, et al. Level of PICALM, a key component of clathrin-mediated endocytosis, is correlated with levels of phosphotau and autophagy-related proteins and is associated with tau inclusions in AD, PSP and Pick disease. *Neurobiol Dis.* 2016;94:32–43.
60. Chen JJ, et al. Compromised function of the ESCRT pathway promotes endolysosomal escape of tau seeds and propagation of tau aggregation. *J Biol Chem.* 2019; 294(50):18952–18966.
61. Yu A, et al. Tau protein aggregates inhibit the protein-folding and vesicular trafficking arms of the cellular proteostasis network. *J Biol Chem.* 2019;294(19):7917–7930.
62. Schaeffer V, et al. Stimulation of autophagy reduces neurodegeneration in a mouse model of human tauopathy. *Brain.* 2012;135(Pt 7):2169–2177.
63. Majumder S, et al. Inducing autophagy by rapamycin before, but not after, the formation of plaques and tangles ameliorates cognitive deficits. *PLoS ONE.* 2011;6(9), e25416.
64. Silva MC, et al. Prolonged tau clearance and stress vulnerability rescue by pharmacological activation of autophagy in tauopathy neurons. *Nat Commun.* 2020;11(1):3258.
65. Kim S, et al. Fisetin stimulates autophagic degradation of phosphorylated tau via the activation of TFEB and Nrf2 transcription factors. *Sci Rep.* 2016;6:24933.
66. Kim YD, et al. Pimozide reduces toxic forms of tau in TauC3 mice via 5′ adenosine monophosphate-activated protein kinase-mediated autophagy. *J Neurochem.* 2017;142(5):734–746.
67. Chen Y, et al. Berberine mitigates cognitive decline in an Alzheimer's disease mouse model by targeting both tau hyperphosphorylation and autophagic clearance. *Biomed Pharmacother.* 2020;121, 109670.
68. Li X, et al. Discovery of nitazoxanide-based derivatives as autophagy activators for the treatment of Alzheimer's disease. *Acta Pharm Sin B.* 2020;10(4):646–666.
69. Sundararajan V, Venkatasubbu GD, Sheik Mohideen S. Investigation of therapeutic potential of cerium oxide nanoparticles in Alzheimer's disease using transgenic Drosophila. *3 Biotech.* 2021;11(4):159.
70. He Z, et al. The protective effect of vanadium on cognitive impairment and the neuropathology of Alzheimer's disease in APPSwe/PS1dE9 mice. *Front Mol Neurosci.* 2020;13:21.
71. Chen Y, et al. Metformin attenuates plaque-associated tau pathology and reduces amyloid-beta burden in APP/PS1 mice. *Alzheimers Res Ther.* 2021;13(1):40.
72. Uddin MN, et al. Strain-specific clearance of seed-dependent tau aggregation by lithium-induced autophagy. *Biochem Biophys Res Commun.* 2021;543:65–71.
73. da Silva LFI, et al. miR-532-3p: a possible altered miRNA in cumulus cells of infertile women with advanced endometriosis. *Reprod BioMed Online.* 2021;42(3):579–588.
74. Nixon RA. Autophagy, amyloidogenesis and Alzheimer disease. *J Cell Sci.* 2007;120(Pt 23):4081–4091.
75. Cortes CJ, La Spada AR. TFEB dysregulation as a driver of autophagy dysfunction in neurodegenerative disease: molecular mechanisms, cellular processes, and emerging therapeutic opportunities. *Neurobiol Dis.* 2019;122:83–93.
76. Chen X, et al. Promoting tau secretion and propagation by hyperactive p300/CBP via autophagy-lysosomal pathway in tauopathy. *Mol Neurodegener.* 2020;15(1):2.

77. Wang Y, et al. Generation of tau aggregates and clearance by autophagy in an inducible cell model of tauopathy. *Neurodegener Dis.* 2010;7(1–3):103–107.
78. Pandit R, Leinenga G, Gotz J. Repeated ultrasound treatment of tau transgenic mice clears neuronal tau by autophagy and improves behavioral functions. *Theranostics.* 2019;9(13):3754–3767.
79. Comerota MM, et al. Near infrared light treatment reduces synaptic levels of toxic tau oligomers in two transgenic mouse models of human tauopathies. *Mol Neurobiol.* 2019;56(5):3341–3355.
80. Baldessarini RJ, Tondo L, Vazquez GH. Pharmacological treatment of adult bipolar disorder. *Mol Psychiatry.* 2019;24(2):198–217.
81. Motoi Y, et al. Lithium and autophagy. *ACS Chem Neurosci.* 2014;5(6):434–442.
82. Severus E, Bauer M, Geddes J. Efficacy and effectiveness of lithium in the long-term treatment of bipolar disorders: an update 2018. *Pharmacopsychiatry.* 2018;51(5):173–176.
83. Zhang D, et al. Lithium promotes recovery of neurological function after spinal cord injury by inducing autophagy. *Neural Regen Res.* 2018;13(12):2191–2199.
84. Moors TE, et al. Therapeutic potential of autophagy-enhancing agents in Parkinson's disease. *Mol Neurodegener.* 2017;12(1):11.
85. Sarkar S, et al. Lithium induces autophagy by inhibiting inositol monophosphatase. *J Cell Biol.* 2005;170(7):1101–1111.
86. Sarkar S, Rubinsztein DC. Inositol and IP3 levels regulate autophagy: biology and therapeutic speculations. *Autophagy.* 2006;2(2):132–134.
87. Sanders DW, et al. Distinct tau prion strains propagate in cells and mice and define different tauopathies. *Neuron.* 2014;82(6):1271–1288.

SECTION VI

Autophagy and pathology in Alzheimer's disease

CHAPTER 10

Granulovacuolar degeneration in neurodegeneration

Kouki Makioka[a], Tsuneo Yamazaki[b], Masayuki Sato[a], Masamitsu Takatama[c], Koichi Okamoto[c], and Yoshio Ikeda[a]
[a]Department of Neurology, Gunma University Graduate School of Medicine, Maebashi, Gunma, Japan
[b]Gunma University School of Health Sciences, Maebashi, Gunma, Japan
[c]Geriatrics Research Institute and Hospital, Maebashi, Gunma, Japan

1. Introduction

The accumulation and deposition of misfolded proteins are common features of neurodegenerative diseases, such as Alzheimer's disease (AD), Parkinson's disease (PD), amyotrophic lateral sclerosis, multiple system atrophy (MSA) and polyglutamine diseases.[1–5] A normal balance between the production and degeneration of cellular protein is required for cell survival, and dysfunction of protein degradation systems has been proposed as one of the causes of neurodegeneration.[6] In AD pathology, impairment of the protein-degradation system including autophagy has been proposed as one of the causes of the accumulation of aberrant proteins.[7] The pathological features of AD included senile plaques, which contain extra cellular deposits of amyloid β and dystrophic neurites, and neurofibrillary tangles (NFTs) containing hyperphosphorylated tau.[1,8] In addition to these two pathological hallmarks, granulovacuolar degeneration (GVD) is increased in AD brain, so GVD is also one of the pathological features of AD.[9,10] In this review, we describe GVD sight of autophagy.

2. Neuropathological features of granulovacuolar degeneration (GVD) in neurons

GVD, which appears in the neuronal cytoplasm, was first described by Simchowicz in 1991.[11] GVD is characterized by the presence of membrane-delineated vacuoles harboring a dense core, the so-called GVD body (GVB) (Fig. 1). GVBs comprise single or multiple granule-containing vacuoles, 3–5 μm in diameter, within the neuronal cytoplasm, and can be observed by hematoxylin and eosin staining, and are argyrophilic in nature.[12]

Autophagy Dysfunction in Alzheimer's Disease and Dementia
https://doi.org/10.1016/B978-0-323-89906-2.00015-0

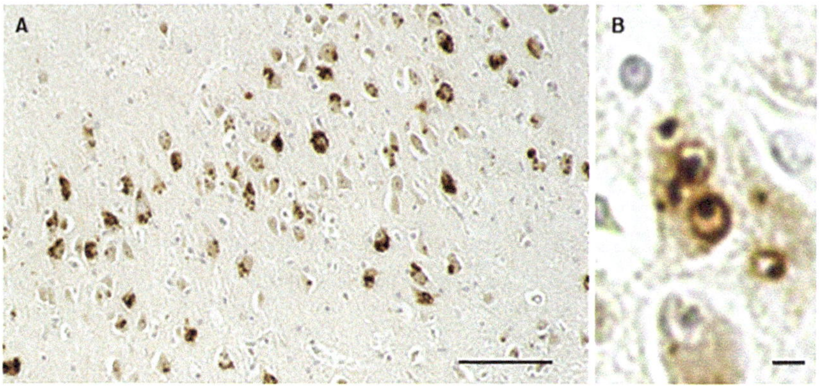

Fig. 1 ***Granulovacuolar degeneration in Alzheimer's disease.*** Immunohistochemical detection of pPERK (Protein kinase R (PKR)-like endoplasmic reticulum kinase) in hippocampus in AD. GVBs have single or multiple granule-containing vacuoles within the neuronal cytoplasm. (Scale bars A: 100 m, B: 10 m.)

GVBs are most commonly observed in the pyramidal cells of the hippocampus, predominantly in the cornu ammonis 1 in AD and aged brain, and the number of cells with GVBs significantly increases in AD cases as compared to healthy individuals.[9,10,13] In addition to AD, GVD is also observed in many other diseases involving tau accumulation, including Pick's disease, Parkinsonism-dementia complex of Guam, Down's syndrome, tuberous sclerosis, and progressive supranuclear palsy.[12,14,15] Thus, GVD is not specific to AD, but is one of the pathological changes associated with tau accumulation.

GVBs coexist with NFTs, and the distribution and spreading pattern of GVD are similar to that of NFT.[16–18] GVBs are positive for tau, including phosphorylated-tau[19–22] and tau kinase.[18,23–25] Moreover, GVBs are frequently observed in pretangle, which means neurons are diffusely stained by tau antibodies.[26,27] Thus, many researchers hypothesized that GVD in the AD brain is associated with NFT formation at an early stage,[23,28–30] but there has been no direct evidence that tau accumulation induces GVBs in neurons. Recently, Wiersma et al. showed that tau accumulation induced GVB-like structures with GVB markers in vitro and in vivo.[28] They observed GVB-like structures with GVB markers in tau-injected mouse brains and in tau-treated primary cultured neuron.[28] Therefore, these results directly show that stress induced by tau accumulation can cause GVD changes. In addition to tauopathy, including AD, GVD is also observed in normal aging brains without AD, with the number of GVBs increasing with age.[10] GVBs might emerge before tau accumulation and their presence might be an early change in neurodegeneration.

3. What kind of organelle are GVBs?

GVBs are small and round organelles in neuronal cells, which are surrounded by membranes and have granules. From this morphology, there are some candidates for the kind of organelle GVBs could be. GVBs are labeled by cytoskeletal proteins, such as tubulin[31] and neurofilament,[21,32,33] along with proteins such as ubiquitin,[34] phosphorylated-TAR-DNA-binding protein 43 (TDP-43),[35] and many organelle markers. In electron microscopic observation, the morphology of the granules is somewhat different in each vacuole, and some vacuoles may contain slight-to-moderate amounts of electron-dense material.[36] These findings indicate that GVBs are not a homogeneous type of organelle and are rather structures with variety. From the varied morphology and presence of markers for different organelles, one can only guess what kind of organelle GVBs are.

Endosomes are a collection of intracellular sorting organelles in eukaryotic cells under normal conditions and are part of the endocytic membrane transport pathway originating from the trans-Golgi network.[37] Immunohistochemical studies have shown that GVBs are positive for endosomal, Golgi apparatus,[37–39] and lysosomal markers.[40,41] In electron microscopic observation, some GVBs have a two-layered membrane, which indicates autophagic vacuoles.[36] Autophagy is initiated when an isolation membrane is created with microtubule-associated light chain 3-II (LC3-II), and the membrane elongates and sequesters cytoplasm and organelles into autophagosome[42] (Fig. 2). Internalized membrane proteins are sorted into intraluminal vesicles of late endosomes (multivesicular body, MVB), and autophagosomes fuse with MVB and lysosomes for digesting the contents (Fig. 2).[43–45] Lysosomes have acidic pH and proteolytic enzymes, and after fusing with lysosomes,[46] the cargo carried by autophagosomes and late endosomes are digested by lysosomal enzymes[45] (Fig. 2). A normal balance between the formation and degradation of cellular proteins is important for homeostasis. Although GVBs have lysosomal proteins, endosomal proteins, Golgi apparatus proteins, and GVBs even have double membranes, indicating an autophagosomal lineage, GVBs are not exactly the same as these organelles that are found under normal conditions in neurons. In addition to markers for these organelle markers, GVBs are positive for the markers of organelles that emerge under stress, including stress granules,

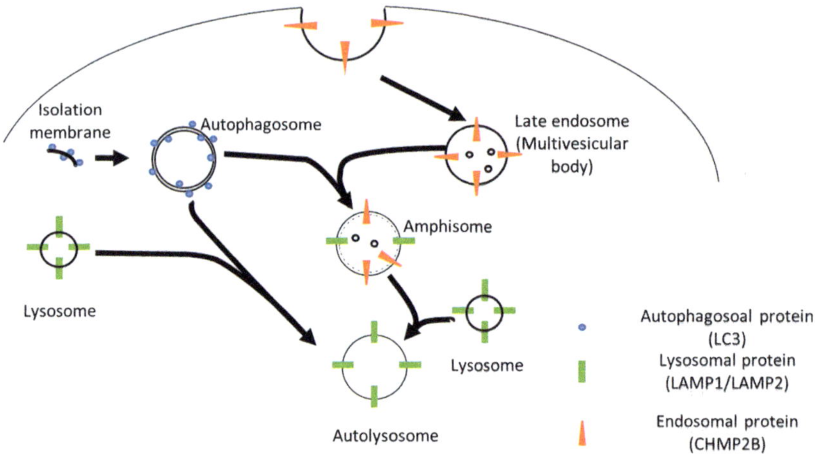

Fig. 2 ***Autophagy-lysosome-endosome pathway in normal condition.*** Autophagy is initiated when an isolation membrane is created with microtubule-associated light chain 3-II (LC3-II), and the membrane elongates and sequesters the cytoplasm and organelles into the autophagosome. Autophagosomes fuse with late endosomes (multivesicular bodies) and lysosomes to digest the contents. Lysosomes have acidic pH and proteolytic enzymes, and after fusing with lysosomes, the cargo carried by autophagosomes and late endosomes are digested by lysosomal enzymes.

necrosome, and unfolded protein response (UPR) (Fig. 3).[47–50] This indicates that GVBs are unique structures that emerge under stress, especially due to tau accumulation in neurons (Fig. 3).

Autophagosomes increase intracellularly under two states: one is when the autophagy is enhanced, while the other happens when degradation in lysosomes is decreased.[51,52] Electron microscopy has revealed that in the AD brain, autophagosomes and autolysosomes are increased, along with LC3-II, which is a marker of autophagosomes.[53] This suggests the possibility of autophagy failure. GVD might be a condition in which the intracellular degradation system via autophagy malfunctions due to tau accumulation, and abnormal proteins cannot be degraded.

4. GVB-like structures in oligodendroglia in multiple system atrophy

GVBs are observed in the cytoplasm of neurons in brains with tauopathy and aged brains, and there have been no previous reports of GVBs in glial cells. We have previously reported structures that were composed of cores and the surrounding outer layer, compatible with GVBs morphology,

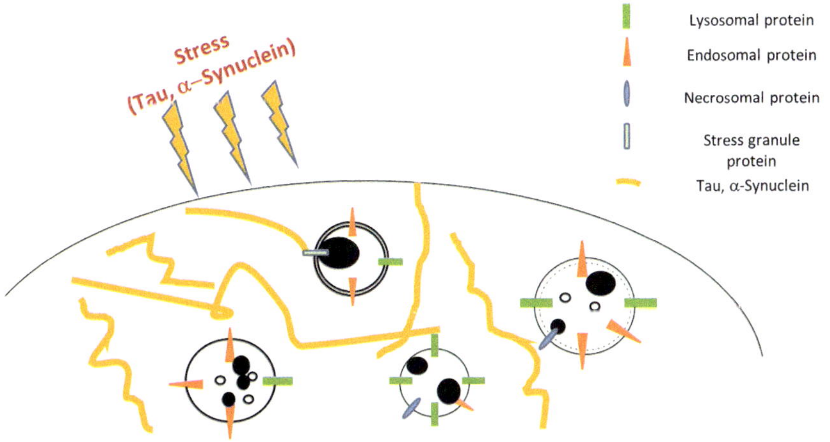

Fig. 3 ***Model of GVBs.***
GVBs are small and round organelles in neuronal cells, which are surrounded by membranes and have granules. GVBs have lysosomal and endosomal proteins, and may be single or double membrane. These results indicate that GVBs are the structures of autophagic and endocytic pathways. GVBs are also positive for the markers of organelles that emerge under stress, including stress granules and the necrosome. The morphology of the granules is somewhat different in each vacuole, and some vacuoles contain slight to moderate amounts of electron-dense material. GVBs are structures that emerge when autophagy is impaired by aggregating proteins, such as tau and α-synuclein, leading to cell death or inclusion body formation.

in glial cells in MSA brains.[54] MSA is a sporadic neurodegenerative disorder characterized by neuronal loss and accompanied by gliosis in the basal ganglia, cerebellum, pons, inferior olivary nuclei, and spinal cord.[55] The histopathological features of MSA are the presence of argyrophilic glial cytoplasmic inclusions (GCIs) with α-synuclein in oligodendroglia, so MSA is one of the α-synucleinopathies, similarly to PD.[56–58] The GVB-like structures in oligodendroglia in MSA also expressed the markers of GVBs in neurons (Fig. 4A and B), including UPR, p-tau, phosphorylated glycogen synthase kinase (pGSK3), and pTDP43.[54] The main components of GCIs are α-synuclein, and GVB-like structures in oligodendroglia, frequently observed in cells that have faint staining of α-synuclein[37] (Fig. 4C). GVB-like structures in oligodendroglia frequently co-localized with tubulin polymerization-prompting protein (TPPP) rather than Ubiquitin.[37] TPPP accumulates in oligodendroglia at the early stage of the formation of GCIs, and ubiquitin at the late stage.[59,60] These results indicate that GVB-like structures emerge at early stages of GCI (Fig. 5). GVBs were also labeled by anti-UPR antibodies in α-synuclein-positive neurons in PD[61] (Fig. 4D);

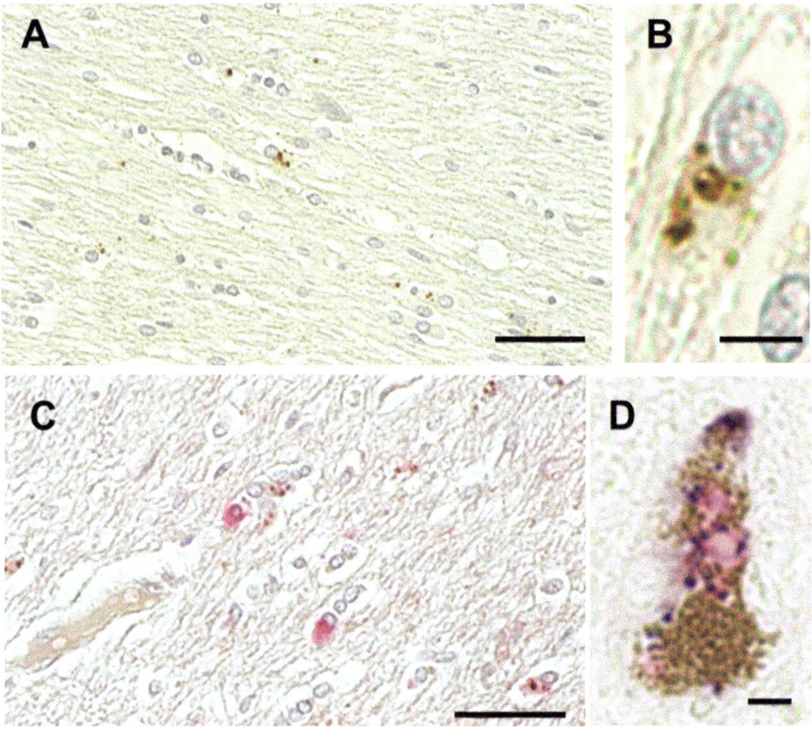

Fig. 4 ***GVBs in α-synucleinopathy.***
Immunohistochemical detection of pPERK in pons in MSA (A, B). pPERK-positive structures were detected as granules in some glial cells. The pPERK-positive structures (*brown* in C) were localized in cells exhibiting α-synuclein labeling (*red* in C), which was frequently but faintly positive. GVBs were detected by anti-pPERK antibodies in α-synuclein-positive neurons in PD (*red*: α-synuclein, purple pPERK) (scale bars A, C: 50 μm, B, D: 10 μm).

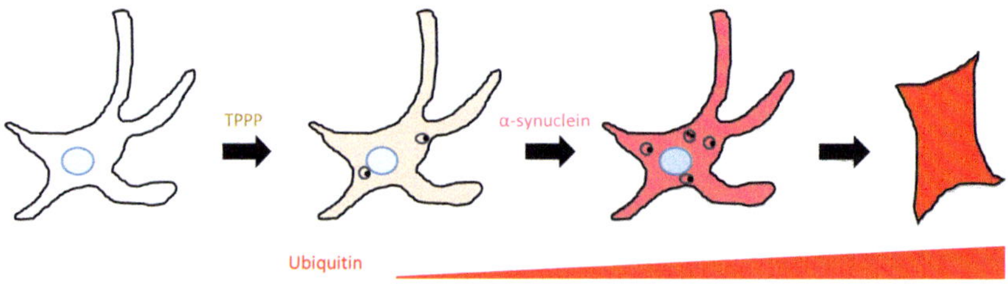

Fig. 5 ***GVBs emerge at early stage of GCIs.***
TPPP accumulates in oligodendroglia at the early stage of the formation of GCIs, and ubiquitin at the late stage. GVB-like structures in oligodendroglia, frequently observed in cells that have faint staining of α-synuclein, and co-localized with tubulin polymerization-prompting protein (TPPP) more than Ubiquitin. GVB-like structures emerge at early stages of GCI.

autophagy is impaired in PD as well as in AD.[62] Therefore, GVBs in PD might indicate a malfunction of autophagy by accumulation of α-synuclein. Thus, GVB-like structures in oligodendroglia emerge from the stress of accumulation of α-synuclein. In MSA, GCIs are LC3-positive, which is a marker for autophagosomes, so it is considered that the dysfunction of autophagy is involved in the pathogenesis.[63,64]

Aggregation of tau in AD and α-synuclein in MSA and PD may cause impaired autophagy pathways, resulting in the appearance of GVB during cell degeneration, leading to inclusion body formation with NFTs, Lewy bodies, and GCIs. Thus, GVD might be in a state where autophagy is impaired by accumulation of proteins such as tau and α-synuclein, leading to cell death or inclusion body formation.

Acknowledgment

This study was supported by grants awarded to T. Yamazaki from the Ministry of Health, Labour and Welfare and the Ministry of Education, Culture, Sports, Science, and Technology of Japan (26461305 and 17K09790).

References

1. Selkoe DJ, Schenk D. Alzheimer's disease: molecular understanding predicts amyloid-based therapeutics. *Annu Rev Pharmacol Toxicol.* 2003;43:545–584.
2. Spillantini MG, Schmidt ML, Lee VM, Trojanowski JQ, Jakes R, Goedert M. Alpha-synuclein in Lewy bodies. *Nature.* 1997;388(6645):839–840.
3. Wijesekera LC, Leigh PN. Amyotrophic lateral sclerosis. *Orphanet J Rare Dis.* 2009;4:3.
4. Gai WP, Power JH, Blumbergs PC, Blessing WW. Multiple-system atrophy: a new alpha-synuclein disease? *Lancet.* 1998;352(9127):547–548.
5. Zoghbi HY, Orr HT. Glutamine repeats and neurodegeneration. *Annu Rev Neurosci.* 2000;23:217–247.
6. Soto C. Unfolding the role of protein misfolding in neurodegenerative diseases. *Nat Rev Neurosci.* Jan 2003;4(1):49–60.
7. Ihara Y, Morishima-Kawashima M, Nixon R. The ubiquitin-proteasome system and the autophagic-lysosomal system in Alzheimer disease. *Cold Spring Harb Perspect Med.* 2012;2(8):a006361. https://doi.org/10.1101/cshperspect.a006361.
8. Mattson MP. Pathways towards and away from Alzheimer's disease. *Nature.* 2004;430(7000):631–639.
9. Ball MJ, Vis CL. Relationship of granulovacuolar degeneration in hippocampal neurones to aging and to dementia in normal-pressure hydrocephalics. *J Gerontol.* 1978;33(6):815–824.
10. Ball MJ, Lo P. Granulovacuolar degeneration in the ageing brain and in dementia. *J Neuropathol Exp Neurol.* 1977;36(3):474–487.
11. Simchowicz T. Histologische studien über die senile dementz. *Histol Histopathol Arbeiten Aber Grosshirnrinde.* 1911;4:267–444.
12. Hirano A, Dembitzer HM, Kurland LT, Zimmerman HM. The fine structure of some intraganglionic alterations. Neurofibrillary tangles, granulovacuolar bodies and

"rod-like" structures as seen in Guam amyotrophic lateral sclerosis and parkinsonism-dementia complex. *J Neuropathol Exp Neurol.* 1968;27(2):167–182.
13. Xu M, Shibayama H, Kobayashi H, et al. Granulovacuolar degeneration in the hippocampal cortex of aging and demented patients—a quantitative study. *Acta Neuropathol.* 1992;85(1):1–9.
14. Steele JC. Progressive supranuclear palsy. Historical notes. *J Neural Transm Suppl.* 1994;42:3–14.
15. Ball MJ, Nuttall K. Topography of neurofibrillary tangles and granulovacuoles in hippocampi of patients with Down's syndrome: quantitative comparison with normal ageing and Alzheimer's disease. *Neuropathol Appl Neurobiol.* 1981;7(1):13–20.
16. Thal DR, Del Tredici K, Ludolph AC, et al. Stages of granulovacuolar degeneration: their relation to Alzheimer's disease and chronic stress response. *Acta Neuropathol.* 2011;122(5):577–589.
17. Braak H, Del Tredici K. Top-down projections direct the gradual progression of Alzheimer-related tau pathology throughout the neocortex. *Adv Exp Med Biol.* 2019;1184:291–303.
18. Ishizawa T, Sahara N, Ishiguro K, et al. Co-localization of glycogen synthase kinase-3 with neurofibrillary tangles and granulovacuolar degeneration in transgenic mice. *Am J Pathol.* 2003;163(3):1057–1067.
19. Bondareff W, Wischik CM, Novak M, Roth M. Sequestration of tau by granulovacuolar degeneration in Alzheimer's disease. *Am J Pathol.* 1991;139(3):641–647.
20. Goedert M, Spillantini MG, Jakes R. Localization of the Alz-50 epitope in recombinant human microtubule-associated protein tau. *Neurosci Lett.* 1991;126(2):149–154.
21. Dickson DW, Ksiezak-Reding H, Davies P, Yen SH. A monoclonal antibody that recognizes a phosphorylated epitope in Alzheimer neurofibrillary tangles, neurofilaments and tau proteins immunostains granulovacuolar degeneration. *Acta Neuropathol.* 1987;73(3):254–258.
22. Das P, Howard V, Loosbrock N, Dickson D, Murphy MP, Golde TE. Amyloid-beta immunization effectively reduces amyloid deposition in FcRgamma−/− knock-out mice. *J Neurosci.* 2003;23(24):8532–8538.
23. Leroy K, Boutajangout A, Authelet M, Woodgett JR, Anderton BH, Brion JP. The active form of glycogen synthase kinase-3beta is associated with granulovacuolar degeneration in neurons in Alzheimer's disease. *Acta Neuropathol.* 2002;103(2):91–99.
24. Augustinack JC, Sanders JL, Tsai LH, Hyman BT. Colocalization and fluorescence resonance energy transfer between cdk5 and AT8 suggests a close association in pre-neurofibrillary tangles and neurofibrillary tangles. *J Neuropathol Exp Neurol.* 2002;61(6):557–564.
25. Pei JJ, Grundke-Iqbal I, Iqbal K, Bogdanovic N, Winblad B, Cowburn RF. Accumulation of cyclin-dependent kinase 5 (cdk5) in neurons with early stages of Alzheimer's disease neurofibrillary degeneration. *Brain Res.* 1998;797(2):267–277.
26. Hondius DC, Koopmans F, Leistner C, et al. The proteome of granulovacuolar degeneration and neurofibrillary tangles in Alzheimer's disease. *Acta Neuropathol.* 2021;141(3):341–358.
27. Nishikawa T, Takahashi T, Nakamori M, et al. Phosphatidylinositol-4,5-bisphosphate is enriched in granulovacuolar degeneration bodies and neurofibrillary tangles. *Neuropathol Appl Neurobiol.* 2014;40(4):489–501.
28. Wiersma VI, van Ziel AM, Vazquez-Sanchez S, et al. Granulovacuolar degeneration bodies are neuron-selective lysosomal structures induced by intracellular tau pathology. *Acta Neuropathol.* 2019;138(6):943–970.
29. Wiersma VI, Scheper W. Granulovacuolar degeneration bodies: red alert for neurons with MAPT/tau pathology. *Autophagy.* 2020;16(1):173–175.
30. Köhler C. Granulovacuolar degeneration: a neurodegenerative change that accompanies tau pathology. *Acta Neuropathol.* 2016;132(3):339–359.

31. Price DL, Altschuler RJ, Struble RG, Casanova MF, Cork LC, Murphy DB. Sequestration of tubulin in neurons in Alzheimer's disease. *Brain Res.* 1986;385(2):305–310.
32. Gheuens J, Cras P, Perry G, et al. Demonstration of a novel neurofilament associated antigen with the neurofibrillary pathology of Alzheimer and related diseases. *Brain Res.* 1991,558(1):43–52.
33. Kahn J, Anderton BH, Probst A, Ulrich J, Esiri MM. Immunohistological study of granulovacuolar degeneration using monoclonal antibodies to neurofilaments. *J Neurol Neurosurg Psychiatry.* 1985;48(9):924–926.
34. Dickson DW, Liu WK, Kress Y, Ku J, DeJesus O, Yen SH. Phosphorylated tau immunoreactivity of granulovacuolar bodies (GVB) of Alzheimer's disease: localization of two amino terminal tau epitopes in GVB. *Acta Neuropathol.* 1993;85(5):463–470.
35. Kadokura A, Yamazaki T, Kakuda S, et al. Phosphorylation-dependent TDP-43 antibody detects intraneuronal dot-like structures showing morphological characters of granulovacuolar degeneration. *Neurosci Lett.* 2009;463(1):87–92.
36. Okamoto K, Hirai S, Iizuka T, Yanagisawa T, Watanabe M. Reexamination of granulovacuolar degeneration. *Acta Neuropathol.* 1991;82(5):340–345.
37. Huotari J, Helenius A. Endosome maturation. *EMBO J.* 2011;30(17):3481–3500.
38. Yamazaki Y, Takahashi T, Hiji M, et al. Immunopositivity for ESCRT-III subunit CHMP2B in granulovacuolar degeneration of neurons in the Alzheimer's disease hippocampus. *Neurosci Lett.* 2010;477(2):86–90.
39. Yan T, Wang L, Gao J, et al. Rab10 phosphorylation is a prominent pathological feature in Alzheimer's disease. *J Alzheimers Dis.* 2018;63(1):157–165.
40. Barrachina M, Maes T, Buesa C, Ferrer I. Lysosome-associated membrane protein 1 (LAMP-1) in Alzheimer's disease. *Neuropathol Appl Neurobiol.* 2006;32(5):505–516.
41. Hassiotis S, Manavis J, Blumbergs PC, et al. Lysosomal LAMP1 immunoreactivity exists in both diffuse and neuritic amyloid plaques in the human hippocampus. *Eur J Neurosci.* 2018;47(9):1043–1053.
42. Asanuma K, Tanida I, Shirato I, et al. MAP-LC3, a promising autophagosomal marker, is processed during the differentiation and recovery of podocytes from PAN nephrosis. *FASEB J.* 2003;17(9):1165–1167.
43. Klumperman J, Raposo G. The complex ultrastructure of the endolysosomal system. *Cold Spring Harb Perspect Biol.* 2014;6(10), a016857.
44. Dunn Jr WA. Studies on the mechanisms of autophagy: formation of the autophagic vacuole. *J Cell Biol.* 1990;110(6):1923–1933.
45. Eskelinen EL. Maturation of autophagic vacuoles in mammalian cells. *Autophagy.* 2005;1(1):1–10.
46. Bellettato CM, Scarpa M. Pathophysiology of neuropathic lysosomal storage disorders. *J Inherit Metab Dis.* 2010;33(4):347–362.
47. Koper MJ, Van Schoor E, Ospitalieri S, et al. Necrosome complex detected in granulovacuolar degeneration is associated with neuronal loss in Alzheimer's disease. *Acta Neuropathol.* 2020;139(3):463–484.
48. Castellani RJ, Gupta Y, Sheng B, et al. A novel origin for granulovacuolar degeneration in aging and Alzheimer's disease: parallels to stress granules. *Lab Invest.* 2011;91(12):1777–1786.
49. Chang RC, Wong AK, Ng HK, Hugon J. Phosphorylation of eukaryotic initiation factor-2alpha (eIF2alpha) is associated with neuronal degeneration in Alzheimer's disease. *Neuroreport.* 2002;13(18):2429–2432.
50. Hoozemans JJ, van Haastert ES, Nijholt DA, Rozemuller AJ, Eikelenboom P, Scheper W. The unfolded protein response is activated in pretangle neurons in Alzheimer's disease hippocampus. *Am J Pathol.* 2009;174(4):1241–1251.
51. Tanida I, Minematsu-Ikeguchi N, Ueno T, Kominami E. Lysosomal turnover, but not a cellular level, of endogenous LC3 is a marker for autophagy. *Autophagy.* 2005;1(2):84–91.

52. Kabeya Y, Mizushima N, Ueno T, et al. LC3, a mammalian homologue of yeast Apg8p, is localized in autophagosome membranes after processing. *EMBO J.* 2000;19(21):5720–5728.
53. Yu WH, Cuervo AM, Kumar A, et al. Macroautophagy—a novel Beta-amyloid peptide-generating pathway activated in Alzheimer's disease. *J Cell Biol.* 2005;171(1):87–98.
54. Makioka K, Yamazaki T, Fujita Y, Takatama M, Nakazato Y, Okamoto K. Involvement of endoplasmic reticulum stress defined by activated unfolded protein response in multiple system atrophy. *J Neurol Sci.* 2010;297(1–2):60–65.
55. Stefanova N, Bücke P, Duerr S, Wenning GK. Multiple system atrophy: an update. *Lancet Neurol.* 2009;8(12):1172–1178.
56. Papp MI, Lantos PL. The distribution of oligodendroglial inclusions in multiple system atrophy and its relevance to clinical symptomatology. *Brain.* 1994;117(Pt. 2):235–243.
57. Papp MI, Kahn JE, Lantos PL. Glial cytoplasmic inclusions in the CNS of patients with multiple system atrophy (striatonigral degeneration, olivopontocerebellar atrophy and Shy-Drager syndrome). *J Neurol Sci.* 1989;94(1–3):79–100.
58. Nakazato Y, Yamazaki H, Hirato J, Ishida Y, Yamaguchi H. Oligodendroglial microtubular tangles in olivopontocerebellar atrophy. *J Neuropathol Exp Neurol.* 1990;49(5):521–530.
59. Wenning GK, Stefanova N, Jellinger KA, Poewe W, Schlossmacher MG. Multiple system atrophy: a primary oligodendrogliopathy. *Ann Neurol.* 2008;64(3):239–246.
60. Abe H, Yagishita S, Amano N, Iwabuchi K, Hasegawa K, Kowa K. Argyrophilic glial intracytoplasmic inclusions in multiple system atrophy: immunocytochemical and ultrastructural study. *Acta Neuropathol.* 1992;84(3):273–277.
61. Hoozemans JJ, van Haastert ES, Eikelenboom P, de Vos RA, Rozemuller JM, Scheper W. Activation of the unfolded protein response in Parkinson's disease. *Biochem Biophys Res Commun.* 2007;354(3):707–711.
62. Park H, Kang JH, Lee S. Autophagy in neurodegenerative diseases: a hunter for aggregates. *Int J Mol Sci.* 2020;21(9):3369. https://doi.org/10.3390/ijms21093369.
63. Malik BR, Maddison DC, Smith GA, Peters OM. Autophagic and endo-lysosomal dysfunction in neurodegenerative disease. *Mol Brain.* 2019;12(1):100.
64. Tanji K, Odagiri S, Maruyama A, et al. Alteration of autophagosomal proteins in the brain of multiple system atrophy. *Neurobiol Dis.* 2013;49:190–198.

CHAPTER 11

Autophagy dysfunction in skeletal myopathies: Inclusion body myositis and Danon disease

Kazuma Sugie
Department of Neurology, Nara Medical University School of Medicine, Kashihara, Nara, Japan

1. Autophagy and autophagic vacuoles

Autophagy is a self-digesting mechanism responsible for the degradation of intracytoplasmic proteins and intercellular organelles in lysosomes, to achieve cell homeostasis and organelle renewal in all the cells.[1] Muscle cells also have this function as cells in the whole body. Lysosomes are recognized as key organelles for the autophagy mechanism. Various hydrolytic enzymes are contained in lysosomes enclosed with membrane structure, and the inside of lysosomes is kept acidic with optimal pH of between 4.5 and 5.0 so that enzymes can efficiently function. Degradation product may be used inside cells or released outside the cells through exocytosis.

Autophagy has two major physiological functions. The first function is homeostatic, where intercellular proteins or organelles that have become unnecessary for the purpose of quality control and depuration in cells are disposed. The second function is autophagy that degrades intracytoplasmic components under stress conditions, including a starved state for continuous energy supply system. Autophagy consists of the following three subtypes: macroautophagy, microautophagy, and chaperone-mediated autophagy.[2] Autophagic vacuoles are thought to form in the process of macroautophagy. First, the isolation membrane, which is a flat membranous sac in cytoplasm, engulfs portions of the proteins and intercellular organelles that are to be degraded while also extending to form the autophagosome. Next, the autophagosome fuses with lysosome to form autolysosome, which is degraded by hydrolytic enzymes contained in the lysosome. Vacuolar structures formed through these processes are collectively called the "autophagic vacuoles".

Autophagy Dysfunction in Alzheimer's Disease and Dementia
https://doi.org/10.1016/B978-0-323-89906-2.00016-2

2. Autophagic vacuolar myopathy

In general, during morphological monitoring of normal muscle tissues, the presence of lysosomes or autophagic vacuoles is rarely observed. Hence for a long time, the role of autophagy in muscle cells remained unclear. The importance of autophagy in muscle cells, however, started to attract attention with the identification of a subgroup of muscle diseases accompanied by many autophagic vacuoles in muscle cells. Myopathies caused by abnormality of autophagy in skeletal muscles are called autophagic vacuolar myopathies, which collectively represent muscle diseases defined by conditions in which autophagic vacuoles present as a muscle pathological feature.[3] Of these, clinicopathological studies in Pompe disease are historically most advanced. As such, the causes of autophagic vacuolar myopathies, except for Pompe disease, lacked clarity until relatively recently. Two novel categories have since been recognized.

Currently, this subgroup can be classified into three subtypes according to its pathomechanical and morphological characteristics (Table 1). Inclusion body myositis (IBM), which is one of the focus of this article, is a representative disease classified into rimmed vacuolar myopathy, and Danon disease, the other focus of this article, is a representative disease classified into AVSF myopathy. The three subtypes are described below.

Table 1 Autophagic vacuolar myopathies.

A. Deficiency of lysosomal enzyme
Pompe disease
B. Secondary lysosomal dysfunctions
Rimmed vacuolar myopathy
✓ **Inclusion body myositis** (IBM)
✓ Distal myopathy with rimmed vacuoles (GNE myopathy)
✓ Inclusion body myopathy with Paget's disease of bone and frontotemporal dementia (IBMPFD)
✓ Oculopharyngeal muscular dystrophy (OPMD)
✓ Oculopharyngodistal myopathy (OPDM)
C. Primary lysosomal dysfunctions
AVSF myopathy
✓ **Danon disease**
✓ X-linked myopathy with excessive autophagy (XMEA)

2.1 Pompe disease

Pompe disease, also known as glycogen storage disease type II, is an autosomal recessive inherited disorder caused by the deficiency of acid alpha-glucosidase (acid maltase, GAA). GAA is an enzyme that is essential in the degradation of glycogen to glucose in lysosomes. Patients with Pompe disease demonstrate muscle weakness of extremities and hypertrophic cardiomyopathy.[4] Muscle pathological findings found excessive glycogen accumulated in lysosomes in muscle cells and observed the formation of characteristic large autophagic vacuoles that occupy the entire cell.[5] Enzyme replacement therapy was introduced in 2006, and its efficacy has been demonstrated so far.[6–8] Nonetheless, insufficient effect or the development of serious adverse reactions in some patients have been reported, and the cause of such conditions has been investigated in various studies including gene therapies.[9]

2.2 Rimmed vacuolar myopathy

The second subtype is a group of diseases with distinctive autophagic vacuoles called rimmed vacuoles. Among rimmed vacuolar myopathies, IBM is a representative nonhereditary disease,[10] listed along with many hereditary diseases such as distal myopathy with rimmed vacuoles (GNE myopathy),[11–13] oculopharyngeal muscular dystrophy,[14] oculopharyngodistal myopathy,[15] inclusion body myopathy with Paget's disease of bone and frontotemporal dementia (IBMPFD) (VCP myopathy),[16] tibial muscular dystrophy, Marinesco-Sjogren syndrome, and Welander distal myopathy. Since all the causative genes identified in these hereditary rimmed vacuolar myopathies are present outside the lysosomes, formation of rimmed vacuoles is considered to be caused by an abnormality of autophagy due to secondary lysosomal dysfunctions. However, as rimmed vacuoles may also be nonspecifically observed in other muscle diseases, careful attention is required in its identification.

2.3 AVSF myopathy

The third subtype is a group of diseases caused by abnormality in the structure and function of lysosomes. Since this subtype is characterized by the presence of specific autophagic vacuoles known as autophagic vacuoles with sarcolemmal features (AVSF), it is referred to as AVSF myopathies.[17] AVSF express virtually all sarcolemmal proteins, in addition to acetylcholinesterase (AChE), on their vacuolar membranes in muscle cells. Danon disease, a representative disease, is caused by primary deficiency of lysosomal-associated membrane protein

2 (LAMP-2).[18,19] Hypertrophic cardiomyopathy is a determinant prognostic factor of this disease, and heart transplantation is still the sole radical treatment. X-linked myopathy with excessive autophagy (XMEA), another representative disease, has recently been revealed to be caused by gene mutations of *vacuolar membrane ATPase activity 21* (*VMA21*), which contributes to acidification inside the lysosomal lumen.[20] Since both causative genes of genetically diagnosable AVSF myopathies are present in lysosomes, AVSF myopathies are thought to be caused by an abnormality of autophagy due to primary lysosomal dysfunctions. AVSF are the pathological finding with extremely high disease specificity unlike other autophagic vacuoles.

3. Inclusion body myositis

3.1 Outline of the disease

Inclusion body myositis (IBM) is the most common idiopathic inflammatory myopathy (IIM) among older persons aged 50 and above in Western countries. In 1967, Chou and colleagues reported the first case that showed severe atrophy and weakness of bilateral quadriceps femoris muscles.[21] Muscle pathology revealed inflammatory cell infiltration in histochemistry and presence of tubular filaments in electron microscopy. An additional characteristic observation was the coexistence of autophagic vacuoles referred to as rimmed vacuoles and deposition of amyloid in muscle cells. While it shares aspects of muscle "degenerative diseases," IBM is classified as IIM along with polymyositis (PM), dermatomyositis (DM), immune-mediated necrotizing myopathy, and antisynthetase syndrome.[22] IBM may not respond to immunological treatments including corticosteroid and/or high-dose immunoglobulin therapy. Therefore, this disease is considered to be a refractory and progressive muscular disease without effective treatments.[23] Accurate diagnosis is pertinent to avoid risk of variable adverse reactions or complications. However, IBM is generally underdiagnosed, and it is sometimes misdiagnosed.

IBM is a sporadic disease in principle, despite reports of IBM in patients' parents, children, or siblings, which suggest a genetic background such as HLA. Recently, genetic abnormalities in *valosin containing protein* (*VCP*) gene have been detected in patients with IBMPFD.[16] This finding has attracted attention in terms of the pathomechanism of IBM, although there is a difference between IBM and IBMPFD; patients with IBMPFD commonly lack inflammation as a pathological feature. Interestingly, the presence of amyloid β protein (Aβ) and phosphorylated tau in rimmed vacuoles has been revealed, suggesting the homology of Alzheimer's disease (AD) and IBM in aspects of the molecular pathogenic mechanisms.

3.2 Clinical features

IBM is a chronic progressive and disabling muscle disease that generally develops from age 50, and the diagnosis is not typically made for 5 years or more after the onset of its first symptoms in many patients. PM and DM dominantly affect females, while IBM dominantly affects males. In patients with IBM, asymmetric muscle weakness and muscle atrophy develop in bilateral quadriceps femoris muscles and flexor muscles of fingers and wrists. Flexor muscles of fingers and wrists are commonly weaker than abductor muscles, and muscle strength of knee extension and ankle dorsiflexion are commonly weaker than that of hip flexion.[22]

A nationwide survey of 146 IBM patients in Japan demonstrated that the ratio of males to females in patients with IBM was 1.4 to 1, indicating a slightly greater number of males.[24] The survey revealed that the age at onset was 64 years old and initial symptoms included difficulty in going upstairs due to weakness of quadriceps femoris muscle found in 74% of the patients surveyed. Dysphagia was observed in 23% of the patients, which is one of factors that have impact on the prognosis. There was significant laterality in the muscle symptoms in 27% of the patient sample. There were no patients with obviously decreased cognitive function. Deep tendon reflexes were normal or mildly decreased. Complications of autoimmune abnormalities such as systemic lupus erythematosus, Sjögren's syndrome, scleroderma, or sarcoidosis were present in approximately 15%, while no increase in the incidence of interstitial lung disease or malignant tumor was demonstrated in patients with IBM, unlike patients with PM or DM. Serum creatine kinase (CK) levels were normal or mildly increased, up to 10 times the upper limit of normal level. The mean value was 511.2 IU/L, and patients whose CK level exceeded 2000 IU/L were rare. Approximately 20% of the patients with IBM were positive for antinuclear antibodies, while they were negative for myositis-specific autoantibodies. Recently, an autoantibody against cytosolic 5′-nucleotidase 1A (cN1A), which is described later, has been identified, and anti-cN1A autoantibodies were detected in approximately one-thirds of the Japanese patients with IBM, suggesting involvement of the immunological pathomechanism.[25]

On skeletal muscle MRI findings, atrophy of flexor digitorum profundus muscles in forearms and quadriceps femoris muscles were especially prominent.[26] In electromyogram, early recruiting, short duration, and low amplitude polyphasic motor unit action potentials (MUAPs) were observed in affected muscles. High amplitude polyphasic MUAPs and increased insertion activity were often observed. Since these findings may lead to a mistaken diagnosis of a neurogenic disease such as amyotrophic lateral sclerosis, careful attention is required. Spontaneous sensory paresthesia is rarely observed, but in a nerve conduction study, sensory nerve abnormalities were noted in around 30% of patients with IBM, partly because of old age or complications.[22] Interestingly, in the survey on 146 IBM patients mentioned above, approximately 20% of the sample were tested positive for human T-lymphotropic virus type I (HTLV-I) or hepatitis C virus (HCV). Previous reports suggest that hypertrophic cardiomyopathy might concurrently exist as a serious complication of IBM.[27]

3.3 Muscle pathological features

Muscle biopsy findings in patients with IBM are characterized by rimmed vacuoles in muscle cells and invasion, and/or encirclement of nonnecrotic muscle cells by monocytes, in addition to lymphocytic inflammation to endomysium (Fig. 1). According to typical pathological findings, the following features can be listed: necrosis and regeneration of muscle cells; monocyte infiltration to nonnecrotic cells mainly containing CD8-positive T cells in immunostaining; overexpression of major histocompatibility complex class-I (MHC-I) in morphologically normal muscle cells; ubiquitin-positive inclusion bodies and amyloid deposition in muscle cells; and relatively higher frequency of the presence of muscle cells stained negative for cyclooxygenase in proportion to age.

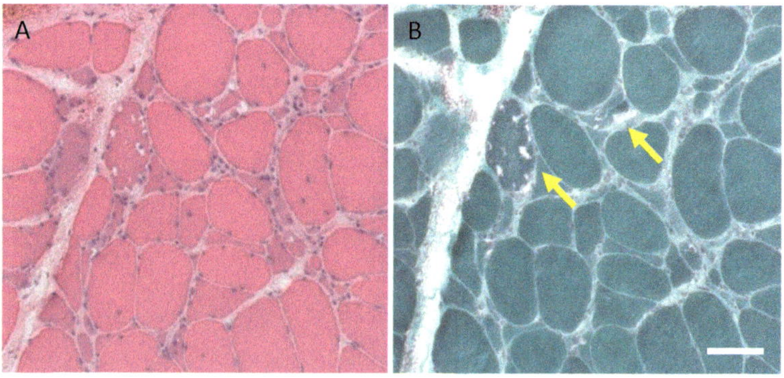

Fig. 1 Muscle pathology of a patient with inclusion body myositis. Muscle cells with rimmed vacuoles (arrows) are scattered with hematoxylin and eosin stain (A) and with modified Gomori-trichrome stain (B). Bar = 50 μm.

Furthermore, the presence of filamentous inclusion bodies of 16–20 mm in size in nucleus and cytoplasm has been confirmed by electron microscopy. Presence of many swollen muscle nuclei and eosinophilic inclusion bodies is observed. Immunostaining shows the presence of emerin, lamin A/C, VCP, histon, TAR DNA-binding protein of 43 kDa (TDP-43), and p62, which are nucleoproteins, inside the inclusion bodies, suggesting that these inclusion bodies are disrupted muscle nuclei. Moreover, Aβ deemed as AD-related proteins can be observed in the inclusion bodies, but its pathological significance has not been identified.[28] In muscle cells with formation of vacuoles, amyloid-like cells of up to 6–10 mm in size accumulate in the cytoplasm. Repeated muscle biopsy may be required in some cases to make a definitive diagnosis of IBM based on pathological features, including electron microscopy findings. Some cases of IBM may potentially be diagnosed as PM because the criteria for muscle biopsy are not satisfied due to sampling errors.

3.4 Pathogenesis and pathomechanism

The pathogenesis and pathomechanism of IBM have not been identified. Whether the primary phenomenon or pathological condition of IBM is inflammation or degeneration has yet to be clarified. The observation of rimmed vacuoles as a muscle pathological feature suggests degeneration or abnormality in the proteolysis pathway.

3.4.1 Mechanism of degeneration: Homology of Alzheimer's disease

Using immunohistochemistry, deposition of Aβ, amyloid precursor protein (APP), phosphorylated tau, prion proteins, apolipoprotein E, α-antichymotrypsin, ubiquitin, and neurofilaments can be observed inside rimmed vacuoles in muscle cells. The similarity between Alzheimer brains and IBM muscles for Aβ accumulation was proposed by Askanas et al..[28] This hypothesis has been supported by findings of muscle degeneration and formation of inclusion bodies in model mice in which muscle-specific overexpression of APP was induced.[29] In addition, it is known that some cytokines released from inflammatory cells increase production of APP. Recently, it has been demonstrated that ultrastructural abnormality in mitochondria occurs prior to formation of inclusion bodies, and inhibition of calcium ion transport to the mitochondria leads to suppression of APP accumulation, suggesting that abnormality in mitochondrial function is an important step in the early stage of the pathogenesis.[30] As Greenberg and colleagues pointed out, however, there are concerns regarding the specificity of antibodies of APP shown in immunostaining.[31] Given that APP accumulation is also observed in the muscle biopsy of patients with PM and DM, further research on disease specificity is necessary.

3.4.2 Abnormality in proteolytic pathway

Homeostasis of muscle cells is presumed to be maintained by a delicate balance between the synthesis and degradation of proteins. Potential disruptions to the proteolysis system or overdegradation by enhanced expression of a certain protein may be assumed. Increased expression of RING finger protein 5 (RNF5), one of ubiquitin E3 ligases, in IBM, and muscle atrophy and formation of inclusion bodies were observed in model mice with overexpression of RNF5.[32] Increased expression of ubiquitin E3 ligase and muscle degeneration and muscle atrophy were also observed in model mice in which defect of autophagy was induced specifically to skeletal muscles.[33] Furthermore, it was reported in drosophilae that autophagy and ubiquitin-proteasome system complementarily function, and the enhancement of histone deacetylase (HDAC) suppresses neurodegeneration, suggesting a potential for the development of a treatment method.[34]

Accumulation of familial ALS-related molecules such as TDP-43 and FUS are also observed in skeletal muscles in patients with IBM. Weihl et al. reported in 2008 the presence of TDP-43-positive intracytoplasmic inclusion bodies in muscles.[35] TDP-43-positive cells are far more frequently observed than rimmed vacuoles and Aβ-positive cells, which are considered to be a frequently observed pathological indicator in IBM. However, this finding is a common pathological change with patients with rimmed vacuolar myopathies, and its disease specificity to IBM is deemed low. P62, which is a ubiquitin-binding protein and observed in brain in patients with AD, shows stainability in muscle cells in patients with IBM at a higher frequency than TDP-43.[36] In addition, even in patients with abnormality in VCP, which is an important protein in proteolytic pathway mentioned earlier, rimmed vacuoles are observed in muscle cells. Abnormality in proteolytic pathway is considered an important mechanism in the pathogenesis of IBM.

3.4.3 Abnormality in immunological system

Another pathological feature in addition to those mentioned above is the presence of dendritic cells around nonnecrotic fibers. These cells are thought to function as antigen-presenting cells that lead to the activation of lymphocytes. In previous reports using microarray and microdissection, clonal proliferation of CD138-positive plasma cells was observed, indicating the involvement of plasma cells. Involvement of virus is also presumed, but it has yet to be demonstrated. In considering the pathogenesis of IBM, it is interesting to note that pathological features similar to those in IBM can be observed in patients with HIV, HTLV-1 infection, or those with sequelae of poliomyelitis, although persistent infection with mumps virus was ruled out.

Clonal proliferation of inflammatory cells indicates a possibility of autoimmune disease mediated by cytotoxic T-cells. As the proportion of muscle fibers that are surrounded by inflammatory cells is higher than that of muscle fibers with rimmed vacuoles or amyloid deposition, the extent of the involvement of inflammation is likely significant. The overexpression of MHC-I and enhancement in proliferation induced by cytokine also indicate the involvement of inflammation. Recently, it has been revealed that the expression of inducible nitric oxide synthase (iNOS) increases in patients with IBM. In addition, an in vitro experiment suggests the possibility that interleukin-1-beta induces iNOS to contribute to Aβ accumulation.[37] The finding of a potential mechanism that can connect inflammation and degeneration is important. However, since this phenomenon is observed even in patients with PM, it may not be considered a disease specific phenomenon. IBM patients also do not respond well to immunosuppressive agents in clinical practice. Thus, IBM is not considered a primary inflammatory disease. The relationship between inflammation and degeneration in IBM is likened to a chicken-and-egg situation, but identifying the process of occurrence of both inflammation and degeneration can potentially lead to the development of new radical treatment methods.

Recently, Greenberg and colleagues' finding of the autoantibody whose target antigen is cN1A that contributes to nucleic acid metabolism and DNA repair in the serum of patients with IBM that reacts to human skeletal muscle extract and roles of the autoantibody in the pathology of IBM have attracted attention.[38,39] Tawara et al. also identified anti-cN1A autoantibodies in 36% of 67 Japanese patients with IBM, which strongly suggests the autoantibodies' involvement in the immunological pathology of IBM.[25]

4. Danon disease

4.1 Outline of the disease

Danon disease (MIM#300257, LAMP-2 deficiency) is an extremely rare disease which was described as a "lysosomal glycogen storage disease with normal acid maltase" by Danon et al. in 1981.[40] In 2000, Nishino et al. identified the *LAMP2* gene on Xq24 as a causative gene of Danon disease and revealed that Danon disease to be caused by a primary deficiency of LAMP-2.[18] In 2002, Sugie et al. first reported that Danon disease showed X-linked dominant inheritance, and hypertrophic cardiomyopathy, myopathy, and mental retardation comprised the clinical manifestation triad of this disease in male patients, whereas cardiomyopathy was a main symptom in female patients.[19] In addition, autophagic vacuoles observed in patients with Danon disease were considerably different from those in other muscular diseases such as Pompe disease and rimmed vacuolar myopathies, which are extremely specific autophagic vacuoles (AVSF) with sarcolemma as a muscle pathological feature.[17,41] In the vacuolar membrane, almost all sarcolemmal proteins including dystrophin and sarcoglycans are expressed, and AChE activities are also observed. Autophagic vacuoles increases with aging, suggesting an accumulative change. LAMP-2 is a component of lysosomal membranes and is thought to protect lysosomal membranes and cytoplasm from proteolytic enzymes in lysosomal lumen. In addition, LAMP-2 has been revealed to contribute to the migration of lysosomes or fusion with autophagosomes.[42] The pathomechanism is different from that of glycogen storage diseases, which are caused by glycolytic enzyme deficiency, and it is thus inappropriate to classify Danon disease as a "glycogen storage disease". In patients with Danon disease, cardiomyopathy is a significant life-threatening factor, and cardiac failure leads to death. This disease can cause a severe arrhythmogenic phenotype that includes a high risk of sudden death. Danon disease has recently attracted attention as a differential diagnosis for hypertrophic cardiomyopathy of unknown cause because symptoms other than cardiomyopathy are mild in the patients.[43–45] Currently, heart transplantation is the sole radical treatment as we await the development of new treatment methods.

4.2 Epidemiology

Since 2000, when causative gene of Danon disease was identified, over 500 patients with genetically confirmed Danon disease have been reported in literature.[46,47] This disease is extremely rare, despite relatively frequent reports of such patients in Japan. In a nationwide survey conducted by Sugie and colleagues, there were 39 patients with genetically confirmed Danon disease in 20 families in Japan (17 male patients and 22 female patients) and at the time of survey, 19 patients in 12 families (8 male patients and 11 female patients) were alive.[48] The cause of death in the other 20 patients were cardiac failure in 19 patients and malignant tumor in one patient. All the patients had cardiomyopathy, and some patients had severe cardiac failure.

Heart transplantation was performed on one patient, while five patients were on the waitlist for the procedure. There was no difference in the distribution of patients with Danon disease among the regions across Japan, and locations of genetic mutations varied according to families. The survey also found patients with features suggesting de novo mutation in 10 families (50%).

4.3 Pathogenesis and pathomechanism

Danon disease is caused by a primary deficiency of LAMP-2, a membrane protein of lysosome. LAMP-2 penetrates through the lysosomal membrane only once.[18,42] Of amino acid sequences in 410 residues, 90% or more of the sequences at N-terminal are present in the lysosomal lumen, and only tens of residues at C-terminal protrude into the cytoplasm. The domain in the lysosomal lumen has four loop structures held by disulfide bonds that are significantly influenced by glycosylation. LAMP-2 comprises approximately 50% of lysosomal membrane together with LAMP-1, which is a homologue on the autosome, and thought to possibly protect lysosomal membrane or cytoplasm from proteolytic enzymes in the lysosomal lumen.[49] In contrast to expression levels of LAMP-1 that are constant, those of LAMP-2 increase in various conditions. Therefore, LAMP-2 is considered to have specific regulated expression and inhibitory function. Based on a study using knockout mice, it was further revealed that LAMP-2 contributes to migration of lysosomes and fusion with autophagosomes. LAMP-2 has three isoforms consisting of LAMP-2A/B/C induced by selective splicing of exon 9.[50] It is known that LAMP-2A is ubiquitously expressed, while LAMP-2B is mainly expressed in cardiac muscle, skeletal muscle, and the brain. Little is known about LAMP-2C. Autophagy that constantly occurs in these regions stops at the final stage (a stage of fusion between autophagosomes and lysosomes) in patients with Danon disease and is considered to be the cause of the symptoms.

4.4 Clinical features

Danon disease shows X-linked dominant inheritance, and the majority of mothers with a male child, a proband, have the symptoms. Clinical symptoms of the patients are relatively homogeneous. The triad of this disease are cardiomyopathy, myopathy, and mental retardation in male patients, while female patients mainly have cardiomyopathy.[19] In male patients, this disease becomes symptomatic in their teens, and patients usually die at around 30 years of age. In female patients, symptoms occur much later than males, from the age of 30, and patients die at around 40 years of age. Cardiomyopathy necessarily develops in both male and female patients and occurs as hypertrophic cardiomyopathy but may progress to a dilated phase at the end stage in some cases. This disease is complicated by cardiac conduction involvement such as Wolff-Parkinson-White syndrome with high probability, and other complications including high left ventricular potentials, abnormal Q wave, atrioventricular block, and atrial fibrillation. Concentric left ventricular hypertrophy with left ventricular dysfunction is observed on echocardiogram in a majority of the patients. Myopathy occurs as muscle weakness and muscle atrophy with a predominance of proximal muscles. Muscle atrophy may involve distal muscles in some cases, but it does not affect daily living activities. However, serum CK levels in male patients typically increase up to around 1000 IU/L and show high levels even during childhood when there are no symptoms, whereas CK levels range from normal to high in female patients. Some patients with high CK level eventually received definitive diagnosis based on results of muscle biopsy. Mental retardation occurs in approximately 60% of patients, although the severity is mild even if it occurs. Brain MRI findings show no remarkable changes. Recently, various complications such as autism, peripheral nerve involvement, short stature, hepatic dysfunction, retinopathy, and small vessel vasculopathy have been reported in some cases.[51,52]

4.5 Muscle pathological features

As a pathological feature, skeletal muscles show mild to moderate variations in muscle cell diameter, without necrotic or regenerative cells. Muscle cells with basophilic granular small vacuoles are scattered with increased acid phosphatase activities (Fig. 2). In addition, AChE activities, which are

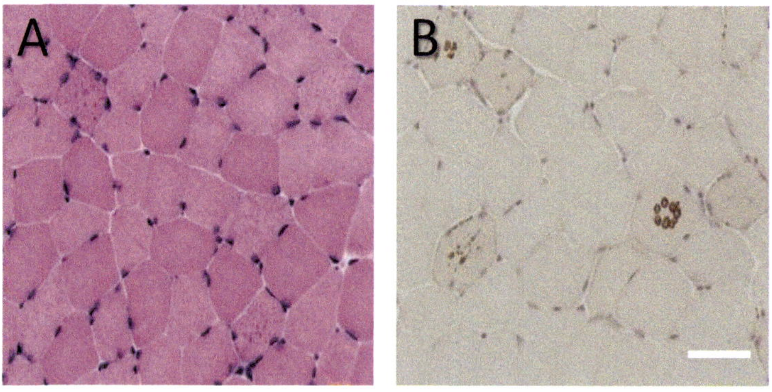

Fig. 2 Muscle pathology of a male patient with Danon disease. Tiny autophagic vacuoles look like basophilic granules in muscle cells with hematoxylin and eosin stain (A). The vacuolar membrane has acetylcholinesterase activity, known as autophagic vacuoles with sarcolemmal features (AVSF) (B). Bar = 50 μm.

typically observed only at neuromuscular junctions, are found in vacuolar membranes that express component proteins of sarcolemma such as dystrophin and sarcoglycans. The autophagic vacuoles with sarcolemmal features have high disease specificity, and we named it AVSF.[17,41] AVSF is thought to be formed through the phenomenon where cells construct extracellular environment inside cells, but little has been established on its mechanism or involvement in the pathogenesis of Danon disease. We confirmed that AVSF increases with aging based on biopsied muscles of patients with Danon disease and knockout mice.[17,53] Some female patients without abnormal pathological findings and young female patients who have AVSF with a rate of 15%–20% have been reported.[43,54] On examination with electron microscope, basal lamina is observed along with the inner part of the vacuolar wall in the autophagic vacuoles containing irregular abnormal structures and glycogen granules. There is also an accumulation of vacuoles without membranes or myeloid bodies in some parts. Pathological features in cardiac muscles, fragmentation, hypertrophy, and vacuolation of cardiac myocytes and increased lipofuscin are observed in addition to the accumulation of autophagic vacuoles.

4.6 Diagnosis and differential diagnosis

A diagnosis of Danon disease is determined based on *LAMP-2* gene analysis and LAMP-2 deficiency by immunohistochemistry or western blotting. As Danon disease is extremely rare, practically, it is difficult to perform *LAMP-2* gene analysis for suspected Danon disease cases based only on clinical symptoms. Muscle biopsy is significant as a practical diagnostic procedure given that biopsied samples in patients with Danon disease show AVSF in muscle cells with very high specificity.[17,41] In female patients, it varies; LAMP-2 proteins in muscle tissues may be defected or normal according to patient. Clinical phenotypes in patients who have mutations in exon 9B that encodes LAMP-2B are known to be relatively mild.

At present, only Danon disease and XMEA are disease types of AVSF myopathies with established disease concepts. XMEA is caused by gene mutations of *VMA21* in Xq28.[20] Other than these disease types, AVSF is also observed in patients with autophagic vacuolar myopathy with late-onset and multiorgan involvement.[55] However, the causative gene or pathomechanism of this disease has been demonstrated. In patients with Danon disease, the severity of myopathy is mild, and the disease may therefore be detected when cardiomyopathy occurs in some cases. Danon disease should be listed as a differential diagnosis for patients who have hypertrophic cardiomyopathy of unknown cause. Since symptoms other than cardiomyopathy are mild, there might have been undiagnosed cases of Danon disease who died of cardiac failure due to cardiomyopathy among individuals from families with the disease. It is hence necessary to perform early examination for latent cardiomyopathy in females without symptoms who are family members of male patients.

4.7 Important differential diagnosis: XMEA

XMEA is a relatively mild myopathy and an extremely rare disease that was first reported by Kalimo et al. in 1988.[56] In 2013, Ramachandran et al. identified *VMA21* gene as a causative gene of XMEA.[20] VMA21 encodes assembly factors of vascular ATPase that contribute to acidification in the lysosomal lumen.[57] In patients' cells, lysosomal pH increases due to the dysfunction of assembly factors, and proteolytic activity decreases, triggering the inactivation of mammalian target of rapamycin (mTOR), which functions as a modulator of cellular metabolism leading to autophagy.

XMEA shows X-linked recessive inheritance, and in male patients with this disease, only myopathy develops as a clinical symptom. Females are carriers and asymptomatic. In infancy, muscle weakness and muscle atrophy in proximal muscles in lower extremities are noticed, but the symptoms are mild. This disease slowly progresses, and distal muscle involvement is extremely mild even at relatively later stages when atrophy in thigh is prominent. Skeletal muscle is the sole affected organ; patients with XMEA have no cardiomyopathy and are intellectually normal unlike Danon disease patients. Our colleagues detected *VMA21* gene mutations in cases of the congenital form of X-linked autophagic vacuolar myopathy[58] and infantile autophagic vacuolar myopathy,[59] and the conditions in these cases were considered allelic disease of XMEA.[60,61] The disease concept of XMEA may be extended in the future. As with Danon disease, AVSF is observed as the muscle pathological feature of biopsied samples.[62] Unlike Danon disease, however, complement complex of C5b-9 called membrane attack complex (MAC) deposits on the surface of muscle cells with vacuoles. This feature is specific to XMEA and not observed in Danon disease. An examination using electron microscope shows reduplication of basal lamina in a part of muscle cell membrane and the accumulation of many autophagic vacuoles between overlapping basal lamina.

A diagnosis of XMEA is determined by *VMA21* gene analysis. This disease is extremely rare, and practically, it is difficult to perform *VMA21* gene analysis for suspected XMEA based only on clinical symptoms. Muscle biopsy is very important as a practical diagnostic procedure as it enables detection of AVSF in muscle cells. There is currently no radical treatment method available, but prognosis in patients with XMEA is good. Muscle symptoms in patients with XMEA are mild, and rehabilitation such as physical therapy is mainly provided as a symptomatic treatment.

4.8 Therapy and prognosis

In patients with Danon disease, cardiomyopathy is a very important determinant of prognosis, and patients die due to cardiac failure.[43–45] This disease can cause a severe arrhythmogenic phenotype that includes a high risk of sudden death. Currently, heart transplantation is the sole radical treatment, and the development of new treatment methods is desired. Its skeletal muscles, symptoms and mental retardation usually have no large impact on daily living activities, and prognosis may significantly improve with successful heart transplantation. At present, patients with Danon disease are commonly treated with pacemaker implantation or implantable cardioverter-defibrillator as a symptomatic treatment. In addition, β-blockers or diuretics, ACE inhibitors, Ca blockers, and anticoagulants are generally used in drug therapy.

5. Conclusion

Although the role of autophagy in muscle tissues remained unknown for a long time, the importance of autophagy in muscle tissues has attracted attention based on the presence of a group of muscle diseases, autophagic vacuolar myopathy, in which many autophagic vacuoles are observed in muscle cells. Its pathomechanism has been gradually clarified by results of clinicopathological or genetic studies. Nevertheless, their rarity meant that the number of patients is small, and there are still only a limited number of diseases for which effective treatment can be found. The detailed identification of the pathomechanism of autophagic vacuolar myopathies such as IBM and Danon disease provides an important clue, not only for the development of treatment methods but also for the identification of the autophagy mechanism in normal muscle tissues. Further explorations in this area are thus expected.

References

1. Mizushima N, Komatsu M. Autophagy: renovation of cells and tissues. *Cell.* 2011;147(4):728–741.
2. Mizushima N, Levine B, Cuervo AM, et al. Autophagy fights disease through cellular self-digestion. *Nature.* 2008;451(7182):1069–1075.

3. Nishino I. Autophagic vacuolar myopathy. *Semin Pediatr Neurol.* 2006;13(2):90–95.
4. Kishnani PS, Beckemeyer AA, Mendelsohn NJ. The new era of Pompe disease: advances in the detection, understanding of the phenotypic spectrum, pathophysiology, and management. *Am J Med Genet C Semin Med Genet.* 2012;160C(1):1–7.
5. Raben N, Hill V, Shea L, et al. Suppression of autophagy in skeletal muscle uncovers the accumulation of ubiquitinated proteins and their potential role in muscle damage in Pompe disease. *Hum Mol Genet.* 2008;17(24):3897–3908.
6. Prater SN, Banugaria SG, DeArmey SM, et al. The emerging phenotype of long-term survivors with infantile Pompe disease. *Genet Med.* 2012;14(9):800–810.
7. Toscano A, Schoser B. Enzyme replacement therapy in late-onset Pompe disease: a systematic literature review. *J Neurol.* 2013;260(4):951–959.
8. Schoser B, Stewart A, Kanters S, et al. Survival and long-term outcomes in late-onset Pompe disease following alglucosidase alfa treatment: a systematic review and meta-analysis. *J Neurol.* 2017;264(4):621–630.
9. Salabarria SM, Nair J, Clement N, et al. Advancements in AAV-mediated gene therapy for Pompe disease. *J Neuromuscul Dis.* 2020;7(1):15–31.
10. Shelly S, Mielke MM, Mandrekar J, et al. Epidemiology and natural history of inclusion body myositis: a 40-year population-based study. *Neurology.* 2021;96(21):e2653–e2661.
11. Nonaka I, Sunohara N, Ishiura S, Satoyoshi E. Familial distal myopathy with rimmed vacuole and lamellar (myeloid) body formation. *J Neurol Sci.* 1981;51(1):141–155.
12. Nishino I, Carrillo-Carrasco N, Argov Z. GNE myopathy: current update and future therapy. *J Neurol Neurosurg Psychiatry.* 2015;86(4):385–392.
13. Nishino I, Noguchi S, Murayama K, et al. Distal myopathy with rimmed vacuoles is allelic to hereditary inclusion body myopathy. *Neurology.* 2002;59(11):1689–1693.
14. Galimberti V, Tironi R, Lerario A, et al. Value of insoluble PABPN1 accumulation in the diagnosis of oculopharyngeal muscular dystrophy. *Eur J Neurol.* 2020;27(4):709–715.
15. Kumutpongpanich T, Ogasawara M, Ozaki A, et al. Clinicopathologic features of oculopharyngodistal myopathy with LRP12 CGG repeat expansions compared with other oculopharyngodistal myopathy subtypes. *JAMA Neurol.* 2021; [online ahead of print].
16. Watts GD, Wymer J, Kovach MJ, et al. Inclusion body myopathy associated with Paget disease of bone and frontotemporal dementia is caused by mutant valosin-containing protein. *Nat Genet.* 2004;36(4):377–381.
17. Sugie K, Noguchi S, Kozuka Y, et al. Autophagic vacuoles with sarcolemmal features delineate Danon disease and related myopathies. *J Neuropathol Exp Neurol.* 2005;64(6):513–522.
18. Nishino I, Fu J, Tanji K, et al. Primary LAMP-2 deficiency causes X-linked vacuolar cardiomyopathy and myopathy (Danon disease). *Nature.* 2000;406(6798):906–910.
19. Sugie K, Yamamoto A, Murayama K, et al. Clinicopathological features of genetically confirmed Danon disease. *Neurology.* 2002;58(12):1773–1778.
20. Ramachandran N, Munteanu I, Wang P, et al. VMA21 deficiency prevents vacuolar ATPase assembly and causes autophagic vacuolar myopathy. *Acta Neuropathol.* 2013;125(3):439–457.
21. Chou SM. Myxovirus-like structures in a case of human chronic polymyositis. *Science.* 1967;158(3807):1453–1455.
22. Amato AA, Barohn RJ. Inclusion body myositis: old and new concepts. *J Neurol Neurosurg Psychiatry.* 2009;80(11):1186–1193.
23. Benveniste O, Guiguet M, Freebody J, et al. Long-term observational study of sporadic inclusion body myositis. *Brain.* 2011;134(Pt11):3176–3184.
24. Suzuki N, Mori-Yoshimura M, Yamashita S, et al. The updated retrospective questionnaire study of sporadic inclusion body myositis in Japan. *Orphanet J Rare Dis.* 2019;14(1):155.

25. Tawara N, Yamashita S, Zhang X, et al. Pathomechanisms of anti-cytosolic 5′-nucleotidase 1A autoantibodies in sporadic inclusion body myositis. *Ann Neurol.* 2017;81(4):512–525.
26. Cox FM, Reijnierse M, van Rijswijk CS, et al. Magnetic resonance imaging of skeletal muscles in sporadic inclusion body myositis. *Rheumatology (Oxford).* 2011;50(6):1153–1161.
27. Inamori Y, Higuchi I, Inoue T, et al. Inclusion body myositis coexisting with hypertrophic cardiomyopathy: an autopsy study. *Neuromuscul Disord.* 2012;22(8):747–754.
28. Askanas V, Engel WK, Nogalska A. Sporadic inclusion-body myositis: a degenerative muscle disease associated with aging, impaired muscle protein homeostasis and abnormal mitophagy. *Biochim Biophys Acta.* 2015;1852(4):633–643.
29. Sugarman MC, Yamasaki TR, Oddo S, et al. Inclusion body myositis-like phenotype induced by transgenic overexpression of beta APP in skeletal muscle. *Proc Natl Acad Sci U S A.* 2002;99(9):6334–6339.
30. Boncompagni S, Moussa CE, Levy E, et al. Mitochondrial dysfunction in skeletal muscle of amyloid precursor protein (APP) overexpressing mice. *J Biol Chem.* 2012;287(24):20534–20544.
31. Salajegheh M, Pinkus JL, Taylor JP, et al. Sarcoplasmic redistribution of nuclear TDP-43 in inclusion body myositis. *Muscle Nerve.* 2009;40(1):19–31.
32. Delaunay A, Bromberg KD, Hayashi Y, et al. The ER-bound RING finger protein 5 (RNF5/RMA1) causes degenerative myopathy in transgenic mice and is deregulated in inclusion body myositis. *PLoS ONE.* 2008;3(2), e1609.
33. Masiero E, Agatea L, Mammucari C, et al. Autophagy is required to maintain muscle mass. *Cell Metab.* 2009;10(6):507–515.
34. Pandey UB, Nie Z, Batlevi Y, et al. HDAC6 rescues neurodegeneration and provides an essential link between autophagy and the UPS. *Nature.* 2007;447(7146):859–863.
35. Weihl CC, Temiz P, Miller SE, et al. TDP-43 accumulation in inclusion body myopathy muscle suggests a common pathogenic mechanism with frontotemporal dementia. *J Neurol Neurosurg Psychiatry.* 2008;79(10):1186–1189.
36. D'Agostino C, Nogalska A, Engel WK, Askanas V. In sporadic inclusion body myositis muscle fibres TDP-43-positive inclusions are less frequent and robust than p62 inclusions, and are not associated with paired helical filaments. *Neuropathol Appl Neurobiol.* 2011;37(3):315–320.
37. Schmidt J, Barthel K, Zschuntzsch J, et al. Nitric oxide stress in sporadic inclusion body myositis muscle fibres: inhibition of inducible nitric oxide synthase prevents interleukin-1beta-induced accumulation of beta-amyloid and cell death. *Brain.* 2012;135(Pt4):1102–1114.
38. Pluk H, van Hoeve BJ, van Dooren SH, et al. Autoantibodies to cytosolic 5′-nucleotidase 1A in inclusion body myositis. *Ann Neurol.* 2013;73(3):397–407.
39. Larman HB, Salajegheh M, Nazareno R, et al. Cytosolic 5′-nucleotidase 1A autoimmunity in sporadic inclusion body myositis. *Ann Neurol.* 2013;73(3):408–418.
40. Danon MJ, Oh SJ, DiMauro S, et al. Lysosomal glycogen storage disease with normal acid maltase. *Neurology.* 1981;31(1):51–57.
41. Sugie K, Nishino I. Lysosomal membrane disorders: LAMP-2 deficiency. In: Rosenberg RN, Pascual JM, eds. *Rosenberg's Molecular and Genetic Basis of Neurological and Psychiatric Disease.* 6th ed. Amsterdam, Nederland: Elsevier; 2020:567–574.
42. Tanaka Y, Guhde G, Suter A, et al. Accumulation of autophagic vacuoles and cardiomyopathy in LAMP-2-deficient mice. *Nature.* 2000;406(6798):902–906.
43. Sugie K, Yoshizawa H, Onoue K, et al. Early onset of cardiomyopathy and intellectual disability in a girl with Danon disease associated with a de novo novel mutation of the LAMP2 gene. *Neuropathology.* 2016;36(6):561–565.
44. Maron BJ, Roberts WC, Arad M, et al. Clinical outcome and phenotypic expression in LAMP2 cardiomyopathy. *JAMA.* 2009;301(12):1253–1259.

45. Miani D, Taylor M, Mestroni L, et al. Sudden death associated with danon disease in women. *Am J Cardiol.* 2012;109(3):406–411.
46. Cenacchi G, Papa V, Pegoraro V, et al. Review: Danon disease: review of natural history and recent advances. *Neuropathol Appl Neurobiol.* 2020;46(4):303–322.
47. Boucek D, Jirikowic J, Taylor M. Natural history of Danon disease. *Genet Med.* 2011;13(6):563–568.
48. Sugie K, Komaki H, Eura N, et al. A nationwide survey on Danon disease in Japan. *Int J Mol Sci.* 2018;19(11):3507.
49. Eskelinen EL. Roles of LAMP-1 and LAMP-2 in lysosome biogenesis and autophagy. *Mol Aspects Med.* 2006;27(5–6):495–502.
50. Endo Y, Furuta A, Nishino I. Danon disease: a phenotypic expression of LAMP-2 deficiency. *Acta Neuropathol.* 2015;129(3):391–398.
51. Schorderet DF, Cottet S, Lobrinus JA, et al. Retinopathy in Danon disease. *Arch Ophthalmol.* 2007;125(2):231–236.
52. Nguyen H, Noguchi S, Sugie K, et al. Small-vessel vasculopathy due to aberrant autophagy in LAMP-2 deficiency. *Sci Rep.* 2018;8(1):3326.
53. Sugie K, Koori T, Yamamoto A, et al. Characterization of Danon disease in a male patient and his affected mother. *Neuromuscul Disord.* 2003;13(9):708–711.
54. Kim H, Cho A, Lim BC, et al. A 13-year-old girl with proximal weakness and hypertrophic cardiomyopathy with Danon disease. *Muscle Nerve.* 2010;41(6):879–882.
55. Kaneda D, Sugie K, Yamamoto A, et al. A novel form of autophagic vacuolar myopathy with late-onset and multiorgan involvement. *Neurology.* 2003;61(1):128–131.
56. Kalimo H, Savontaus ML, Lang H, et al. X-linked myopathy with excessive autophagy: a new hereditary muscle disease. *Ann Neurol.* 1988;23(3):258–265.
57. Forgac M. Vacuolar ATPases: rotary proton pumps in physiology and pathophysiology. *Nat Rev Mol Cell Biol.* 2007;8(11):917–929.
58. Yan C, Tanaka M, Sugie K, et al. A new congenital form of X-linked autophagic vacuolar myopathy. *Neurology.* 2005;65(7):1132–1134.
59. Yamamoto A, Morisawa Y, Verloes A, et al. Infantile autophagic vacuolar myopathy is distinct from Danon disease. *Neurology.* 2001;57(5):903–905.
60. Munteanu I, Ramachandran N, Ruggieri A, et al. Congenital autophagic vacuolar myopathy is allelic to X-linked myopathy with excessive autophagy. *Neurology.* 2015;84(16):1714–1716.
61. Munteanu I, Kalimo H, Saraste A, Nishino I, Minassian BA. Cardiac autophagic vacuolation in severe X-linked myopathy with excessive autophagy. *Neuromuscul Disord.* 2017;27(2):185–187.
62. Kurashige T, Takahashi T, Yamazaki Y, et al. Elevated urinary β2 microglobulin in the first identified Japanese family afflicted by X-linked myopathy with excessive autophagy. *Neuromuscul Disord.* 2013;23(11):911–916.

SECTION VII

Autophagy and other disorders causing dementia

CHAPTER 12

Autophagy in Lewy body diseases and multiple system atrophy

Yasuo Miki, Kunikazu Tanji, and Koichi Wakabayashi
Department of Neuropathology, Institute of Brain Science, Hirosaki University Graduate School of Medicine, Hirosaki, Aomori, Japan

1. Introduction

The management of abnormal proteins is key to cellular survival. While cells are young and healthy enough to clear abnormal proteins, they are disposed of by protein quality control systems, including autophagy and the ubiquitin-proteasome system.[1,2] However, the ability to degrade proteins declines with age, leading to the accumulation of undegraded proteins in cells and ultimately cell death. In addition, some genetic mutations that accelerate the synthesis of misfolded proteins or impair clearance systems can also affect cellular activities. Aging is the primary risk factor for most neurodegenerative diseases that have one pathological feature in common—aggregation of abnormal proteins in cells.[3–5] In this review, we focus on the alterations of autophagy that occur in synucleinopathies, which are characterized by the presence of abnormal α-synuclein aggregates in neurons or glial cells, including Parkinson's disease (PD), dementia with Lewy bodies (DLB), and multiple system atrophy (MSA).[4,5] We also discuss the current challenges and future directions of autophagy research.

2. Autophagy

Autophagy is the major cellular mechanism responsible for maintaining the environment of cells by degrading damaged organelles and misfolded proteins.[1,6] In addition, autophagy plays a pivotal role in the maintenance of amino acid pooling, development, clearance of intracellular microbes, and regulation of immunity.[7] There are three different forms of autophagy that each uses a different route for delivering substrates to lysosomes: microautophagy, chaperon-mediated autophagy, and macroautophagy.[1,6,8,9] In microautophagy, lysosomes invaginate by themselves and directly engulf

Autophagy Dysfunction in Alzheimer's Disease and Dementia
https://doi.org/10.1016/B978-0-323-89906-2.00008-3

small portions of the cytoplasm.[8] Chaperon-mediated autophagy is a selective process that recognizes the KFERQ-like motif by lysosomal-associated membrane protein 2A (LAMP-2A).[9] On the other hand, macroautophagy (hereafter referred to as autophagy) is the main autophagic pathway whereby substrates are delivered to the lysosome through a double membrane-bound vesicle known as an autophagosome. Autophagosomes then fuse with lysosomes to form autolysosomes.[1,6]

The process of autophagy includes: (i) induction and phagophore formation, (ii) elongation and autophagosome formation, and (iii) lysosome fusion and degradation, all of which are tightly regulated by essential core autophagic proteins (Fig. 1). In this review, we focus on the essential autophagic proteins that we have investigated in previous studies. Under basal conditions the mammalian target of rapamycin (mTOR) inhibits the function of the Unc-51-like kinase (ULK) 1/ULK2-ATG13-ATG101-200 kDa FAK family kinase-interacting (FIP200) complex.[10–16] Inhibition of mTOR in turn leads to the activation of this complex, initiating the formation of phagophores.[17] Next, the class III phosphatidylinositol 3-kinase (PI3KC3/VPS34), VPS15, Beclin1, activating molecule in beclin1-regulated autophagy (AMBRA1), and ATG14 complex are activated. This PI3KC3 complex

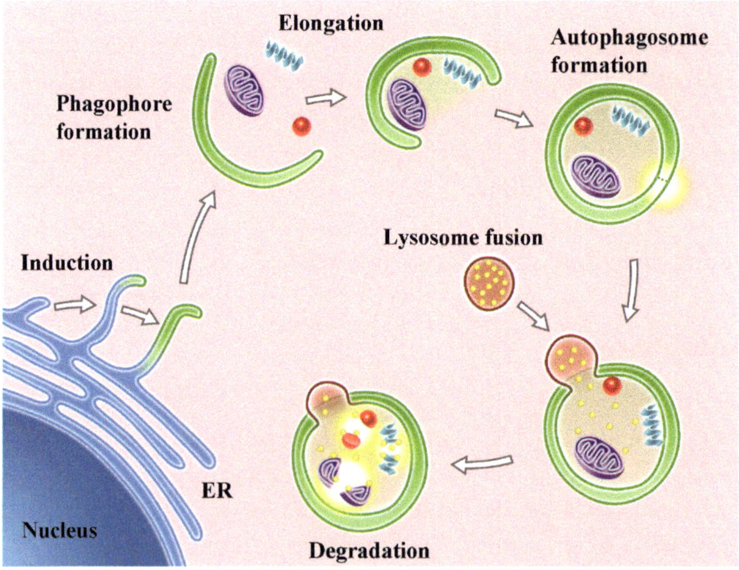

Fig. 1 Schematic overview of autophagy. The processes of autophagy include induction, phagophore formation, elongation, autophagosome formation, lysosome fusion, and degradation. ER: endoplasmic reticulum.

is then recruited to the endoplasmic reticulum (ER), giving rise to phagophore genesis in a specialized ER domain called omegasomes.[18–21] For phagophore elongation, ATG5 conjugates ATG12 to form a complex with ATG16L. This complex induces curvature into the growing phagophore through the recruitment of microtubule-associated protein light chain 3 (LC3), a mammalian homolog of yeast Atg8. After autophagy is activated the C-terminal amino acid 22 of LC3 is cleaved by ATG4 to generate LC3-I in the cytoplasm. After LC3-I is transferred to ATG3 through ATG7, phosphatidyl ethanolamine is conjugated to LC3-I, followed by the conversion of LC3-I to LC3-II, the lapidated form of LC3.[22–24] Importantly, LC3-II gets involved in phagophore closure and stays on the inner membrane of the autophagosome until engulfed proteins are degraded in the lysosomes; thus LC3-II has been utilized as a marker of autophagy activity. Furthermore, when substrates are engulfed by a phagophore, various proteins can serve as membrane cargo receptors, including p62, neighbor of BRCA1 gene (NBR1) or optineurin, which combine aggregated proteins and LC3-II.[1] In addition to LC3, at least three Atg8 homologs exist in mammals which participate in several parts of autophagy, including interacting with ULK1 to connect the ULK1 complex with phagophore, phagophore closure, or autophagosome-lysosome fusion. These homologs include: (i) γ-aminobutyric acid type A receptor-associated protein (GABARAP); (ii) GABARAP-like 1 (GABARAPL1); and (iii) Golgi-associated ATPase enhancer of 16 kDa (GATE-16).[25,26] After a series of reactions with these autophagic proteins, autophagosomes fuse with lysosomes with the help of lysosomal membrane glycoprotein (LAMP-1 or -2) and small GTPase Rab7.[27–29] The inner vesicles are then degraded by lysosomal acid hydrolases. As described above the delicate control of autophagy is based on cross talk among many autophagic proteins. Therefore when this complex orchestrated autophagic process is disturbed due to aging, genetic mutations, or other reasons, abnormal proteins can accumulate in cells and affect cellular activities.

3. Lewy body diseases

3.1 Clinical features of Lewy body diseases

PD and DLB are collectively known as Lewy body diseases based on the presence of Lewy bodies and Lewy neurites in the central and peripheral autonomic nervous systems.[4,30] Based on these pathological hallmarks, PD and DLB are considered to be on the same pathological spectrum; however, these conditions clinically differ in the time of onset of parkinsonism and dementia.[31,32] Individuals with PD develop bradykinesia, a resting tremor, rigidity, and/or postural instability. Dementia is also a common feature, but it often occurs in the later stages of PD. On the other hand, DLB primarily causes progressive dementia, including fluctuations in cognitive function with prominent executive dysfunction and visual hallucinations. The term DLB is applied when dementia and parkinsonism begin concurrently or when dementia starts within 1 year of motor symptoms. Furthermore, parkinsonism in DLB tends to be mild, if present at all, reflecting less severe neuronal loss in the substantia nigra of individuals with DLB compared with PD cases with dementia.[33,34] In addition to these motor symptoms, patients with PD and DLB can develop various nonmotor symptoms such as autonomic dysfunction, cognitive impairment, rapid eye movement behavior disorder, sleep difficulties, depression, apathy, or pain.[31,35]

3.2 Pathological features of Lewy body diseases

The pathological features of Lewy body diseases include loss of dopaminergic neurons in the substantia nigra (Fig. 2A–D) and the presence of Lewy bodies and Lewy neurites that consist of insoluble fibrillary aggregates of phosphorylated α-synuclein.[4,36,37] While the exact physiological functions of α-synuclein remain elusive, α-synuclein exists predominantly in the presynaptic nerve terminals and is believed to regulate synaptic plasticity, vesicle trafficking, and subsequent neurotransmitter release.[38–40] In Lewy body diseases the self-assembly of α-synuclein into Lewy bodies occurs though the oligomeric form of α-synuclein and protofibrils.[32] Lewy bodies in the brainstem, referred to as brainstem-type Lewy bodies, show eosinophilic cores with peripheral halos in hematoxylin and eosin staining (Fig. 2E). Lewy bodies in the cortex (cortical Lewy bodies) are also eosinophilic, but often irregular, and their structure is poorly defined without a conspicuous halo or central core (Fig. 2F). Studies using immunohistochemistry have revealed that both cortical and brainstem-type Lewy bodies are strongly immunoreactive for phosphorylated α-synuclein, especially in the halo and peripheral portions of the core of brainstem-type Lewy bodies (Fig. 3A, B).

3.3 Pathological staging systems for Lewy body diseases

Two staging systems (Braak PD stage and McKeith's criteria) are commonly used to evaluate Lewy body pathologies.[41–43] For example, Braak et al. described a pathological staging system for PD that begins with α-synuclein pathology in the dorsal vagal nucleus and the olfactory bulb (stage 1) and then progresses to the pontine tegmentum (stage 2). Braak PD stage 1 or

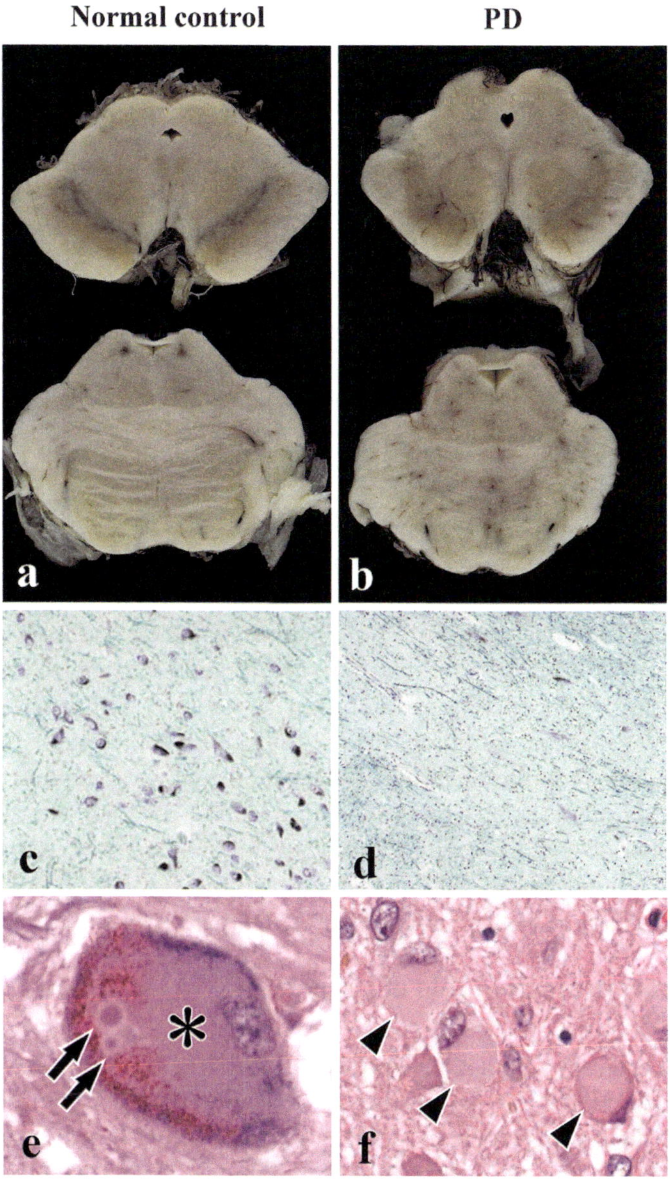

Fig. 2 Pathological features of Parkinson's disease. Gross appearance of the midbrain and pons of a control case (A) and a case of Parkinson's disease (PD) (B) showing depigmentation of the substantia nigra and locus coeruleus in PD (B). Compared with normal control (C), loss of dopaminergic neurons is evident in PD (D). (E) Two Lewy bodies (arrows) and a pale body, a precursor of Lewy bodies (asterisk) in a dopaminergic neuron in the substantia nigra of PD. (F) Cortical Lewy bodies in neurons in the temporal cortex of PD. (C, D) Klüver-Barrera(KB); (E, F) Hematoxylin and eosin (HE) staining.

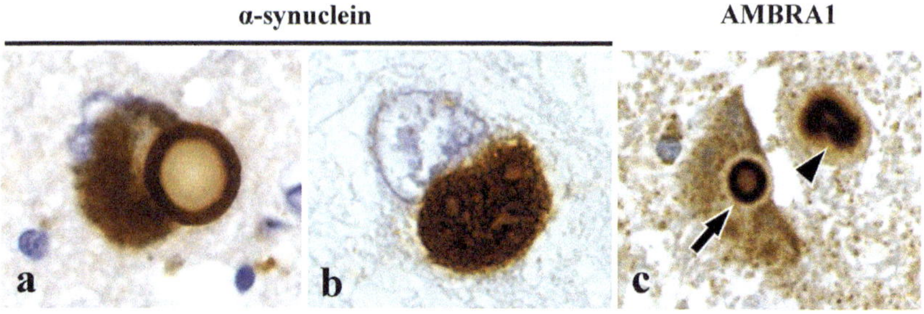

Fig. 3 Immunoreactivity of Lewy bodies. Brainstem-type (A) and cortical Lewy bodies (B) immunoreactive for phosphorylated α-synuclein. (C) Lewy body (arrow) and Lewy neurite (arrowhead) immunopositive for activating molecule in beclin1-regulated autophagy (AMBRA1).

2 corresponds to an incidental Lewy body disease. Individuals with incidental Lewy body disease do not manifest any symptoms of parkinsonism or dementia. In stage 3, α-synuclein further ascends through the midbrain and neostriatum, before reaching the mesocortex and allocortex (stage 4). Only when Lewy pathologies reach stage 3 or 4 do individuals with such pathologies develop parkinsonism. Finally, α-synuclein pathology spreads to the neocortex at stages 5 and 6, where cognition starts to decline in addition to the presence of parkinsonism symptoms.[41] On the other hand, McKeith et al. set the pathological criteria for DLB based on the pattern of Lewy pathology in the brainstem, limbic and neocortical regions, with the recent addition of the olfactory bulb and amygdala-predominant categories. The likelihood that a patient's pathological findings are associated with the clinical manifestations of DLB is assessed according to the severity of Lewy body pathology as well as concomitant Alzheimer's pathology.[42,43]

3.4 Neuroprotective role of Lewy bodies in Lewy body diseases

The role of Lewy bodies in neurodegeneration is still an open topic of discussion. However, it has been hypothesized that α-synuclein oligomers or protofibrils can be detrimental to neurons, whereas fibrillar aggregates of α-synuclein may instead be neuroprotective.[44–46] In fact, a previous study found that the more α-synuclein aggregates there were, the less toxic α-synuclein became in a Drosophila model of PD.[47] As we will discuss later in more detail, many autophagosomal and autophagic adaptor proteins are found in Lewy bodies.[48,49] Furthermore, most Lewy bodies are reactive for histone deacetylase 6 (HDAC6), which helps in the transport of abnormal proteins via the microtubule network to form aggresomes.[50] These findings suggest that neurons harboring Lewy bodies may be dying, as the excessive amount of toxic α-synuclein likely overwhelms cellular damage control. However, forming Lewy bodies in and of itself can be an aggresome-related process that confines damaged, mutated, or cytotoxic proteins to Lewy bodies.

3.5 α-Synuclein as a substrate for autophagy

The clearance of abnormal α-synuclein depends heavily on protein quality control systems, including autophagy. Importantly, multiple studies have reported α-synuclein as a substrate for autophagy.[51–54] To understand the relationship between α-synuclein clearance and β-glucocerebrosidase activity, Rothaug et al. investigated the function of the lysosomal integral membrane protein type-2 (LIMP-2) in PD. They found that LIMP-2-deficient mice exhibited autophagic and lysosomal deficits, which developed into neurological symptoms in an α-synuclein dosage-dependent manner.[51] Another study by Winslow et al. demonstrated that overexpression of wild-type α-synuclein impaired autophagy by inhibiting Rab1a, a GTPase required for binding with optineurin. Notably, α-synuclein disturbed autophagy through Rab1, resulting in the mislocalization of ATG9 and a subsequent decrease in the formation of omegasomes.[52] In addition, autophagy is primarily impaired in some familial forms of PD. Furthermore, mutations in the lysosomal membrane protein ATP13A2 are known to reduce lysosomal degradation in human fibroblasts from patients with Kufor-Rakeb syndrome (PARK9), characterized by early onset parkinsonism, dementia, and pyramidal tract degeneration.[53,54] Finally, mutations in *VPS35*, the causative gene of autosomal dominant, late-onset PD (PARK17), impair autophagy by perturbing trafficking of ATG9.[55]

3.6 Impairment of autophagy in the brain of individuals with Lewy body diseases

Given the role of autophagy in clearing abnormal proteins, we examined multiple autophagic proteins in the brains of individuals with Lewy body diseases. First, we confirmed the upregulation of proteins associated with autophagy induction and phagophore formation (ULK1, ULK2, Beclin1, and VPS34) that occur after transfection of α-synuclein and synphilin-1 in HEK293 cells. We then investigated the localization and levels of these proteins in Lewy body diseases, finding that ULK1, ULK2, VPS34, and AMBRA1 were incorporated into brainstem-type Lewy bodies (Fig. 3C). In addition, western blot analysis using brain lysates showed specific bands of ULK1, ULK2, Beclin-1, or VPS34 in the soluble or insoluble fraction of patients with DLB, but not in normal controls. However, unlike the findings in the cellular model of PD, upregulation of these proteins in the brains of patients with PD or DLB was only partial.[49] Next, we attempted to understand how wild-type or abnormal α-synuclein can interact with autophagic proteins. Notably, abnormal α-synuclein showed a nine times stronger affinity toward AMBRA1 compared with that of wild-type α-synuclein.[56] Taken together, these results suggest that autophagy can be induced in Lewy body diseases. However, its capacity to degrade proteins may be suboptimal in cases of abnormal α-synuclein, which may prevent the function of autophagic proteins by sequestering autophagic proteins from their original positions. We also investigated changes in the downstream parts of autophagic proteins associated with elongation and autophagosome formation (LC3, GABARAP, and GATE-16) as well as autophagic adaptor proteins (NBR1 and p62). As expected, these proteins were found in brainstem-type and/or cortical Lewy bodies.[48,57,58] Importantly, α-synuclein pathology was enhanced by deletion of p62 in mice expressing α-synuclein. In the same study, p62-deficient mice showed increased mRNA and protein levels of NBR1.[57] Thus p62 and NBR1 can serve as adaptor proteins for recognizing α-synuclein. Finally, we compared the protein levels of LC3-II between DLB and normal controls, revealing significantly increased protein levels of LC3-II in patients with DLB.[48] Contrary to our expectations the level of GABARAP was decreased in DLB.[48] Our extensive investigations covering the upstream as well as downstream parts of autophagy clearly suggest that the accumulation of abnormal α-synuclein in Lewy body diseases is inextricably intertwined with dysregulation of autophagy. That is, abnormal α-synuclein accumulates in cells, and its accumulation may impair autophagy, or vice versa. The alterations in autophagic proteins in the brains of patients with Lewy body diseases are summarized in Table 1.

3.7 Impairment of autophagy in PD—Is it a primary or secondary event?

A simple question arose from our earlier studies using autopsied brains: does the dysregulation of autophagy reflect a primary aspect of Lewy body diseases or a secondary feature as a result of extensive neurodegeneration? To that end, we studied autophagy of peripheral blood mononuclear cells in early- to middle-stage PD cases and controls.[58] Protein

Table 1 Alterations of autophagy in Lewy body diseases versus MSA.

	Immunohistochemistry			**Protein levels (Western blotting)**		
	Lewy body diseases		**MSA**	**Lewy body diseases**		
	Brainstem-type LBs	**Cortical LBs**	**GCIs**	**PD**	**DLB**	**MSA**
ULK1	+	−	−			Increased
ULK2	+	−	−			Increased
VPS34	+	−	−		Increased	
Beclin1	+	−	−	Increased		
AMBRA1	+	+	+			Increased
p62	+	+	+			Increased
NBR1	+	+	+			Increased
LC3	+	+	+			
GABARAP	−	+	−		Decreased	Decreased
GATE-16	+	±	+			Decreased

Abbreviations: −, immunonegative; +, immunoreactive; AMBRA1, activating molecule in beclin1-regulated autophagy; DLB, dementia with Lewy bodies; GABARAP, γ-aminobutyric-acid type A receptor associated protein; GATE-16, Golgi-associated ATPase enhancer of 16 kDa; GCI, glial cytoplasmic inclusion; LB, Lewy body; LC3, microtubule-associated protein light chain 3; MSA, multiple system atrophy; NBR1, neighbor of BRCA1 gene; PD, Parkinson's disease; ULK1, Unc-51-like kinase 1;VPS34, class III phosphatidyl inositol 3-kinase.
Wakabayashi K, Yoshimoto M, Tsuji S, Takahashi H (1998), αSynuclein immunoreactivity in glial cytoplasmic inclusions in multiple system atrophy. Neurosci Lett 249:180–182.

levels of ULK1, VPS34, Beclin-1, and AMBRA1 increased in parallel with α-synuclein oligomers, whereas the mRNA levels of these proteins, except for ULK1, were downregulated in patients with PD compared with normal controls. In addition, the oligomeric form of α-synuclein was positively correlated with the clinical severity of PD and negatively correlated with cognitive function as well as early-phase cardiac uptake of meta-iodobenzylguanidine (MIBG). Importantly, autophagy in peripheral blood mononuclear cells does not directly reflect the conditions of autophagy in the brains of patients with PD. However, our study suggests that dysregulation of autophagy can occur concurrently with the accumulation of α-synuclein or in a relatively early phase of PD.[59] Recently, Niimi et al. also compared serum lysosomal acid hydrolase activities of patients with PD ($N = 68$) with that of controls ($N = 32$) and patients with Parkinsonian syndrome (N = 32).[59] β-Galactosidase activity was dysregulated in patients with PD,[60] corroborating our hypothesis that impairment of autophagy could be one of the primary events in PD. Further study is warranted to vilify this hypothesis.

4. MSA

4.1 Clinical and pathological features of MSA

MSA is an adult-onset, sporadic, neurodegenerative disease that manifests with levodopa unresponsive parkinsonism, cerebellar ataxia, and/or autonomic failure.[61,62] The average disease duration to death is 9 years after symptom onset. Patients with MSA are classified into two clinical phenotypes based on the predominant motor symptoms: a parkinsonian variant (MSA-P) related to striatonigral degeneration (SND) (Fig. 4A–H) and a cerebellar variant (MSA-C) associated with olivopontocerebellar atrophy (OPCA) (Fig. 5A–F).[62,63] In addition to the core clinical features of MSA, individuals with MSA can also develop specific signs or symptoms called red flag features that assist in its clinical diagnosis.[64,65] For example, rapid eye movement sleep behavior disorder can precede years before motor onset in not only Lewy body diseases but also MSA. Furthermore, recent clinicopathological studies have reported that up to 37% of patients with pathologically proven MSA develop mild to moderate cognitive impairments in life.[66,67] The pathological hallmarks of MSA include the widespread occurrence of abnormal α-synuclein aggregates in oligodendrocytes and, to a lesser extent, in neurons (Fig. 6A, B).[5,68,69] Despite the predominant pathological involvement of either MSA-P or MSA-C, both SN and OPC systems are often equally affected, especially in MSA cases with a long disease duration.[63]

4.2 Impairment of autophagy in the brain of individuals with MSA

With the aim of understanding alterations in autophagy in MSA, we investigated the brains of individuals with MSA.[56] Similar to Lewy body diseases, autophagic proteins associated with multiple steps of autophagy (AMBRA1, LC3, and GATE-16) were found in oligodendrocytes that harbor abnormal α-synuclein called glial cytoplasmic inclusions (Fig. 6C, D). While Beclin1 and VPS34 were increased in PD and DLB, respectively, different autophagic proteins associated with autophagy induction (ULK1, ULK2, and AMBRA1) were upregulated in MSA. By contrast the protein level of GABARAP was significantly decreased in both DLB and MSA.[70] Thus impaired autophagy can occur along with, or as a result of, α-synuclein accumulation in the brain of patients with Lewy body diseases or MSA. However, the profiles of autophagy dysregulation may differ between Lewy body diseases and MSA (Table 1).

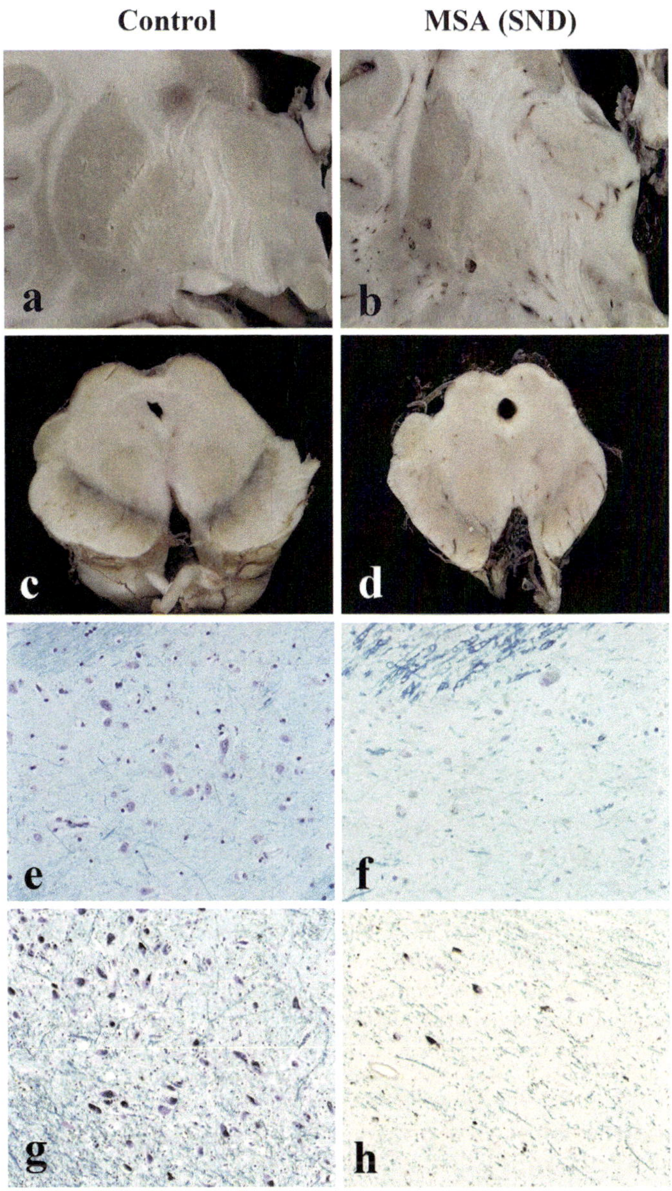

Fig. 4 Pathological features of multiple system atrophy (striatonigral degeneration). (A–D) Gross appearance of the basal ganglia and midbrain of a control case (A, C) and a case of multiple system atrophy (MSA) (striatonigral degeneration (SND)) (B, D) showing *gray*-brown discoloration and atrophy of the putamen and depigmentation of the substantia nigra in an MSA case. (Compared to a control case (E, G)) Histological appearance of the putamen and substantia nigra revealing almost complete loss of neurons (F, H, in an MSA case). (E–H) KB staining.

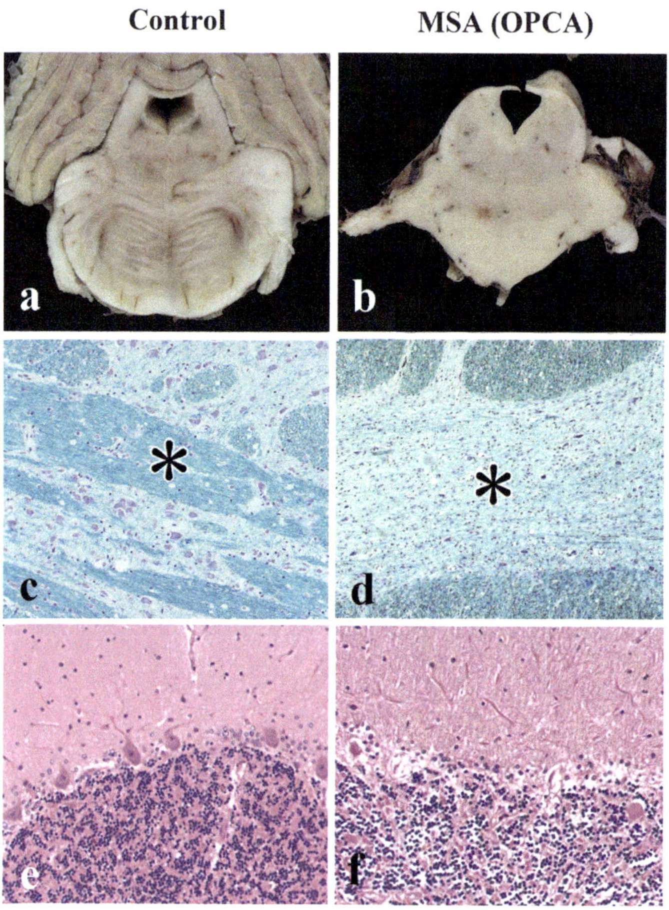

Fig. 5 Pathological features of MSA (olivopontocerebellar atrophy). Gross appearance of the pons of a control case (A) and a case of MSA (olivopontocerebellar atrophy (OPCA)) showing severe atrophy of the pontine base of MSA case (B). Compared with a control case (C, E), loss of transverse fibers of the pons (D) and cerebellar Purkinje cells (F) is evident in a patient with MSA. (C, D) KB; (E, F) HE staining.

4.3 Impairment of autophagy in MSA—Is it a primary or secondary event?

In our earlier study, we revealed that autophagy can be dysregulated concurrently or along with the progression of PD. Thus impaired autophagy may be a primary aspect of Lewy body disease. To date, there have been no reports of autophagy in the peripheral blood mononuclear cells of patients with MSA. However, Monzio et al. generated induced pluripotent stem cell-derived dopaminergic neurons from MSA cases and compared

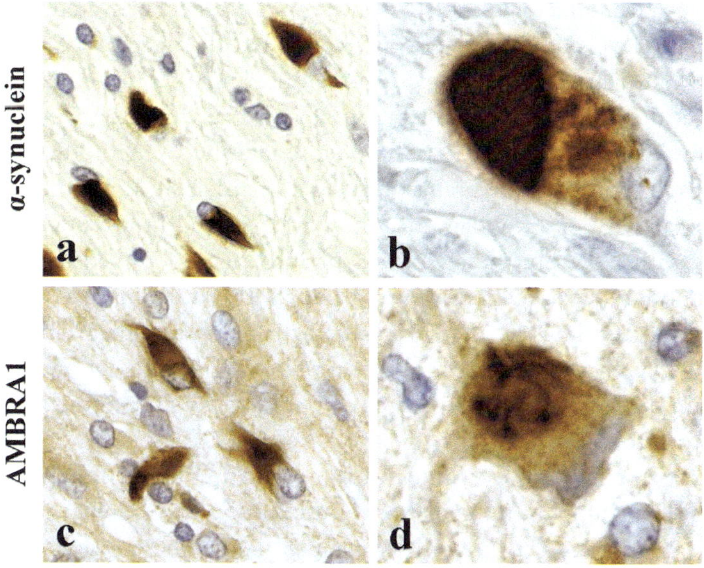

Fig. 6 Immunoreactivity of glial cytoplasmic inclusions and neuronal cytoplasmic inclusions. Glial cytoplasmic inclusions (A) and a neuronal cytoplasmic inclusion (B) in a patient with MSA immunoreactive for phosphorylated α-synuclein. Glial cytoplasmic inclusions (C) and a neuronal cytoplasmic inclusion (D) immunopositive for AMBRA1.

the activity of autophagy with that of normal controls.[71] Interestingly, autophagy flux analysis of LC3-II, the lapidated form of LC3, revealed decreased autophagy activity in MSA. In addition, lysosomal enzyme activity was significantly impaired in patients with MSA compared with that in normal controls.[71] Thus, similar to Lewy body diseases, autophagy may also be impaired at an early stage of MSA.

4.4 Current challenges and future directions in autophagy of synucleinopathies

While previous studies have led to a better understanding of autophagy in the pathogenesis of synucleinopathies, we have yet to unveil the complete picture of autophagy due to several critical limitations. First, to evaluate autophagy in patients with synucleinopathies, we often use formalin-fixed, paraffin-embedded, or frozen tissues. We then attempt to understand how autophagy is affected during the course of the illness by examining the immunoreactivity and/or altered protein levels of autophagic proteins, including LC3-II. However, alterations of autophagy in these types of tissues may simply reflect their levels at the time of autopsy and not alterations of autophagy over time. More importantly, increases in the level of LC3-II, or even the appearance of autophagosomes, do not reflect activated autophagy. In fact, the levels of LC3-II can also be increased as a result of the inhibition of autophagosome clearance. Therefore it may be more reasonable to interpret the activation or inactivation of autophagy by measuring multiple autophagic proteins. Second, we have not established a method for measuring autophagy flux in the brains of living humans. Instead, we infer that the baseline activity of autophagy may be compromised in the brains of individuals with PD from the alterations of autophagy observed in peripheral blood mononuclear cells. However, it goes without saying that we need to confirm our findings by measuring the activity of autophagy in the brain directly. Therefore establishing direct monitoring systems of autophagy in the brain of living humans is critically important for developing a drug that activates autophagy.

Notwithstanding these limitations in evaluating the activity of autophagy in living humans, upregulation of autophagy to enhance degradation of aggregates has great potential for ameliorating these disease processes. For this purpose, there are several strategies to activate autophagy, including activation of the upstream part of autophagy or activation of a specific part of autophagy that is impaired in PD or DLB. It is also reasonable to activate a key molecule implicated in several processes of autophagy or to activate multiple key proteins simultaneously. Nevertheless, an autophagic protein can be likened to a cog in a large machine. Even if we activate a single autophagic protein, "this single cog" may only spin in the autophagy machinery. In this sense, activating multiple key autophagic proteins may be more efficient in boosting autophagy. In fact, multiple studies have reported autophagy activation as a therapeutic strategy for neurodegenerative diseases.[72,74] Taken together, we may be about to experience a new era of treatment for neurodegenerative diseases, including synucleinopathies.

5. Conclusions

Recent studies have only begun to uncover the impairment of autophagy that occurs in synucleinopathies, thus novel frontiers lie ahead, and despite several critical limitations, modulating autophagy remains a promising therapeutic strategy.

Acknowledgments

The authors wish to express our gratitude to all patients, who bequeathed their brains, and their family members. Without their understanding, none of our studies would have been possible.

References

1. Damme M, Suntio T, Saftig P, Eskelinen EL. Autophagy in neuronal cells: general principles and physiological and pathological functions. *Acta Neuropathol.* 2015;129(3):337–362.
2. Galves M, Rathi R, Prag G, Ashkenazi A. Ubiquitin signaling and degradation of aggregate-prone proteins. *Trends Biochem Sci.* 2019;44(10):872–884.
3. Hou Y, Dan X, Babbar M, et al. Ageing as a risk factor for neurodegenerative disease. *Nat Rev Neurol.* 2019;15(10):565–581.
4. Spillantini MG, Schmidt ML, Lee VM, Trojanowski JQ, Jakes R, Goedert M. Alpha-synuclein in Lewy bodies. *Nature.* 1997;388(6645):839–840.
5. Wakabayashi K, Yoshimoto M, Tsuji S, Takahashi H. αSynuclein immunoreactivity in glial cytoplasmic inclusions in multiple system atrophy. *Neurosci Lett.* 1998,249(2–3):180–182.
6. Boya P, Reggiori F, Codogno P. Emerging regulation and functions of autophagy. *Nat Cell Biol.* 2013;15(7):713–720.
7. Mizushima N, Yoshimori T, Levine B. Methods in mammalian autophagy research. *Cell.* 2010;140(3):313–326.
8. Li WW, Li J, Bao JK. Microautophagy: lesser-known self-eating. *Cell Mol Life Sci.* 2012;69(7):1125–1136.
9. Kaushik S, Cuervo AM. Chaperone-mediated autophagy: a unique way to enter the lysosome world. *Trends Cell Biol.* 2012;22(8):407–417.
10. Chan EY, Longatti A, McKinght NC, Tooze SA. Kinase-inactivated ULK proteins inhibit autophagy via their conserved C-terminal domains using an Atg13-dependent mechanism. *Mol Cell Biol.* 2009;29(1):157–171.
11. Ganley IG, Lam DH, Wang J, Ding X, Chen S, Jiang X. ULK1. ATG13. FIP200 complex mediates mTOR signaling and is essential for autophagy. *J Biol Chem.* 2009;284(18):12297–12305.
12. Hara T, Takamura A, Kishi C, et al. FIP200, a ULK-interacting protein, is required for autophagosome formation in mammalian cells. *J Cell Biol.* 2008;181(3):497–510.
13. Hosokawa N, Hara T, Kaizuka T, et al. Nutrient-dependent mTORC1 association with the ULK1-Atg13-FIP200 complex required for autophagy. *Mol Biol Cell.* 2009;20(7):1981–1991.
14. Hosokawa N, Sasaki T, Iemura S, Natsume T, Hara T, Mizushima N. Atg101, a novel mammalian autophagy protein interacting with Atg13. *Autophagy.* 2009;5(7):973–979.
15. Lee EJ, Tournier C. The requirement of uncoordinated 51-like kinase 1 (ULK1) and ULK2 in the regulation of autophagy. *Autophagy.* 2011;7(7):689–695.
16. Mercer CA, Kaliappan A, Dennis PB. A novel, human Atg13 binding protein, Atg101, interacts with ULK1 and is essential for macroautophagy. *Autophagy.* 2009;5(5):649–662.
17. Jung CH, Jun CB, Ro SH, et al. ULK-Atg13-FIP200 complexes mediate mTOR signaling to the autophagy machinery. *Mol Biol Cell.* 2009;20(7):1992–2003.
18. Matsunaga K, Saitoh T, Tabata K, et al. Two Beclin 1-binding proteins, Atg14L and Rubicon, reciprocally regulate autophagy at different stages. *Nat Cell Biol.* 2009;11(4):385–396.
19. Sun Q, Fan W, Zhong Q. Regulation of Beclin1 in autophagy. *Autophagy.* 2009;5(5):713–716.
20. Matsunaga K, Morita E, Saitoh T, et al. Autophagy requires targeting of the PI3-kinase complex via Atg14L. *J Cell Biol.* 2010;190(4):511–521.
21. Kabeya Y, Mizushima N, Ueno T, et al. LC3, a mammalian homologue of yeast Apg8p, is localized in autophagosome membranes after processing. *EMBO J.* 2000;19(21):5720–5728.
22. Fujita N, Itoh T, Omori H, Fukuda M, Noda T, Yoshimori T. The Atg16L complex specifies the site of LC3 lipidation for membrane biogenesis in autophagy. *Mol Biol Cell.* 2008;19(5):2092–2100.
23. Itakura E, Mizushima N. Characterization of autophagosome formation site by a hierarchical analysis of mammalian Atg proteins. *Autophagy.* 2010;6(6):764–776.

24. Kabeya Y, Mizushima N, Yamamoto A, Oshitani-Okamoto S, Ohsumi Y, Yoshimori T. LC3, GABARAP and GATE16 localize to autophagosomal membrane depending on form-II formation. *J Cell Sci*. 2004;117(Pt. 13):2805–2812.
25. Chakrama FZ, Seguin-Py S, Le Grand JN, et al. GABARAPL1 (GEC1) associates with autophagic vesicles. *Autophagy*. 2010;6(4):495–505.
26. González-Polo RA, Boya P, Pauleau AL, et al. The apoptosis/autophagy paradox: autophagic vacuolization before apoptotic death. *J Cell Sci*. 2005;118(Pt. 14):3091–3102.
27. Hubert V, Peschel A, Langer B, Gröger M, Rees A, Kain R. LAMP-2 is required for incorporating syntaxin-17 into autophagosomes and for their fusion with lysosomes. *Biol Open*. 2016;5(10):1516–1529.
28. Bento CF, Puri C, Moreau K, Rubinsztein DC. The role of membrane-trafficking small GTPases in the regulation of autophagy. *J Cell Sci*. 2013;126(Pt. 5):1059–1069.
29. Baba M, Nakajo S, Tu PH, et al. Aggregation of α-synuclein in Lewy bodies of sporadic Parkinson's disease and dementia with Lewy bodies. *Am J Pathol*. 1998;152(4):879–884.
30. Wlaker Z, Possin KL, Boeve BF, Aarsland D. Lewy body dementias. *Lancet*. 2015;386(10004):1683–1697.
31. Wakabayashi K, Tanji K, Odagiri S, Miki Y, Mori F, Takahashi H. The Lewy body in Parkinson's disease and related neurodegenerative disorders. *Mol Neurobiol*. 2013;47(2):495–508.
32. Burn DJ, Rowan EN, Minett T, et al. Extrapyramidal features in Parkinson's disease with and without dementia and dementia with Lewy bodies: a cross-sectional comparative study. *Mov Disord*. 2003;18(8):884–889.
33. Tsuboi Y, Dickson DW. Dementia with Lewy bodies and Parkinson's disease with dementia: are they different? *Parkinsonism Relat Disord*. 2005;11(Suppl. 1):S47–S51.
34. Hansen D, Ling H, Lashley T, Holton JL, Warner TT. Review: clinical, neuropathological and genetic features of Lewy body dementias. *Neuropathol Appl Neurobiol*. 2019;45(7):635–654.
35. Lewy FH. Paralysis agitans. I. Pathologische Anatomie. In: Lewandowsky M, ed. *Hundbuch der Neurologie III*. Berlin: Springer; 1912:920–933.
36. Tretiakoff MC. *Contribution a l'étude de l'anatomie pathologique de Locus Niger de Soemmerling*. Université de Paris; 1919.
37. Kahle PJ, Neumann M, Ozmen L, et al. Subcellular localization of wild-type and Parkinson's disease-associated mutant alpha-synuclein in human and transgenic mouse brain. *J Neurosci*. 2000;20(17):6365–6373.
38. Fortin DL, Troyer MD, Nakamura K, Kubo S, Anthony MD, Edwards RH. Lipid rafts mediates the synaptic localization of alpha-synuclein. *J Neurosci*. 2004;24(30):6715–6723.
39. Clayton DF, George JM. Synucleins in synaptic plasticity and neurodegenerative disorders. *J Neurosci Res*. 1999;58(1):120–129.
40. Braak H, Del Tredici K, Rüb U, de Vos RA, Jansen Steur EN, Braak E. Staging of brain pathology related to sporadic Parkinson's disease. *Neurobiol Aging*. 2003;24(2):197–211.
41. McKeith IG, Dickson DW, Lowe J, et al. Consortium on DLB, Diagnosis and management of dementia with Lewy bodies: third report of the DLB Consortium. *Neurology*. 2005;65(12):1863–1872.
42. McKeith IG, Boeve BF, Dickson DW, et al. Diagnosis and management of dementia with Lewy bodies: fourth consensus report of the DLB Consortium. *Neurology*. 2017;89(1):88–100.
43. Lashuel HA, Petre BM, Wall J, et al. Alpha-synuclein, especially the Parkinson's disease-associated mutants, forms pore-like annular and tubular protofibrils. *J Mol Biol*. 2002;322(5):1089–1102.
44. Ding TT, Lee SJ, Rochet JC, Lansbury Jr PT. Annular α-synuclein protofibrils are produced when spherical protofibrils are incubated in solution or bound to brain-derived membranes. *Biochemistry*. 2002;41(32):10209–10217.

45. Lashuel HA, Hartley D, Petre BM, Walz T, Lansbury Jr PT. Neurodegenerative disease: amyloid pores from pathogenic mutations. *Nature.* 2002;418(6895):291.
46. Chen L, Feany MB. α-Synuclein phosphorylation controls neurotoxicity and inclusion formation in a Drosophila model of Parkinson disease. *Nat Neurosci.* 2005;8(5):657–663.
47. Tanji K, Mori F, Kakita A, Takahashi H, Wakabayashi K. Alteration of autophagosomal proteins (LC3, GABARAP and GATE-16) in Lewy body disease. *Neurobiol Dis.* 2011;43(3):690–697.
48. Miki Y, Tanji K, Mori F, et al. Alteration of upstream autophagy-related proteins (ULK1, ULK2, Beclin1, VPS34 and AMBRA1) in Lewy body disease. *Brain Pathol.* 2016;26(3):359–370.
49. Miki Y, Mori F, Tanji K, Kakita A, Takahashi H, Wakabayashi K. Accumulation of histone deacetylase 6, an aggresome-related protein, is specific to Lewy bodies and glial cytoplasmic inclusions. *Neuropathology.* 2011;31(6):561–568.
50. Rothaug M, Zunke F, Mazzulli JR, et al. LIMP-2 expression is critical for β-glucocerebrosidase activity and α-synuclein clearance. *Proc Natl Acad Sci U S A.* 2014;111(43):15573–15578.
51. Winslow AR, Chen CW, Corrochano S, et al. α-Synuclein impairs macroautophagy: implications for Parkinson's disease. *J Cell Biol.* 2010;190(6):1023–1037.
52. Dehay B, Ramirez A, Martinez-Vicente M, et al. Loss of P-type ATPase ATP13A2/PARK9 function induces general lysosomal deficiency and leads to Parkinson disease neurodegeneration. *Proc Natl Acad Sci U S A.* 2012;109(24):9611–9616.
53. van Veen S, Martin S, Van den Haute C, et al. ATP13A2 deficiency disturbs lysosomal polyamine export. *Nature.* 2020;578(7795):419–424.
54. Zavodszky E, Seaman MN, Moreau K, et al. Mutation in VPS35 associated with Parkinson's disease impairs WASH complex association and inhibits autophagy. *Nat Commun.* 2014;5:3828.
55. Miki Y, Tanji K, Mori F, et al. AMBRA1, a novel α-synuclein-binding protein, is implicated in the pathogenesis of multiple system atrophy. *Brain Pathol.* 2018;28(1):28–42.
56. Tanji K, Odagiri S, Miki Y, et al. p62 deficiency enhances α-synuclein pathology in mice. *Brain Pathol.* 2015;25(5):552–564.
57. Odagiri S, Tanji K, Mori F, Kakita A, Takahashi H, Wakabayashi K. Autophagic adapter protein NBR1 is localized in Lewy bodies and glial cytoplasmic inclusions and is involved in aggregate formation in α-synucleinopathy. *Acta Neuropathol.* 2012;124(2):173–186.
58. Miki Y, Shimoyama S, Kon T, et al. Alteration of autophagy-related proteins in peripheral blood mononuclear cells of patients with Parkinson's disease. *Neurobiol Aging.* 2018;63:33–43.
59. Niimi Y, Ito S, Mizutani Y, et al. Altered regulation of serum lysosomal acid hydrolase activities in Parkinson's disease: a potential peripheral biomarker? *Parkinsonism Relat Disord.* 2019;61:132–137.
60. Wenning GK, Geser F, Krismer F, et al. The natural history of multiple system atrophy: a prospective European cohort study. *Lancet Neurol.* 2013;12(3):264–274.
61. Gilman S, Wenning GK, Low PA, et al. Second consensus statement on the diagnosis of multiple system atrophy. *Neurology.* 2008;71(9):670–676.
62. Ozawa T, Paviour D, Quinn NP, et al. The spectrum of pathological involvement of the striatonigral and olivopontocerebellar systems in multiple system atrophy: clinicopathological correlations. *Brain.* 2004;127(Pt 12):2657–2671.
63. Miki Y, Foti SC, Asi YT, et al. Improving diagnostic accuracy of multiple system atrophy: a clinicopathological study. *Brain.* 2019;142(9):2813–2827.
64. Miki Y, Tsushima E, Foti SC, et al. Identification of multiple system atrophy mimicking Parkinson's disease or progressive supranuclear palsy. *Brain.* 2021;144(40):1138–1151.

65. Koga S, Aoki N, Uitti RJ, et al. When DLB, PD, and PSP masquerade as MSA: an autopsy study of 134 patients. *Neurology.* 2015;85(5):404–412.
66. Miki Y, Foti SC, Hansen D, et al. Hippocampal α-synuclein pathology correlates with memory impairment in multiple system atrophy. *Brain.* 2020;143(6):1798–1810.
67. Papp MI, Kahn JE, Lantos PL. Glial cytoplasmic inclusions in the CNS of patients with multiple system atrophy (striatonigral degeneration, olivopontocerebellar atrophy and Shy-Drager syndrome). *J Neurol Sci.* 1989;94(1–3):79–100.
68. Papp MI, Lantos PL. Accumulation of tubular structures in oligodendroglial and neuronal cells as the basic alteration in multiple system atrophy. *J Neurol Sci.* 1992;107(2):172–182.
69. Tanji K, Odagiri S, Maruyama A, et al. Alteration of autophagosomal proteins in the brain of multiple system atrophy. *Neurobiol Dis.* 2013;49:190–198.
70. Monzio Compagnoni G, Kleiner G, Samarani M, et al. Mitochondrial dysregulation and impaired autophagy in iPSC-derived dopaminergic neurons of multiple system atrophy. *Stem Cell Rep.* 2018;11(5):1185–1198.
71. Djajadikerta A, Keshri S, Pavel M, Prestil R, Ryan L, Rubinsztein DC. Autophagy induction as a therapeutic strategy for neurodegenerative diseases. *J Mol Biol.* 2020;432(8):2799–2821.
72. Kocak M, Erdi SE, Jorba G, et al. Targeting autophagy in disease: established and new strategies. *Autophagy.* 2022;18(3):473–495.

CHAPTER 13

Autophagy and Huntington's disease

Tatsuo Mano[a] and Atsushi Iwata[b]
[a]Department of Neurology, The University of Tokyo Hospital, Tokyo, Japan
[b]Department of Neurology, Tokyo Metropolitan Geriatric Medical Center Hospital, Tokyo, Japan

1. Introduction

Huntington's disease (HD) is a progressive neurodegenerative disorder that is clinically characterized by severe motor dysfunction, including chorea, cognitive dysfunction, and psychiatric symptoms.[1] HD is inherited in a dominant form, and its genetic cause has been identified as the abnormal elongation of a trinucleotide repeat of CAG (cytosine, adenine, and guanine) in the huntingtin (*Htt*) gene.[1,2] Onset is typically during midlife, and the disease duration can span multiple decades. Neurodegeneration starts from the medium spiny neurons of the caudate-putamen and then spreads to the cortex, cerebellum, thalamus, and white matter, causing neurological symptoms.[3] The abnormal expansion of CAG repeats decodes a long polyglutamine (polyQ) tract close to the amino terminus of the Htt protein. This elongated polyQ stretch has a propensity to aggregate[4–6]; thus, polyQ aggregates form intranuclear and intracytoplasmic neuronal inclusions, which are one of the most important pathological hallmarks of HD.[7–10] Protein aggregation due to abnormal polyQ elongation has been confirmed in HD mice models, including R6/2 mouse model, transgenically containing N-terminally truncated mutant huntingtin protein (mHTT) with CAG repeat expansion within the *Htt* gene exon 1,[8,9,11] and recently in the neurons generated from human HD fibroblasts.[12] While intraneuronal inclusions are observed in at most 10% of cortical, striatal, and thalamic neurons in adult HD[3], various other neurons contain smaller oligomeric species; this abnormal accumulation of the mutation Htt is considered to contribute to the pathogenesis of HD. Only polyQ expression in Htt exon 1 can reproduce the HD pathology, and abolishing the mutant Htt in symptomatic animals can mitigate neurological deficits[13–17]; therefore, it is clear that the toxicity of the elongated polyQ stretch is critical to the pathomechanism of HD.

Autophagy Dysfunction in Alzheimer's Disease and Dementia
https://doi.org/10.1016/B978-0-323-89906-2.00001-0

Meanwhile, the expression of Htt itself is essential for life, and dysfunction of HTT due to abnormally elongated polyQ also impairs neuronal function. Htt is a large, 350-kDa protein dominated by the HEAT (Huntingtin, EF3, PP2A, and TOR1) motif, indicative of its function as a scaffold protein. Htt is involved in various cellular processes, including transcriptional regulation, nucleo-cytoplasmic shuttling, mitochondrial dynamics, vesicle trafficking, synaptic function, and antiapoptotic activity.[18–21] Elimination of *Htt* gene expression results in embryonic lethality between embryonic days 8.5 to 9.5 in mice.[22–24] Subsequent genetic studies investigating the postnatal elimination of *Htt* gene expression found that the loss of Htt can have consequences both within and outside the brain, even when the expression is eliminated in adults.[25–27] Thus, considering the neuronal enrichment of Htt expression, partial loss of Htt function is also attributed to the pathomechanism of HD. Neurodevelopmental alterations have been recently reported in patients carrying this disease-causing mutation even before the motor dysfunction.[28,29]

In this review, we discuss how autophagy contributes to the elimination of aggregated Htt to mitigate its toxicity, as well as the physiological role of Htt in synaptic maintenance via autophagy.

2. Macroautophagy in HD

In animal and cellular models, the expression of mutant Htt has been shown to induce the accumulation of aggregated Htt. Interestingly, a reduction in the expression of mutant Htt decreases the amount of Htt aggregation and mitigates motor and cognitive functions even after the manifestation of neurological symptoms.[13–16] These studies suggested that mutant Htt pathology is in equilibrium between aggregation and degradation, and neuronal cells can degrade the aggregated Htt to eliminate its toxicity. The lysosome-mediated degradation pathway plays a critical role in the clearance of aggregated proteins,[30–33] in this process, histone deacetylase 6 (HDAC6) binds the ubiquitinated aggregates to microtubules to contribute to the incorporation of aggregates to the autophagy-lysosome pathway.[34] Autophagy pathways in mammalian cells are classified into three major types: macroautophagy, microautophagy, and chaperone-mediated autophagy. Among these, macroautophagy processes relatively larger substrates, such as organelles and protein aggregates.[35] Several studies have shown that other pathways can also degrade protein aggregates[36–38]; however, macroautophagy is considered to be a robust degradation pathway of protein aggregates.[39–41]

In macroautophagy, cargos are captured into a multilamellar structured organelle, the autophagosome. While the formation of autophagosomes is mediated by autophagy-related (*ATG*) genes,[42] several adaptor proteins selectively capture protein aggregates to incorporate them into the autophagosome.[43,44] Several adaptor proteins, including p62/Sequestesome-1,[39,45–47] neighbor of BRCA1 (NBR1),[47,48] optineurin,[49] and autophagy-linked FYVE protein (Alfy)[41,50] are involved in the degradation of aggregated proteins. These adaptor proteins selectively sequester their cargos and bind to autophagy-related genes including the mammalian orthologs of the yeast protein, Atg8, which are attached to the autophagosome membrane to capture adaptor-recognized cargos.[46] For example, p62 contains a Phox and Bem1p (PB1) domain, an ATG8/LC3-interacting region (LIR) motif and an ubiquitin-associated (UBA) domain,[51–53] where it recognizes polyubiquitinated cargo via its UBA domain and binds to ATG8/LC3 via its LIR domain. By this process, p62 selectively connect the polyubiquitinaed cargo and the autophagosome menbrane. While these adaptor proteins share common functions to bring the nascent autophagic membrane to the aggregates and stabilize the autophagosome membrane to the cargo, they interact with different Atg8 families, suggestive of their functional difference. Although the relevance of the different interactions between these adaptor proteins and Atg8 homologs remains unclear, several studies have investigated the different functions of p62 and Alfy, and we discuss in the following sections.

2.1 Alfy

Alfy plays an important role in the clearance of aggregated proteins, as expected from its function as an adaptor protein connecting the aggregated proteins to the autophagosome. Alfy can bind to the p62-positive inclusions[54] and the Atg-5-12-16L complex[41] to help conjugate Atg8 proteins to the autophagosome membrane. Interestingly, Alfy does not bind to the LC3 protein, which p62 binds as discussed below, instead, it binds to other Atg8 homologs, such as the gamma aminobutyric acid type A receptor-associated protein.[55]

The depletion of Alfy expression in the cells expressing mutant Htt was followed by the increased accumulation of aggregated Htt, and the increased expression of Alfy decreased the amount of aggregated Htt, suggesting the role of Alfy in the clearance of aggregated proteins.[41,56,57] Furthermore, the increase in aggregated Htt by the abolished expression of Alfy leads to neurodegeneration.[58] Thus, Alfy protects against the toxicity of the abnormally elongated stretch of polyQ by eliminating Htt aggregates. Considering that Alfy facilitates the degradation of preformed aggregates of Htt, enhancing its function might be a potential target for the development of therapeutics.

2.2 p62

p62 is a selective autophagy receptor and mediates clearance of polyubiquitinated inclusions. While the depletion of Alfy accelerates the accumulation of aggregates and the neurodegenerative process, interestingly, the loss of p62 led to partial neuroprotection in mouse models of HD, i.e., the R6/2 and HD109QG transgenic models.[59] Similar to the depletion of Alfy, the depletion of p62 also enhances the cytosolic aggregation of Htt; however, it leads to a significant decrease in the number of nuclear inclusions. Based on the in vitro assay, it is considered that the abnormally elongated polyQ stretch is supposedly the minimal unit for the initiation of aggregation, and the recognition of these aggregates by an adaptor protein facilitates the degradation of the aggregates. However, the above-mentioned study revealed the maladaptive behavior of p62 to sequester Htt aggregates within the nucleus and hinder its degradation by autophagy. The depletion of p62 reduces the sequestration of Htt aggregates within the nucleus and mitigates the toxicity of the abnormally elongated polyQ in the nucleus, leading to partial neuroprotection. However, the degradation of Htt aggregates in the cytosol is suppressed by the depletion of p62, resulting in polyQ toxicity, which can be attributed to the partial neuroprotection. While the levels of Htt aggregates were increased in the cytosol upon depletion of p62, the decreased amount of Htt aggregates in the nucleus suggests that they are degraded by autophagy independent of p62 adaptation in the nucleus. For instance, other adaptor proteins can compensate for the loss offunction of p62, or Htt aggregates can be captured and degraded in a nonadaptor-mediated manner. Another possibility is that the function or binding proteins are different in the cytosol and nucleus. Similar to p62, the inhibition of histone deacetylase 3 (HDAC3), which directly binds to Htt and suppresses its neurotoxicity, has differential effects on cytoplasmic and nuclear Htt aggregates, suggesting a different mode of regulation of Htt degradation.[60] Taken together, Htt aggregates are degraded by macroautophagy, which is regulated by several adaptor proteins. The physiological significance of the different adaptor proteins with differential behavior in the cytosol and nucleus should be investigated in more detail in the future.

2.3 NBR1

NBR1 is a selective autophagy receptor containing PB1, LIR, and UBA domains, and its structure and function are similar to p62.[48,52,61] NBR1 itself functions as a selective autophagy receptor and also interact with p62 to support sequestration of protein aggregates.[52] Compared to p62 to bind to polyubiquitinated proteins, NBR1 also binds to monoubiquitin rather than polyubiqutinated protein preferentially.[62] As mentioned above, mHTT forms nuclear inclusions with p62 and reduces cytololic p62. Howerver, even in the late stage of HD, NBR1 accumulates in the cytosol and fill in for p62 function, suggesting that NBR1 can compensate the pathogenic dysfunction of p62 in the late stage of HD.[63,64]

2.4 Optineurin

Optineurin (OPTN) acts as an adaptor protein in selective autophagy and facilitates cytosolic and intranuclear mHTT clearance.[65–67] OPTN recruits Atg5-Atg12-Atg16L complex to the site of autophagosome initiation,[68] and the lack of OPTN causes defets in trafficking and autopagy leading to the deterioration of proteostasis.[69–71] Consistent these studies, overexpression of OPTN increases selective autophagy to reduce mHTT aggregation.[72] Mutations within OPTN ubiquitin-binding domain prevent its interaction with mHTT aggregation in a dominant-negative manner, suggestive of its oligomeliation in selective autophagy.[72,73]

2.5 Tollip

Toll-interacting protein (Tollip) functions as a cargo receptor protein in selective autophagy and have ubiquitin/Atg8-binding CUE domain to promote protein aggregates clearance.[74] It is found in nuclear inclusions of HD mice model suggestive of its involvement in clearance of mHTT aggregation,[75] and another study demonstrated that overexpression of Tollip increased aggregate clearance.[76]

3. Physiological function of Htt in autophagy

Although mHTT is widely expressed in a variety of cells, its functional abnormalities in neurons have been the focus of various studies. In adult-onset HD, where polyQ elongation is relatively short, inclusion bodies are observed in various cells at a low frequency; however, in juvenile-onset HD with a long polyQ elongation, inclusion bodies accumulate in neurons, especially in frontal cortical neurons,[77] suggesting that autophagy is more rate-limiting in neurons than in glial cells. In the early stage of the disease, autophagosomes are formed at the tips of neuronal axons and mature as they are transported to the cell bodies of neurons, which is thought to underlie the basis for the importance of autophagy in reducing polyQ neurotoxicity at axonal tips.[78–81] Considering that the degree of formation of aggregates and degeneration of neurons differs among the subtypes of neurons and brain regions, it is necessary to clarify the role of autophagy in each of these subtypes. Furthermore, in addition to aggregate formation, functional and morphological abnormalities have also been reported in glial cells, including astrocytes, microglia, and oligodendrocytes in the brains of human patients with HD and mouse models of HD.[82–85] Despite these functional abnormalities, the mechanism by which autophagy contributes to the degradation of protein aggregrates in these cell types remains unclear.

As reviewed above, polyQ toxicity is important in the pathogenesis of HD, and previous studies have shown that the autophagic processing of Htt aggregates contributes to the reduction in polyQ toxicity. If polyQ toxicity can explain the pathogenesis of HD, the gene dosage of abnormal *Htt* should be considered important; however, the fact that even heterozygous patients are not necessarily more affected than heterozygous patients[86,87] suggests that a pure gain of toxicity of the polyQ stretch is not the only explanation, and that polyQ-induced disruption of the native function of Htt may also contribute to the pathogenesis. A complete loss of function of Htt results in embryonic lethality, suggesting that a partial loss of function can be a potential pathomechanism.

Htt is a scaffold protein that plays an important role as an integrator of transport along the cellular cytoskeleton in endocytosis, endosomal motility, and axonal transport to regulate the dynamics of trafficking and transport.[88–95] Among various functions, Htt has structural similarities to selective autophagy proteins Atg11, Atg23, and Vac8 and interacts with Atg11-binding partners[96,97] to contribute to selective autophagy. The loss of the polyQ stretch in Htt enhances its autophagic capacity in neuronal cells and longevity in mice.[96] In agreement with this study, the loss of Htt or its interactor, HAP1 (Huntington-associated protein 1), impaired retrograde transport of autophagosomes containing selective cargo, leading to protein accumulation.[95,98] Htt interacts with p62 to facilitate the interaction between p62 and LC3B, and binds to the autophagy initiator ULK1 (Unc-51 like autophagy activating kinase 1) by releasing it from negative regulation by the mammalian target of rapamycin (mTOR).[99] These data suggest that Htt plays a significant role in selective autophagy, and the depletion of its function can contribute to the accumulation of abnormal proteins leading to HD pathology.

Furthermore, Htt dysfunction impairs the autophagic turnover of synaptic proteins, thereby reducing synaptic plasticity. To maintain synaptic plasticity, where a large number of proteins are packed in a highly organized manner, proteins regularly undergo assembly and disassembly processes.[100] Synaptic proteins and synaptic vesicles are locally incorporated into autophagosomes at synapses, transported retrogradely along with the axon, and degraded with autophagosomes fused with lysosomes.[101–109] A decrease in autophagic activity disrupts the rapid turnover of synaptic components, leading to impairments in learning and memory by reducing synaptic plasticity.[102,110,111] Htt interacts with dynein, dynactin, and kinesin through Htt-associated proteins, contributes to the retrograde transport of synaptic vesicles, and interacts with Htt-associated protein 40 to regulate endosomal trafficking.[112–118] Thus, a dysfunction of Htt can result in the disruption of vesicle trafficking, leading to reduced synaptic plasticity even without Htt aggregates. Previous observational studies have provided clear evidence that cognitive dysfunction in HD can occur in the earliest stage and even precede motor symptoms,[119–122] and neuropathological changes at both pre- and postsynaptic sites have been reported in the early stages of HD.[123,124] Additionally, polyQ aggregates also impede Htt function and autophagic efficiency, leading to a further reduction in the turnover of synaptic proteins.

4. Autophagy: A therapeutic target

In HD, autophagy is responsible for the degradation of aggregates; thus, the activation of autophagy has been considered as a therapeutic strategy. Activation of autophagy by rapamycin, an mTOR inhibitor, has been examined in a mouse model of HD and has been shown to decrease Htt aggregates and reduce toxicity.[31,125] Moreover, it has been shown that autophagy can be activated by the modulation of intracellular calcium, which is independent of the mTOR pathway. In cell and Drosophila models, Htt aggregates are reduced by activating autophagy through intracellular calcium modulation. Additionally, trehalose and metformin have been shown to have beneficial effects by activating autophagy via the calcium-induced activation of AMP-activated protein kinase in mice.[126–134]

The global activation of autophagy may be effective in reducing Htt aggregates in the brain at the model level, but it should be noted that it activates autophagy not only in the brain but also in the whole body. Since muscle wasting is a crucial issue especially in the advanced stage of HD[135], it should be taken into consideration that increased muscle catabolism due to enhanced autophagy may aggravate muscle wasting.

Considering that there is no significant abnormality in the expression of genes related to autophagy in HD, the selective enhancement of autophagy against Htt aggregates may constitute a potential therapeutic target.[135] In the brains of patients with HD, adaptor proteins such as Alfy and optineurin are decreased; thus, Htt aggregates can be reduced by enhancing the recognition of cargo by adaptor proteins and inducing their incorporation into the autophagosome.

Therefore, autophagy is a potential therapeutic target. The global activation of autophagy can reduce Htt aggregates in the brain and induce therapeutic effects in animal models, but more brain-specific induction of autophagy or enhanced recognition of Htt aggregates by adaptor proteins would be more desirable as a therapeutic strategy in the future.

5. Conclusion

In HD, autophagy is a rate-limiting factor in the degradation of polyQ aggregates. In this process, polyQ aggregates are recognized by several adaptor proteins and recruited to autophagosomes, which then fuse to the lysosome to degrade the cargo. Similar to p62 and Alfy, each adaptor protein has a unique function in cargo recognition and transport; however, its biological significance remains unclear. To date, therapeutic strategies for the global activation of autophagy have been successful in cellular and animal models, but the unnecessary activation of autophagy may cause undesirable effects. Therefore, a deeper understanding of the physiological roles of each adaptor protein is required to achieve the exclusive elimination of polyQ aggregates in the brain. Moreover, the length of the polyQ stretch in cellular and mouse models is much longer than that in humans with HD; thus, it is important to develop a suitable model with a polyQ length closer to that of humans with HD, to bridge the gap between models and human patients.

References

1. Sturrock A, Leavitt BR. The clinical and genetic features of Huntington disease. *J Geriatr Psychiatry Neurol.* 2010;23(4):243–259.
2. The Huntington's Disease Collaborative Research Group. A novel gene containing a trinucleotide repeat that is expanded and unstable on Huntington's disease chromosomes. *Cell.* 1993;72(6):971–983.

3. Vonsattel JPG, Keller C, Cortes Ramirez EP. Huntington's disease—neuropathology. In: Weiner WJ, Tolosa E, eds. *Handb Clin Neurol*. vol. 100; 2011:83–100.
4. Perutz MF, Johnson T, Suzuki M, Finch JT. Glutamine repeats as polar zippers: their possible role in inherited neurodegenerative diseases. *Proc Natl Acad Sci U S A*. 1994;91(12):5355–5358.
5. Scherzinger E, Lurz R, Turmaine M, et al. Huntingtin-encoded polyglutamine expansions form amyloid-like protein aggregates in vitro and in vivo. *Cell*. 1997;90(3):549–558.
6. Bäuerlein FJB, Saha I, Mishra A, et al. In situ architecture and cellular interactions of PolyQ inclusions. *Cell*. 2017;171(1):179–187.e10.
7. Roizin L. Neuronal nuclear-cytoplasmic changes in Huntington's chorea: electron microscope investigations. In: *Advances in Nueorlogy 23. Huntingtons Disease*; 1979. Published online https://ci.nii.ac.jp/naid/10008323980/
8. Davies SW, Turmaine M, Cozens BA, et al. Formation of neuronal intranuclear inclusions underlies the neurological dysfunction in mice transgenic for the HD mutation. *Cell*. 1997;90(3):537–548.
9. DiFiglia M, Sapp E, Chase KO, et al. Aggregation of huntingtin in neuronal intranuclear inclusions and dystrophic neurites in brain. *Science*. 1997;277(5334):1990–1993.
10. Gutekunst CA, Li SH, Yi H, et al. Nuclear and neuropil aggregates in Huntington's disease: relationship to neuropathology. *J Neurosci*. 1999;19(7):2522–2534.
11. Mangiarini L, Sathasivam K, Seller M, et al. Exon 1 of the HD gene with an expanded CAG repeat is sufficient to cause a progressive neurological phenotype in transgenic mice. *Cell*. 1996;87(3):493–506.
12. Victor MB, Richner M, Olsen HE, et al. Striatal neurons directly converted from Huntington's disease patient fibroblasts recapitulate age-associated disease phenotypes. *Nat Neurosci*. 2018;21(3):341–352.
13. Yamamoto A, Lucas JJ, Hen R. Reversal of neuropathology and motor dysfunction in a conditional model of Huntington's disease. *Cell*. 2000;101(1):57–66.
14. Régulier E, Trottier Y, Perrin V, Aebischer P, Déglon N. Early and reversible neuropathology induced by tetracycline-regulated lentiviral overexpression of mutant huntingtin in rat striatum. *Hum Mol Genet*. 2003;12(21):2827–2836.
15. Harper SQ, Staber PD, He X, et al. RNA interference improves motor and neuropathological abnormalities in a Huntington's disease mouse model. *Proc Natl Acad Sci U S A*. 2005;102(16):5820–5825.
16. DiFiglia M, Sena-Esteves M, Chase K, et al. Therapeutic silencing of mutant huntingtin with siRNA attenuates striatal and cortical neuropathology and behavioral deficits. *Proc Natl Acad Sci U S A*. 2007;104(43):17204–17209.
17. Kordasiewicz HB, Stanek LM, Wancewicz EV, et al. Sustained therapeutic reversal of Huntington's disease by transient repression of huntingtin synthesis. *Neuron*. 2012;74(6):1031–1044.
18. Harjes P, Wanker EE. The hunt for huntingtin function: interaction partners tell many different stories. *Trends Biochem Sci*. 2003;28(8):425–433.
19. Schulte J, Littleton JT. The biological function of the Huntingtin protein and its relevance to Huntington's disease pathology. *Curr Trends Neurol*. 2011;5:65–78.
20. Reddy PH, Shirendeb UP. Mutant huntingtin, abnormal mitochondrial dynamics, defective axonal transport of mitochondria, and selective synaptic degeneration in Huntington's disease. *Biochim Biophys Acta*. 2012;1822(2):101–110.
21. Saudou F, Humbert S. The biology of huntingtin. *Neuron*. 2016;89(5):910–926.
22. Duyao M, Ambrose C, Myers R, et al. Trinucleotide repeat length instability and age of onset in Huntington's disease. *Nat Genet*. 1993;4(4):387–392.
23. Nasir J, Floresco SB, O'Kusky JR, et al. Targeted disruption of the Huntington's disease gene results in embryonic lethality and behavioral and morphological changes in heterozygotes. *Cell*. 1995;81(5):811–823.

24. Zeitlin S, Liu JP, Chapman DL, Papaioannou VE, Efstratiadis A. Increased apoptosis and early embryonic lethality in mice nullizygous for the Huntington's disease gene homologue. *Nat Genet*. 1995;11(2):155–163.
25. Dragatsis I, Levine MS, Zeitlin S. Inactivation of Hdh in the brain and testis results in progressive neurodegeneration and sterility in mice. *Nat Genet*. 2000;26(3):300–306.
26. Wang G, Liu X, Gaertig MA, Li S, Li X-J. Ablation of huntingtin in adult neurons is nondeleterious but its depletion in young mice causes acute pancreatitis. *Proc Natl Acad Sci U S A*. 2016;113(12):3359–3364.
27. Dietrich P, Johnson IM, Alli S, Dragatsis I. Elimination of huntingtin in the adult mouse leads to progressive behavioral deficits, bilateral thalamic calcification, and altered brain iron homeostasis. *PLoS Genet*. 2017;13(7), e1006846.
28. Kerschbamer E, Biagioli M. Huntington's disease as neurodevelopmental disorder: altered chromatin regulation, coding, and non-coding RNA transcription. *Front Neurosci*. 2015;9:509.
29. Lee JK, Conrad A, Epping E, et al. Effect of trinucleotide repeats in the Huntington's gene on intelligence. *EBioMedicine*. 2018;31:47–53.
30. Johnston JA, Ward CL, Kopito RR. Aggresomes: a cellular response to misfolded proteins. *J Cell Biol*. 1998;143(7):1883–1898.
31. Ravikumar B, Duden R, Rubinsztein DC. Aggregate-prone proteins with polyglutamine and polyalanine expansions are degraded by autophagy. *Hum Mol Genet*. 2002;11(9):1107–1117.
32. Iwata A, Christianson JC, Bucci M, et al. Increased susceptibility of cytoplasmic over nuclear polyglutamine aggregates to autophagic degradation. *Proc Natl Acad Sci U S A*. 2005;102(37):13135–13140.
33. Yamamoto A, Cremona ML, Rothman JE. Autophagy-mediated clearance of huntingtin aggregates triggered by the insulin-signaling pathway. *J Cell Biol*. 2006;172(5):719–731.
34. Iwata A, Riley BE, Johnston JA, Kopito RR. HDAC6 and microtubules are required for autophagic degradation of aggregated huntingtin. *J Biol Chem*. 2005;280(48):40282–40292.
35. Lamark T, Johansen T. Aggrephagy: selective disposal of protein aggregates by macroautophagy. *Int J Cell Biol*. 2012;2012. https://doi.org/10.1155/2012/736905.
36. Juenemann K, Schipper-Krom S, Wiemhoefer A, Kloss A, Sanz Sanz A, Reits EAJ. Expanded polyglutamine-containing N-terminal huntingtin fragments are entirely degraded by mammalian proteasomes. *J Biol Chem*. 2013;288(38):27068–27084.
37. Schipper-Krom S, Juenemann K, Jansen AH, et al. Dynamic recruitment of active proteasomes into polyglutamine initiated inclusion bodies. *FEBS Lett*. 2014;588(1):151–159.
38. Hjerpe R, Bett JS, Keuss MJ, et al. UBQLN2 mediates autophagy-independent protein aggregate clearance by the proteasome. *Cell*. 2016;166(4):935–949.
39. Bjorkoy G, Lamark T, Brech A, et al. p62/SQSTM1 forms protein aggregates degraded by autophagy and has a protective effect on huntingtin-induced cell death. *J Cell Biol*. 2005;171(4):603–614.
40. Filimonenko M, Stuffers S, Raiborg C, et al. Functional multivesicular bodies are required for autophagic clearance of protein aggregates associated with neurodegenerative disease. *J Cell Biol*. 2007;179(3):485–500.
41. Filimonenko M, Isakson P, Finley KD, et al. The selective macroautophagic degradation of aggregated proteins requires the PI3P-binding protein Alfy. *Mol Cell*. 2010;38(2):265–279.
42. Mizushima N, Yoshimori T, Ohsumi Y. The role of Atg proteins in autophagosome formation. *Annu Rev Cell Dev Biol*. 2011;27:107–132.
43. Dikic I. Proteasomal and autophagic degradation systems. *Annu Rev Biochem*. 2017;86:193–224.
44. Gatica D, Lahiri V, Klionsky DJ. Cargo recognition and degradation by selective autophagy. *Nat Cell Biol*. 2018;20(3):233–242.

45. Shin J. P62 and the sequestosome, a novel mechanism for protein metabolism. *Arch Pharm Res.* 1998;21(6):629–633.
46. Pankiv S, Clausen TH, Lamark T, et al. p62/SQSTM1 binds directly to Atg8/LC3 to facilitate degradation of ubiquitinated protein aggregates by autophagy. *J Biol Chem.* 2007;282(33):24131–24145.
47. Lamark T, Kirkin V, Dikic I, Johansen T. NBR1 and p62 as cargo receptors for selective autophagy of ubiquitinated targets. *Cell Cycle.* 2009;8(13):1986–1990.
48. Kirkin V, Lamark T, Sou Y-S, et al. A role for NBR1 in autophagosomal degradation of ubiquitinated substrates. *Mol Cell.* 2009;33(4):505–516.
49. Korac J, Schaeffer V, Kovacevic I, et al. Ubiquitin-independent function of optineurin in autophagic clearance of protein aggregates. *J Cell Sci.* 2013;126(Pt 2):580–592.
50. Simonsen A, Birkeland HCG, Gillooly DJ, et al. Alfy, a novel FYVE-domain-containing protein associated with protein granules and autophagic membranes. *J Cell Sci.* 2004;117(Pt 18):4239–4251.
51. Johansen T, Lamark T. Selective autophagy mediated by autophagic adapter proteins. *Autophagy.* 2011;7(3):279–296.
52. Lamark T, Perander M, Outzen H, et al. Interaction codes within the family of mammalian Phox and Bem1p domain-containing proteins. *J Biol Chem.* 2003;278(36):34568–34581.
53. Vadlamudi RK, Joung I, Strominger JL, Shin J. p62, a phosphotyrosine-independent ligand of the SH2 domain of p56lck, belongs to a new class of ubiquitin-binding proteins. *J Biol Chem.* 1996;271(34):20235–20237.
54. Clausen TH, Lamark T, Isakson P, et al. p62/SQSTM1 and ALFY interact to facilitate the formation of p62 bodies/ALIS and their degradation by autophagy. *Autophagy.* 2010;6(3):330–344.
55. Lystad AH, Ichimura Y, Takagi K, et al. Structural determinants in GABARAP required for the selective binding and recruitment of ALFY to LC3B-positive structures. *EMBO Rep.* 2014;15(5):557–565.
56. Finley KD, Edeen PT, Cumming RC, et al. Blue cheese mutations define a novel, conserved gene involved in progressive neural degeneration. *J Neurosci.* 2003;23(4):1254–1264.
57. Dragich JM, Kuwajima T, Hirose-Ikeda M, et al. Autophagy linked FYVE (Alfy/WDFY3) is required for establishing neuronal connectivity in the mammalian brain. *Elife.* 2016;5. https://doi.org/10.7554/eLife.14810.
58. Eenjes E, Dragich JM, Kampinga HH, Yamamoto A. Distinguishing aggregate formation and aggregate clearance using cell-based assays. *J Cell Sci.* 2016;129(6):1260–1270.
59. Kurosawa M, Matsumoto G, Kino Y, et al. Depletion of p62 reduces nuclear inclusions and paradoxically ameliorates disease phenotypes in Huntington's model mice. *Hum Mol Genet.* 2015;24(4):1092–1105.
60. Mano T, Suzuki T, Tsuji S, Iwata A. Differential effect of HDAC3 on cytoplasmic and nuclear huntingtin aggregates. *PLoS One.* 2014;9(11), e111277.
61. Odagiri S, Tanji K, Mori F, Kakita A, Takahashi H, Wakabayashi K. Autophagic adapter protein NBR1 is localized in Lewy bodies and glial cytoplasmic inclusions and is involved in aggregate formation in α-synucleinopathy. *Acta Neuropathol.* 2012;124(2):173–186.
62. Walinda E, Morimoto D, Sugase K, Konuma T, Tochio H, Shirakawa M. Solution structure of the ubiquitin-associated (UBA) domain of human autophagy receptor NBR1 and its interaction with ubiquitin and polyubiquitin. *J Biol Chem.* 2014;289(20):13890–13902.
63. Rué L, López-Soop G, Gelpi E, Martínez-Vicente M, Alberch J, Pérez-Navarro E. Brain region- and age-dependent dysregulation of p62 and NBR1 in a mouse model of Huntington's disease. *Neurobiol Dis.* 2013;52:219–228.
64. Nicot A-S, Lo Verso F, Ratti F, et al. Phosphorylation of NBR1 by GSK3 modulates protein aggregation. *Autophagy.* 2014;10(6):1036–1053.

65. Li F, Xu D, Wang Y, et al. Structural insights into the ubiquitin recognition by OPTN (optineurin) and its regulation by TBK1-mediated phosphorylation. *Autophagy.* 2018;14(1):66–79.
66. Mori F, Tanji K, Toyoshima Y, et al. Optineurin immunoreactivity in neuronal nuclear inclusions of polyglutamine diseases (Huntington's, DRPLA, SCA2, SCA3) and intranuclear inclusion body disease. *Acta Neuropathol.* 2012;123(5):747–749.
67. Schwab C, Yu S, McGeer EG, McGeer PL. Optineurin in Huntington's disease intranuclear inclusions. *Neurosci Lett.* 2012;506(1):149–154.
68. Bansal M, Moharir SC, Sailasree SP, et al. Optineurin promotes autophagosome formation by recruiting the autophagy-related Atg12-5-16L1 complex to phagophores containing the Wipi2 protein. *J Biol Chem.* 2018;293(1):132–147.
69. Sahlender DA, Roberts RC, Arden SD, et al. Optineurin links myosin VI to the Golgi complex and is involved in Golgi organization and exocytosis. *J Cell Biol.* 2005;169(2):285–295.
70. Tumbarello DA, Waxse BJ, Arden SD, Bright NA, Kendrick-Jones J, Buss F. Autophagy receptors link myosin VI to autophagosomes to mediate Tom1-dependent autophagosome maturation and fusion with the lysosome. *Nat Cell Biol.* 2012;14(10):1024–1035.
71. Paulus JD, Link BA. Loss of optineurin in vivo results in elevated cell death and alters axonal trafficking dynamics. *PLoS One.* 2014;9(10), e109922.
72. Shen W-C, Li H-Y, Chen G-C, Chern Y, Tu P-H. Mutations in the ubiquitin-binding domain of OPTN/optineurin interfere with autophagy-mediated degradation of misfolded proteins by a dominant-negative mechanism. *Autophagy.* 2015;11(4):685–700.
73. Gao J, Ohtsubo M, Hotta Y, Minoshima S. Oligomerization of optineurin and its oxidative stress- or E50K mutation-driven covalent cross-linking: possible relationship with glaucoma pathology. *PLoS ONE.* 2014;9(7), e101206.
74. Lu K, Psakhye I, Jentsch S. A new class of ubiquitin-Atg8 receptors involved in selective autophagy and polyQ protein clearance. *Autophagy.* 2014;10(12):2381–2382.
75. Doi H, Mitsui K, Kurosawa M, Machida Y, Kuroiwa Y, Nukina N. Identification of ubiquitin-interacting proteins in purified polyglutamine aggregates. *FEBS Lett.* 2004;571(1–3):171–176.
76. Lu K, Psakhye I, Jentsch S. Autophagic clearance of polyQ proteins mediated by ubiquitin-Atg8 adaptors of the conserved CUET protein family. *Cell.* 2014;158(3):549–563.
77. Jansen AHP, van Hal M, Op den Kelder IC, et al. Frequency of nuclear mutant huntingtin inclusion formation in neurons and glia is cell-type-specific. *Glia.* 2017;65(1):50–61.
78. Hollenbeck PJ. Products of endocytosis and autophagy are retrieved from axons by regulated retrograde organelle transport. *J Cell Biol.* 1993;121(2):305–315.
79. Maday S, Holzbaur ELF. Autophagosome assembly and cargo capture in the distal axon. *Autophagy.* 2012;8(5):858–860.
80. Maday S, Wallace KE, Holzbaur ELF. Autophagosomes initiate distally and mature during transport toward the cell soma in primary neurons. *J Cell Biol.* 2012;196(4):407–417.
81. Maday S, Holzbaur ELF. Autophagosome biogenesis in primary neurons follows an ordered and spatially regulated pathway. *Dev Cell.* 2014;30(1):71–85.
82. Myers RH, Vonsattel JP, Paskevich PA, et al. Decreased neuronal and increased oligodendroglial densities in Huntington's disease caudate nucleus. *J Neuropathol Exp Neurol.* 1991;50(6):729–742.
83. Huang B, Wei W, Wang G, et al. Mutant huntingtin downregulates myelin regulatory factor-mediated myelin gene expression and affects mature oligodendrocytes. *Neuron.* 2015;85(6):1212–1226.
84. Jin J, Peng Q, Hou Z, et al. Early white matter abnormalities, progressive brain pathology and motor deficits in a novel knock-in mouse model of Huntington's disease. *Hum Mol Genet.* 2015;24(9):2508–2527.

85. Phillips OR, Joshi SH, Squitieri F, et al. Major superficial white matter abnormalities in Huntington's disease. *Front Neurosci*. 2016;10:197.
86. Wexler NS, Young AB, Tanzi RE, et al. Homozygotes for Huntington's disease. *Nature*. 1987;326(6109):194–197.
87. Squitieri F, Gellera C, Cannella M, et al. Homozygosity for CAG mutation in Huntington disease is associated with a more severe clinical course. *Brain*. 2003;126(Pt 4):946–955.
88. Gauthier LR, Charrin BC, Borrell-Pagès M, et al. Huntingtin controls neurotrophic support and survival of neurons by enhancing BDNF vesicular transport along microtubules. *Cell*. 2004;118(1):127–138.
89. Gunawardena S, Goldstein LSB. Polyglutamine diseases and transport problems: deadly traffic jams on neuronal highways. *Arch Neurol*. 2005;62(1):46–51.
90. Her L S, Goldstein LSB. Enhanced sensitivity of striatal neurons to axonal transport defects induced by mutant huntingtin. *J Neurosci*. 2008;28(50):13662–13672.
91. Caviston JP, Holzbaur ELF. Huntingtin as an essential integrator of intracellular vesicular trafficking. *Trends Cell Biol*. 2009;19(4):147–155.
92. Power D, Srinivasan S, Gunawardena S. In-vivo evidence for the disruption of Rab11 vesicle transport by loss of huntingtin. *Neuroreport*. 2012;23(16):970–977.
93. Liot G, Zala D, Pla P, Mottet G, Piel M, Saudou F. Mutant Huntingtin alters retrograde transport of TrkB receptors in striatal dendrites. *J Neurosci*. 2013;33(15):6298–6309.
94. Zala D, Hinckelmann M-V, Yu H, et al. Vesicular glycolysis provides on-board energy for fast axonal transport. *Cell*. 2013;152(3):479–491.
95. Wong YC, Holzbaur ELF. The regulation of autophagosome dynamics by huntingtin and HAP1 is disrupted by expression of mutant huntingtin, leading to defective cargo degradation. *J Neurosci*. 2014;34(4):1293–1305.
96. Zheng S, Clabough EBD, Sarkar S, Futter M, Rubinsztein DC, Zeitlin SO. Deletion of the huntingtin polyglutamine stretch enhances neuronal autophagy and longevity in mice. *PLoS Genet*. 2010;6(2), e1000838.
97. Steffan JS. Does Huntingtin play a role in selective macroautophagy? *Cell Cycle*. 2010;9(17):3401–3413.
98. Ochaba J, Lukacsovich T, Csikos G, et al. Potential function for the Huntingtin protein as a scaffold for selective autophagy. *Proc Natl Acad Sci U S A*. 2014;111(47):16889–16894.
99. Rui Y-N, Xu Z, Patel B, et al. Huntingtin functions as a scaffold for selective macroautophagy. *Nat Cell Biol*. 2015;17(3):262–275.
100. Vijayan V, Verstreken P. Autophagy in the presynaptic compartment in health and disease. *J Cell Biol*. 2017;216(7):1895–1906.
101. Zhai B, Shang X, Fu J, Li F, Zhang T. Rapamycin relieves anxious emotion and synaptic plasticity deficits induced by hindlimb unloading in mice. *Neurosci Lett*. 2018;677:44–48.
102. Yan J, Porch MW, Court-Vazquez B, Bennett MVL, Suzanne ZR. Activation of autophagy rescues synaptic and cognitive deficits in fragile X mice. *Proc Natl Acad Sci U S A*. 2018;115(41):E9707–E9716.
103. Nikoletopoulou V, Sidiropoulou K, Kallergi E, Dalezios Y, Tavernarakis N. Modulation of autophagy by BDNF underlies synaptic plasticity. *Cell Metab*. 2017;26(1):230–242.e5.
104. Zhang H, Shang Y, Xiao X, Yu M, Zhang T. Prenatal stress-induced impairments of cognitive flexibility and bidirectional synaptic plasticity are possibly associated with autophagy in adolescent male-offspring. *Exp Neurol*. 2017;298:68–78.
105. Shehata M, Matsumura H, Okubo-Suzuki R, Ohkawa N, Inokuchi K. Neuronal stimulation induces autophagy in hippocampal neurons that is involved in AMPA receptor degradation after chemical long-term depression. *J Neurosci*. 2012;32(30):10413–10422.
106. Hoffmann-Conaway S, Brockmann MM, Schneider K, et al. Parkin contributes to synaptic vesicle autophagy in Bassoon-deficient mice. *Elife*. 2020;9, e56590.

107. Matsuda S, Miura E, Matsuda K, et al. Accumulation of AMPA receptors in autophagosomes in neuronal axons lacking adaptor protein AP-4. *Neuron.* 2008;57(5):730–745.
108. Rowland AM, Richmond JE, Olsen JG, Hall DH, Bamber BA. Presynaptic terminals independently regulate synaptic clustering and autophagy of GABAA receptors in Caenorhabditis elegans. *J Neurosci.* 2006;26(6):1711–1720.
109. Khan MM, Strack S, Wild F, et al. Role of autophagy, SQSTM1, SH3GLB1, and TRIM63 in the turnover of nicotinic acetylcholine receptors. *Autophagy.* 2014;10(1):123–136.
110. Glatigny M, Moriceau S, Rivagorda M, et al. Autophagy is required for memory formation and reverses age-related memory decline. *Curr Biol.* 2019;29(3):435–448.e8.
111. Hylin MJ, Zhao J, Tangavelou K, et al. A role for autophagy in long-term spatial memory formation in male rodents. *J Neurosci Res.* 2018;96(3):416–426.
112. Zala D, Hinckelmann M-V, Saudou F. Huntingtin's function in axonal transport is conserved in Drosophila melanogaster. *PLoS ONE.* 2013;8(3), e60162.
113. Twelvetrees AE, Yuen EY, Arancibia-Carcamo IL, et al. Delivery of GABAARs to synapses is mediated by HAP1-KIF5 and disrupted by mutant huntingtin. *Neuron.* 2010;65(1):53–65.
114. McGuire JR, Rong J, Li S-H, Li X-J. Interaction of Huntingtin-associated protein-1 with kinesin light chain: implications in intracellular trafficking in neurons. *J Biol Chem.* 2006;281(6):3552–3559.
115. Li SH, Gutekunst CA, Hersch SM, Li XJ. Interaction of huntingtin-associated protein with dynactin P150Glued. *J Neurosci.* 1998;18(4):1261–1269.
116. Engelender S, Sharp AH, Colomer V, et al. Huntingtin-associated protein 1 (HAP1) interacts with the p150Glued bubunit of dynactin. *Hum Mol Genet.* 1997;6(13):2205–2212.
117. Caviston JP, Ross JL, Antony SM, Tokito M, Holzbaur ELF. Huntingtin facilitates dynein/dynactin-mediated vesicle transport. *Proc Natl Acad Sci U S A.* 2007;104(24):10045–10050.
118. Pal A, Severin F, Lommer B, Shevchenko A, Zerial M. Huntingtin-HAP40 complex is a novel Rab5 effector that regulates early endosome motility and is up-regulated in Huntington's disease. *J Cell Biol.* 2006;172(4):605–618.
119. Scahill RI, Zeun P, Osborne-Crowley K, et al. Biological and clinical characteristics of gene carriers far from predicted onset in the Huntington's disease Young adult study (HD-YAS): a cross-sectional analysis. *Lancet Neurol.* 2020;19(6):502–512.
120. Paulsen JS, Hayden M, Stout JC, et al. Preparing for preventive clinical trials: the Predict-HD study. *Arch Neurol.* 2006;63(6):883–890.
121. Paulsen JS, Smith MM, Long JD. PREDICT HD investigators and coordinators of the Huntington Study Group. Cognitive decline in prodromal Huntington Disease: implications for clinical trials. *J Neurol Neurosurg Psychiatry.* 2013;84(11):1233–1239.
122. Tabrizi SJ, Scahill RI, Owen G, et al. Predictors of phenotypic progression and disease onset in premanifest and early-stage Huntington's disease in the TRACK-HD study: analysis of 36-month observational data. *Lancet Neurol.* 2013;12(7):637–649.
123. DiProspero NA, Chen E-Y, Charles V, Plomann M, Kordower JH, Tagle DA. Early changes in Huntington's disease patient brains involve alterations in cytoskeletal and synaptic elements. *J Neurocytol.* 2004;33(5):517–533.
124. Sapp E, Penney J, Young A, Aronin N, Vonsattel JP, DiFiglia M. Axonal transport of N-terminal huntingtin suggests early pathology of corticostriatal projections in Huntington disease. *J Neuropathol Exp Neurol.* 1999;58(2):165–173.
125. Berger Z, Ravikumar B, Menzies FM, et al. Rapamycin alleviates toxicity of different aggregate-prone proteins. *Hum Mol Genet.* 2006;15(3):433–442.
126. Sarkar S. Regulation of autophagy by mTOR-dependent and mTOR-independent pathways: autophagy dysfunction in neurodegenerative diseases and therapeutic application of autophagy enhancers. *Biochem Soc Trans.* 2013;41(5):1103–1130.

127. Criollo A, Maiuri MC, Tasdemir E, et al. Regulation of autophagy by the inositol trisphosphate receptor. *Cell Death Differ.* 2007;14(5):1029–1039.
128. Zhang L, Yu J, Pan H, et al. Small molecule regulators of autophagy identified by an image-based high-throughput screen. *Proc Natl Acad Sci U S A.* 2007,104(48):19023–19028.
129. Williams A, Sarkar S, Cuddon P, et al. Novel targets for Huntington's disease in an mTOR-independent autophagy pathway. *Nat Chem Biol.* 2008;4(5):295–305.
130. Johnson CW, Melia TJ, Yamamoto A. Modulating macroautophagy: a neuronal perspective. *Future Med Chem.* 2012;4(13):1715–1731.
131. Tanaka M, Machida Y, Niu S, et al. Trehalose alleviates polyglutamine-mediated pathology in a mouse model of Huntington disease. *Nat Med.* 2004;10(2):148–154.
132. Ma TC, Buescher JL, Oatis B, et al. Metformin therapy in a transgenic mouse model of Huntington's disease. *Neurosci Lett.* 2007;411(2):98–103.
133. Sarkar S, Davies JE, Huang Z, Tunnacliffe A, Rubinsztein DC. Trehalose, a novel mTOR-independent autophagy enhancer, accelerates the clearance of mutant huntingtin and α-synuclein. *J Biol Chem.* 2007;282(8):5641–5652.
134. Jiang W, Wei W, Gaertig MA, Li S, Li X-J. Therapeutic effect of berberine on Huntington's disease transgenic mouse model. *PLoS One.* 2015;10(7), e0134142.
135. Hodges A, Strand AD, Aragaki AK, et al. Regional and cellular gene expression changes in human Huntington's disease brain. *Hum Mol Genet.* 2006;15(6):965–977.

SECTION VIII

Drug discovery in Alzheimer's disease by modulating autophagy

CHAPTER 14

Drug discovery in Alzheimer's disease by regulating autophagy

Tadanori Hamano[a,b,c,d] **and Yoshinori Endo**[a,b]
[a]Second Department of Internal Medicine, Faculty of Medical Sciences, University of Fukui, Fukui, Japan
[b]Department of Neurology, University of Fukui Hospital, Fukui, Japan
[c]Department of Aging and Dementia (DAD), University of Fukui, Fukui, Japan
[d]Life Science Innovation Center, University of Fukui, Fukui, Japan

1. Introduction

Typical pathological findings of Alzheimer's disease (AD) are senile plaques (SPs) consisting of amyloid β protein (Aβ) and neurofibrillary tangles (NFT) consisting of tau protein. Aβ and tau protein cause neuronal damage and synaptic dysfunction.[1,2] Excessive phosphorylation of tau is an important step to form NFT in patients with AD.[3,4] Inhibition of the formation, or removal of Aβ or tau is the major therapeutic strategy to prevent or treat AD.[5] Autophagy is one of the major organelles by which cells break down abnormal proteins. Removal of Aβ and tau may also be achieved through autophagy.[6,7] Regulating the activity of autophagy can be an important strategy in treating AD or slowing the progression of AD. The autophagy pathway is shown in Fig. 1, and the autophagy modulators are shown in Fig. 2 and Table 1.[4,8–110] Autophagy inducers are mainly composed of two subgroups that act directly on mechanistic target of rapamycin complex (mTORC) and other pathways to activate autophagy. In this chapter, autophagy-activating drugs considered to be useful for AD treatment are outlined. Note that some autophagy stimulants were demonstrated to have considerable promise as AD therapeutics in clinical trials (Table 1).

2. Drugs that induce autophagy

2.1 Direct mTORC inhibitors

2.1.1 Rapamycin

The mTORC signaling pathway is started by nutrients and growth factors and regulates autophagy[105] (Fig. 1). Rapamycin (sirolimus) is a selective inhibitor of mTORC, which promotes autophagy. Rapamycin is also known to act through direct binding to FKBP12. A diet including rapamycin was reported

Autophagy Dysfunction in Alzheimer's Disease and Dementia
https://doi.org/10.1016/B978-0-323-89906-2.00002-2

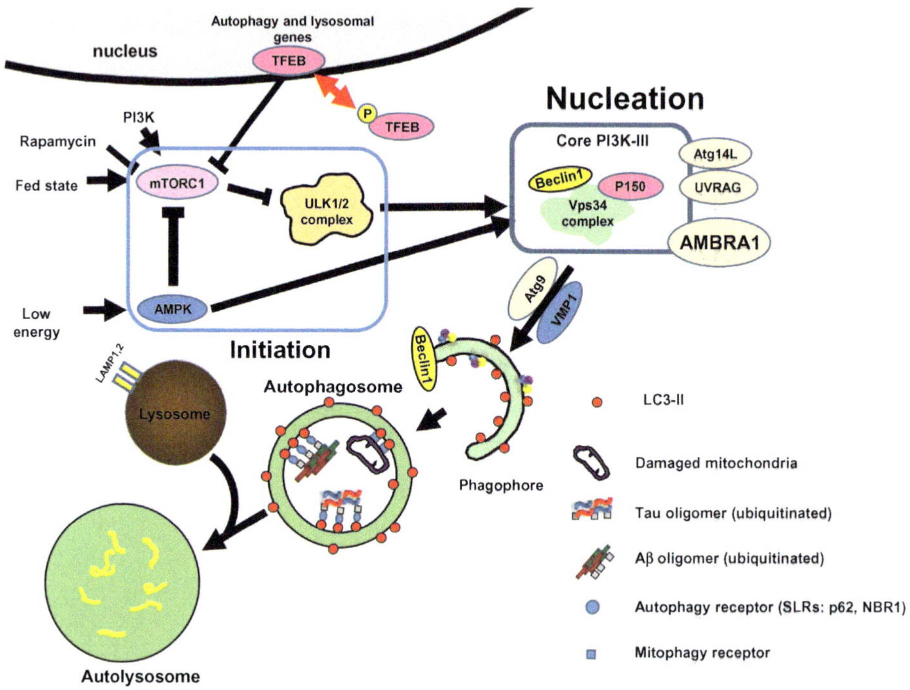

Fig. 1 Scheme of autophagy flux. Initiation. In response to nutrient deprivation, AMPK phosphorylates mTORC1, releasing its inhibition on ULK, and activates the initiation complex. Downregulation of mTORC1 leads to the translocation of TFEB to the nucleus, where TFEB becomes upregulated, and various autophagy- and lysosome-related genes are translated. ULK phosphorylates AMBRA-1, which promotes the interaction of Beclin1 with p150 and PI3K-III (Vps34) in the PI3K-III complexes. UVRAG and Atg14L bind differently to form separate PI3K-III-complex populations in cells. Mitochondria can be directly bound by mitochondrial receptors without poly-ubiquitination. In aggrephagy, poly-ubiquitinated protein aggregates (Aβ oligomer and tau oligomer) are engulfed by the phagophore via SLRs (sequestosome 1/p62-like receptors), which include p62 or NBR1. Maturation. Once the loaded phagophore is closed and the outer fraction of LC3-II is removed, the completed autophagosome fuses with the lysosome via a bonding process involving syntaxin members, and LAMP-1 and -2 creating an autolysosome. Abbreviations: *AMPK*, 5′ adenosine monophosphate-activated protein kinase; *PI3K-III*, class III phosphatidylinositol 3-kinase; *LC3-II*, microtubule-associated protein light chain 3-II; *LAMP-1, LAMP-2*, lysosome associated membrane protein 1, 2; *SLRs*, sequestosome 1/p62-like receptors; *FYCO1*, FYVE and coiled-coil domain containing 1; *p62*, p62/sequestosome-1; *NBR1*, neighbor of BRCA1 gene 1; *TFEB*, transcription factor EB. *(Revised from Int J Mol Sci. 2021;22(14):7475.)*

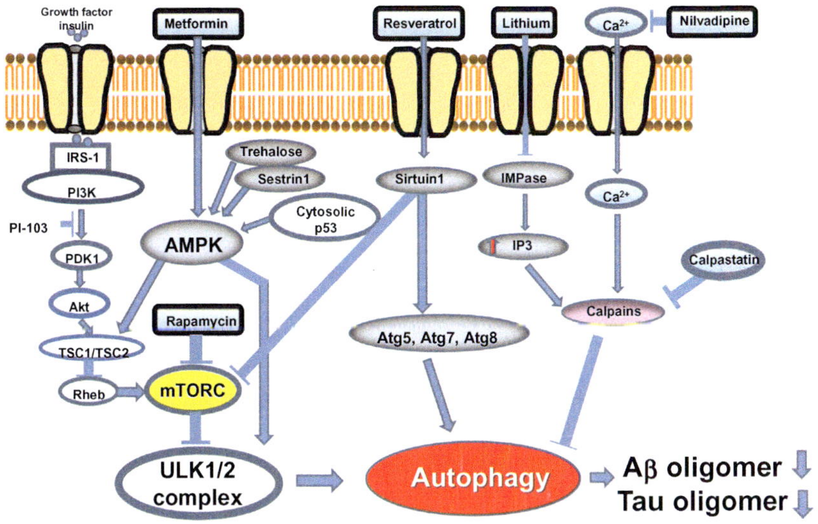

Fig. 2 Autophagy upregulating drugs. Two signaling pathways, mTORC1-dependent and -independent, are involved in autophagy regulation. Abbreviations: *mTORC1*, mechanistic target of rapamycin 1; *IRS-1*, insulin receptor substrate-1; *PI3K*, phosphoinositide 3-kinase; *SIRT1*, sirtuin 1. *(Revised from* Curr Alzheimer Res. *2018;15:1283–1296 and* Int J Mol Sci. *2021;22(14):7475.)*

to prolong the life of animals.[106] Aging is a most important factor in the etiology of AD, especially for the development of AD therapeutics.[107] Human studies suggested the involvement of mTORC signaling in AD.[108,109] Blockade of mTORC signaling by rapamycin may mitigate cognitive impairment and reduce Aβ deposition by increasing autophagy in the AD mouse brain.[8]

Furthermore, overexpressing mTORC or not expressing mTORC may affect the balance of phospho-tau/nonphospho-tau in the cytosol and different cell compartments, which may promote tau deposition. mTORC activation significantly increases cytosolic tau and its relocalization to exocytotic vesicles, which are not associated with exosomes. This suggests that mTORC is involved in controlling the distribution of tau to organelles in cells and initiating the secretion of tau from cells to extracellular organs.[110] Rapamycin promotes Aβ degradation and reduces Aβ deposition. It also functions to reduce tau.[8,9] Indeed, rapamycin reduced NFT in P301S transgenic mice.[10] Rapamycin also reduced phosphorylated tau in zebrafish expressing A152T tau.[11] Rapamycin ameliorated memory disturbances in AD model mice[9] and improved cell viability treated with Aβ42.[12] Furthermore, rapamycin was proven to improve neuroinflammation, restore blood vessel and metabolic functions in the brain, and ameliorate cognitive dysfunction in many in vitro and in vivo AD settings.[13,39]

Table 1 Drugs possibly effective for Alzheimer's disease by activating autophagy.

Category Subcategory Autophagy target drugs	Other actions	Experimental effects	Clinical trial	References
mTORC dependent autophagy inducer				
Rapamycin	Direct mTORC1 inhibitor	Improving learning and memory and reduced Aβ and tau pathology in AD model mouse Increased viability of cells against Aβ42	Promising results with not significant decrease in β galactosidase (NCT01649960)	8–14
Rapamycin analogs				
CCI-779 (temsirolimus)	Direct mTORC1 inhibitor	Improved Aβ clearance and cognition in mouse model of AD Reduced tau and rescue motor dysfunction in P301L tau mouse	Not tested	15–17
RAD001 (everolimus)	Direct mTORC1 inhibitor			18
AP23573, selective ATP-competitive small molecule				
PP242, OSI-027, AZD2014, AZD8055	mTORC1/2 inhibitor			19,20
Other mTORC1 dependent autophagy inducer				
Methylene blue	mTORC inhibitor	Inhibits tau oligomerization in organic hippocampus slice/neuron and P301L tau mice	Finished 4 phase III clinical trials NCT01689246 NCT01689233 NCT 01626378 Ongoing NCT02380573	21–23

Curcumin	PI3K/Akt/mTORC inhibitor Induces Hsp70 involved in CMA			24,25
Latrepirdine	mTORC inhibitor	Reduces Aβ peptides in the brains of mice	Promising results	26,27
β-Asarone	mTORC dependent	Decreases Aβ42 in hippocampus improves memory in animal model of AD.	Not tested	28,29
Oleuropein Aglycone (polyphenol)	mTORC dependent	Counter acts amyloid aggregation and toxicity		30–32
Omega-3 fatty acids	Akt/mTORC1 pathway inhibitor Disruption of Beclin1-bcl2 binding	Decreases tau, improves cognitive function	Ambiguous results on cognitive function in AD.	33–35
Other autophagy modulating drugs				
AMPK activator				
Metformin	PP2A agonists	Reduced levels of Aβ Decreased tau phosphorylation	Already in clinical use for diabetes patients Improve cognitive function	36–38
Trehalose	AMPK activator mTORC independent	Inhibits aggregation on both Aβ40 and tau Reduces their cytotoxicity in vitro Protects against cognitive dysfunction Decreases hippocampal Aβ	not tested	39–43

Continued

Table 1 Drugs possibly effective for Alzheimer's disease by activating autophagy—cont'd

Category				
Subcategory				
Autophagy target drugs	**Other actions**	**Experimental effects**	**Clinical trial**	**References**
Resveratrol	AMPK activator	Long term treatment rescued memory loss and reduced Aβ in AD mouse model	Well tolerated and biomarker change	44–49
	Atg upregulation		Aβ40 ↑	
AICAR	AMPK activator	Reduced inflammatory response to Aβ Decreased tau phosphorylation Restored mitochondrial function Improved cognitive function	Not tested	50,51
Arctigenin	AMPK activator			52
Statins	AMPK/mTORC1 pathway	Reduced Aβ plaque and toxicity	Beneficial effects for MCI patients	53–57
		Decreased tau phosphorylation	Improves cognitive function in AD	
IMPase, IP_3 inhibitor				
Lithium	Iinositol and IP_3 reducer, GSK3β inhibitor	Induces autophagy and inhibits GSK3β mediated tau phosphorylation	Beneficial on cognition in MCI and AD.	39,58–63
			Currently AD phase II clinical trials, NCT02129348	
Carbamazepine	Phosphoinositide cycle inhibitor		Ambiguous results on behavioral symptoms related to AD	64,65

Imidazoline receptor agonists (cAMP reduction)				
Clonidine, rilmenidine		Accelerated clearance kinetics of A152T-tau		11
TFEB activation				
Flubendazole Bromhexine	TFEB activator	Reduced accumulation of phosphorylated tau in N2a cells	Not tested	66
Flavonol	TEFB activator			
Fisetin		Autophagic degradation of phosphorylated tau in 293T cells expressing tau and GSK3	Currently in AD phase II clinical trial, NCT02380573	67,68
Curcumin analog (C1)	TFEB activator	Reduced accumulation of phosphorylated tau	Multiple trial completed or currently in AD phase II clinical trials	69
Enhancing lysosome/autolysosome acidification, reducing autophagosome accumulation				
Nicotinamide			Currently Early Phase 1 clinical trial NCT04430517	70–73
Activating JNK1/Bcl-2/PI3K class III pathway				
L-NAME		Improves cognitive function	Not tested	39
Reduction of intracellular Calcium levels				
Calcium channel blockers		Enhances Aβ degradation Decreases tau, ameliorates neuroinflammation Improves cognitive function	Ambiguous results for cognitive functions in AD patients	39,74

Continued

Table 1 Drugs possibly effective for Alzheimer's disease by activating autophagy—cont'd

Category Subcategory Autophagy target drugs	Other actions	Experimental effects	Clinical trial	References
Nitrendipine				39
Nilvadipine				39
VPS34 activation				
Memantine	mTORC dependent		Already in clinical use for AD patients Beneficial for AD with good tolerance	75–77
Atg dependent pathway				
SMER28	Increasing autophagosome synthesis and enhancing the clearance of autophagy substrates	Reduced Aβ peptide in gamma secretase independent manner		78
Spermidine	Preventing Beclin1 cleavage: Upregulation of Atgs expression, EP300 inhibition	Reduced Aβ secretion	Beneficial for memory performance	79–81
		Improves cognitive function		
Tetrahydrohyperforin	ATG5 dependent pathway	Prevents memory impairment and physiological dysfunction including tau hyperphosphorylation or disturbed turnover of Aβ plaque	not tested	82,83
Elevating ATP60V01				

Rifampicin		Suppressed deposition of tau oligomers, highly phosphorylated tau, synaptic loss, and microglial activity in animal model of AD		84,85
Tau acetylation inhibition				
P300/CBP inhibitor SMDC37892	Atg5,7,8,12 inhibitor	Reduces tau acetylation, dampens tau secretion		86
Cilostazol	Beclin1, Atg5, LC3, cathepsin ↑ mTORC, P300 ↓	Reduces tau acetylation and phosphorylation	Clinically approved for vascular disease. Currently phase II clinical trial (NCT02491268)	87,88
		Reduces Aβ in N2a cells with Swedish mutation		
Multiple mechanisms				
Vitamin D	Ca^{2+} ↑ ATG16L1, LC3B induction	Reduces Aβ generation	Ambiguous results on cognitive function in AD	89,90
		Enhances Aβ degradation Ameliorates neuro-inflammation		
ABL1 inhibitor				
Nilotinib, bosutinib	Tyrosine kinase inhibitor	Degraded phosphorylated tau	Currently in AD phase II clinical trial NCT02947893	91
Unknown				
GTM-1 (derivative of quinolone)	non-mTORC	Rescued cognitive dysfunction and Aβ pathology in mouse model of AD		65,92

Continued

Table 1 Drugs possibly effective for Alzheimer's disease by activating autophagy.—cont'd

Category Subcategory Autophagy target drugs	Other actions	Experimental effects	Clinical trial	References
Bexarotene		Restores defective mitochondrial respiration in P301L-tau mutation		93
Lonafenib	Falnesyl transferase inhibitor	Prevent Rhes mediated tau accumulation	Already in clinical use for cancer patients	94
Rho-ROCK inhibitor (H1152, Y-27632, and fasudil)		Reduction of phosphorylated and oligomeric tau in cell line (M17D), and P301L tau mice	Already in clinical use for subarachnoidal hemorrhage patients	95
Gene therapy				
MicroRNA (miRNA-132/212)		Increased clearance of tau in AD mice		96,97
siRNA (calpain knockdown)	Inhibitor of calpain	Ameliorated tau induced toxicity in drosophila		98
Optogenetic TFEB inducer	TFEB inducer	Reduced accumulation of phosphorylated tau in N2a cells and iPSC from AD patient	Not tested	99
AAV-TFEB	TFEB inducer			100
Flag TFEB transgenic mouse		Reduced PHF-tau, restored learning ability		101

In rapamycin clinical trials, a correlation between some aging markers and physical performance was noted, but frailty was not improved (NCT01649960). Furthermore, the reduction in expression of β galactosidase, a cellular aging marker by rapamycin, was not significant.[14] In summary, drugs acting on the mTORC pathway are expected to be effective for autophagy-related AD treatment.

2.1.2 Rapamycin analogs

Other drugs inhibiting mTORC in a similar manner to rapamycin also demonstrated protective effects against AD. Everolimus ameliorated the impairment of cognition and mood in AD mice.[18] Temsirolimus (CCI-779) induced Aβ clearance via autophagy and reversed memory loss in mouse models of AD.[15] Temsirolimus also reduced tau accumulation[16] and improved motor dysfunction in tau mutant mice.[17] ATP competing mTORC inhibitors OI-027 and AZD2014 significantly reduced phosphorylated tau and aggregated tau in iPSC-derived neuron models, which was more effective than rapamycin.[19] As a result, the vulnerability of tau-mediated neurotoxic stress was reduced. The authors found that mTORC inhibition and autophagy upregulation led directly to tau clearance.[19] ATP competitive mTORC1/2 blocker PP242 also reduced tau.[20]

2.2 Other mTORC-dependent autophagy inducers

2.2.1 Methylene blue

Methylene blue induced autophagy and attenuated sarkosyl insoluble tau accumulation via mTORC1 suppression in cultured brain slices from a mouse model of frontotemporal dementia (FTD) and in cell culture models.[21] Gauther et al. reported a clinical trial of methylene blue, Leucomethylthioninium Bis (LMTM). Eight hundred and ninety-one patients were assigned as follows: 357 to control, 268 to 75 mg of LMTM two times per day, and 266 to 125 mg of LMTM two times per day. The primary analyses revealed no effects of either dose on primary outcomes (ADAS-Cog score decline) compared with control. Gastrointestinal and urinary adverse events were observed with LMTM treatment. The primary analysis was not positive and did not support the use of LMTM as an add-on therapy for participants with mild to moderate AD.[22] As a subanalysis, mild AD participants ($n = 800$) were randomly assigned to 100 mg of LMTM two times per day or 4 mg two times per day. Patients treated with LMTM as monotherapy at either dose tested consistently performed better than those receiving the same dose as an add-on to several acetylcholinesterase inhibitors or memantine. Furthermore, the participants administered 100 mg of LMTM two times per day as monotherapy performed better than those administered 4 mg twice a day.[23]

2.2.2 Curcumin

Curcumin was reported to reduce the dimeric tau in the soluble fraction in human tau (hTau) mice through autophagy activation.[24] The authors assumed that this was due to an increase in membrane-bound Hsc70 by curcumin, which is required for the degradation of tau in CMA.[24] Wang et al. reported that curcumin inhibits the PI3K/Akt/mTOR pathway.[25]

2.2.3 Latrepirdine

Latrepirdine stimulates autophagy in an mTOR- and Atg5-dependent manner and reduces the amount of intracellular amyloid precursor protein (APP) metabolites containing Aβ in the mouse brain.[26] Meta-analysis revealed that although latrepirdine did not increase adverse events compared with placebo, it did not affect cognitive function, or the activity of daily living (ADL) score in mild-to-moderate AD patients, although there was a modest improvement in behavior.[27]

2.2.4 β-asarone

β-asarone is an ether found *Acori graminei*.[28] β-asarone treatment reduces Aβ42 levels in the hippocampus, possibly through mTORC-dependent autophagy, and reverses memory in APP/PS1 transgenic mouse models of AD.[29]

2.2.5 Oleuropein aglycone (polyphenol)

Oleuropein aglycone is a polyphenol in olives and induces autophagy through the mTORC pathway.[30,31] Autophagy regulation is reported to be one of the mechanisms by which oleuropein aglycone prevent the aggregation and toxicity of Aβ.[32]

2.2.6 Omega-3 fatty acids

Omega-3 fatty acids inhibit Akt/mTORC1 signaling and induce autophagy by blocking the binding of Beclin1 and Bcl-2.[89] Omega-3 fatty acids were also reported to be effective in several preclinical models of AD, but the results are inconsistent.[33,34] Long-term fish oil (FO) supplementation can ameliorate AD pathology. Even a short period of FO intake restored the phospholipid composition, and it significantly reduced SPs in 5xFAD brains if administered in the early phase of AD pathology.[35]

2.3 AMPK activator

2.3.1 Metformin

Metformin, an adenosine 5′-monophosphate-activated protein kinase (AMPK) activator, reduced tau accumulation.[36] Metformin is also a protein phosphatase 2A agonist, which reduces phosphorylation levels of tau. Clinical trials suggested that protein phosphatase 2A agonists, including metformin, can inhibit excessive phosphorylation of tau.[37] Mice studies also provided similar results.[38] On the other hand, in db/db mice, metformin did not affect spatial learning, memory disturbances, or long-term hyperglycemia.[38]

2.3.2 Trehalose

Two-sugar trehalose upregulates autophagy via the AMPK/mTORC1 pathway. It was also reported that trehalose activates autophagy by increasing the expression of Beclin1[39] (Fig. 1). Trehalose reduced endogenous tau in primary neurons.[40] It should be noted that phosphorylation of tau in the AD brain had little effect on its degradation via autophagy. In addition, autophagy upregulation suppressed tau aggregation and reduced its cytotoxicity.[40] Trehalose also reduced insoluble tau in P301S mice.[41] Trehalose inhibited both Aβ40 and tau aggregation and reduced cytotoxicity in vitro.[40,42] Trehalose also induced tau degradation in neurons and inhibited their aggregation. Furthermore, disturbed cognitive and learning functions recovered in APP/PS1 transgenic mice.[43]

2.3.3 Resveratrol

Resveratrol, a polyphenol derived from grapes, and its derivatives are used as dietary supplements to reduce blood pressure. Resveratrol reduced the accumulation of extracellular Aβ peptides by inducing autophagy via the AMPK/mTORC1 pathway (Fig. 1).[44] Resveratrol is known to reduce Aβ toxicity and ameliorate memory dysfunction, mainly in animal models of AD.[45] Of note, resveratrol was also reported to activate autophagy epigenetically by promoting Atgs expression. This may be related to the extended life span in drosophila, *Caenorhabditis elegans* (*C. elegans*), and yeast.[46] A recently published phase II clinical trial exploring the effects of resveratrol on AD revealed that it was well tolerated, but AD biomarkers, including plasma Aβ40 levels were higher in the treatment group than in controls.[47] On the other hand, long-term resveratrol therapy ameliorated cognitive dysfunction and reduced Aβ levels in AD mouse models.[48] Antiinflammatory effects were also demonstrated in patients with AD.[49] However, further studies are needed to interpret these effects.

2.3.4 5-Aminoimidazole-4-carboxamide-1-beta-4-ribofuranoside (AICAR)

Activating AMPK directly with 5-aminoimidazole-4-carboxamide-1-beta-4-ribofuranoside (AICAR) induced autophagy through mTORC1 inhibition. AICAR was reported to reduce inflammatory responses by Aβ in astrocytes, reduce tau phosphorylation, repair mitochondrial function, and rescue spatial learning in animal models of AD.[50,51] In contrast, other studies revealed that AICAR promoted tau phosphorylation in neurons, especially in the cytosol, because AMPK is a tau kinase.[111]

2.3.5 Arctigenin

Arctigenin, a polyphenols extracted from *Arctium lappa*, is another autophagy regulator that inhibited Aβ production and memory dysfunction in mouse models of AD.[52] These effects were considered to be via mTORC- and AMPK-dependent autophagy activation.[52]

2.3.6 Statins

Autophagy induction through the AMPK/mTORC1 pathway is induced by statins.[53] Statin promote autophagy by inhibiting Beclin1-Bcl-2 binding. Stain prevented oxidative stress and cognitive dysfunction in AD mouse models[54] and were reported to reduce the Aβ load,[55] phosphorylated tau levels,[56] and neurotoxicity in vitro and in vivo models. Regarding the relationship between statins and dementia, meta-analysis demonstrated ambiguous effects in patients with dementia, but it is expected to prevent the progression from mild cognitive impairment (MCI) to dementia.[57]

2.4 IMPase, IP_3 inhibitors

2.4.1 Lithium

Lithium, the most extensively investigated autophagy modulator in patients with AD, was reported to activate autophagy and reduce tau accumulation in an mTORC1-independent manner, i.e., via inositol monophosphatase (IMPase) inhibition.[58] Clinical studies revealed that lithium improves AD and that this effect may be related to its mTOR-independent autophagy-promoting ability.[59,60] Lithium reduces inositol and inositol-1, 4,5-triphosphate (IP_3) levels by affecting the phosphoinositide cycle and induces autophagy. Lithium has been investigated in multiple AD-related studies and was demonstrated to promote Aβ degradation, reduce SPs formation, and tau phosphorylation, and mitigate neurodegenerating processes, mainly in mouse and rat models of AD.[39,60,61] Lithium is effective for cognitive function in MCI and AD patients and is used in a current Phase IV clinical trial.[62] In a meta-analysis of clinical studies on AD, lithium significantly reduced cognitive decline compared with placebo, with no significant adverse events.[63]

2.5 Carbamazepine

Carbamazepine was originally developed as a drug to treat epilepsy.[112] Two studies reported that autophagy induction by carbamazepine also protects against cognitive decline and reduces Aβ in the mouse model of AD brain.[64,65] Carbamazepine and other anticonvulsant valproic acid activate autophagy by reducing inositol and IP_3 levels through the phosphoinositide cycle, similar to lithium.

2.6 Imidazoline receptor agonists (reducing cAMP)

2.6.1 Clonidine and rilmenidine

As autophagy inducers, the imidazoline receptor agonists, clonidine, and rilmenidine, were effective in zebrafish models harboring A152T tau.[11]

2.7 Drugs targeting TFEB

2.7.1 About TFEB

Transcription factor EB (TFEB) is a master activator of several lysosomal genes (ATP6V1H, ATP6V0E1, ATP6V1E1, and LAMP1), genes associated with lysosomal biosynthesis and function (CSTB, M6PR, IGF2R2009, and CTSD), and autophagy genes (SQSTM1, LC3-II (microtubule associated protein light chain3-II), UVRAG, and ATG9B)[113,114] (Fig. 1). TFEB is expressed in neuron, microglia, and astrocyte.

Several studies revealed that TFEB induction can reduce the tau burden and ameliorate both behavioral and synaptic functions in in vitro and in vivo models. Injections of adeno-associated virus (AAV)-TFEB into the lateral ventricles of both hemispheres on day 0 (P0) of the rTg4510 mouse brain reduced the NFTs and improved behavioral and synaptic deficiencies compared with wild-type controls after 1 month.[100] TFEB specifically degraded soluble and aggregated tau by up-regulating autophagy proteins (LC3-II and p62) and lysosomal proteins (LAMP1 and CTSD). Especially, phosphatase and tensin homologs (PTEN) is a direct target of TFEB that is required for TFEB-dependent aberrant tau clearance.[100] The overexpression TFEB in P301S tauopathy mice upregulated autophagy, mitigated neuronal loss, reduced PHF tau, ameliorated age-related accumulation of lipofuscin granules, and recovered spatial, working, and reference memory.[101]

2.7.2 Flubendazole and bromhexine

Two FDA-approved drugs, flubendazole and bromhexine, function as autophagy inducers by promoting the nuclear transition of TFEB by disrupting dynamic microtubules, thereby inactivating mTORC and dissociating from lysosomes.[66] The effects of bromhexine can be attributed to ambroxol, a metabolite associated with TFEB. Of note, both drugs removed the highly phosphorylated tau (S202 and S396/S404) in N2a cells transfected with 3R0N-T231D/S235D phosphorylation. When Beclin1 was knocked down, extensively phosphorylated tau was accumulated. Thus, the removal of phosphorylated tau by these two drugs was hypothesized to depend on autophagy.

2.7.3 Fisetin

Fisetin, an organic flavonoid found in fruits and vegetables, can reduce pathological tau, and ameliorated learning and memory impairment in APPswe/PS1dE9 transgenic AD mice. Fisetin treatment reduced S262 and S396/S404 extensively phosphorylated tau in mouse and rat primary cortical neurons.[67] Kim et al. reported that fisetin upregulated TFEB.[68] It also activated autophagy by increasing the levels of AMPK phosphorylation and sirtuin 1 (SIRT1) expression.

2.7.4 Curcumin analog

A new small molecule curcumin analog C1 (C1) was identified as a TFEB activator.[69] C1 directly binds TFEB and activates it thereby promoting autophagy and lysosomal activity. It reduced APP, APP C-terminal fragments (CTF-β/α), Aβ peptides, and tau aggregates in 5xFAD, P301S, and 3xTg-AD mice, respectively, and ameliorated synaptic and memory dysfunction.[69]

2.7.5 Optogenetic TFEB inducer

Binder et al. optimized a light induction system to promote TFEB expression in various cell-culture models of tauopathy. Blue light (465 nm) irradiation was suggested to promote TFEB expression, its migration to the nucleus, and subsequent activation.[99] This further induces autophagy and lysosomal gene expression, which ultimately removes pathological tau from neuronal cells and AD patient-derived human iPSC-neurons.

2.8 Increased lysosome/autolysosome acidification and reduce autophagosome accumulation

2.8.1 Nicotinamide

Long-term therapy with nicotinamide (vitamin B3/PP) ameliorated cognitive dysfunction and Aβ and tau pathologies in mouse models of AD. The effects of nicotinamide are likely mediated by increased acidification of lysosomes or autophagosomes, which reduced autophagosome adhesion.[70,71] Nicotinamide increased NAD^+, increased mitochondrial resistance to oxidative stress, and promoted autophagy-lysosomal processes.[70] Nicotinamide activity also depends on its ability to induce the degradation of β-site APP cleaving enzyme 1 (BACE1). Clinical trials revealed that nicotinamide is safe, but not effective against cognitive dysfunction in patients with AD.[72] However, the anti-AD activity of nicotinamide is still being explored, and further studies are currently underway.[73]

2.9 JNK1/Bcl-2/PI3K activator

2.9.1 L-NAME

L-NAME is a nonselective inhibitor of nitric oxide synthase (NOS) that promotes autophagosome formation by inducing the JNK1/Bcl-2/PI3K Class III pathway. Memory impairment by Aβ in rats was improved,[39] but other studies suggest that it exacerbates AD pathology in APP/PS1 mice.[115]

2.10 Calcium channel antagonist

2.10.1 Nitrendipine, and nilvadipine

Calcium channel antagonists are known as autophagy inducers that disable the autophagy inhibitory effects of intracellular calcium signaling. Several calcium channel blockers (nitrendipine, nilvadipine, and isradipine) were tried as potential anti-AD drugs and confirmed to reduce Aβ levels and tau aggregates in vitro and in animal AD models and to improve cognitive dysfunction in AD mouse models.[39] Isradipine treatment in a triple transgenic animal model of AD was well tolerated. Isradipine reduced tau accumulation and upregulated autophagy function in the brain.[74]

2.11 VPS34 activator

2.11.1 Memantine

The NMDA (*N*-methyl-D-aspartic acid) receptor antagonist memantine is commonly applied to treat moderate to severe AD. Metaanalysis revealed that memantine is well tolerated and effective for AD.[75] This effects of memantine may be mediated, at least in part, by autophagy in either an mTORC-dependent or -independent manner.[76] The possibility of autophagy activation via VPS34 activation by memantine has been noted.[77]

2.12 Atg-dependent pathway

2.12.1 SMER28

SMER28, increases autophagy through the Atg5-dependent pathway and reduces the level of Aβ peptides independently of γ-secretase.[78]

2.12.2 Spermidine

Spermidine is another poorly examined inducer of autophagy, which decreases with age, suggesting a possible association between decreased endogenous spermidine levels and deterioration associated with aging.[79] Spermidine has been demonstrated to induce autophagy with neuroprotective action in multiple neurodegenerative diseases, including AD, by preventing Beclin1 cleavage by caspase-3, increasing Atg expression, and inhibiting the endogenous autophagy repressor EP300.[79–81] Spermidine also inhibited Aβ aggregation in vitro and ameliorated age-related memory impairment in vivo. Of note, like resveratrol, spermidine was suggested to extend life span in vivo and possibly in humans.[79,80]

2.12.3 Tetrahydrohyperforin

Tetrahydrohyperforin is a derivative of hyperforin, the active ingredient of *Hypericum perforatum*. Tetrahydrohyperforin ameliorated memory dysfunction and physiological impairments such as excessive phosphorylation of tau or disturbed metabolic turnover of SPs in animal models of AD.[82,116] One of its beneficial effects is associated with autophagy-related activity, mainly APP clearance via the Atg5-dependent pathway.[83]

2.13 ATP6V0A1 enhancer

2.13.1 Rifampicin

Tau mice administered rifampicin exhibited dose-dependently decreased tau oligomers, highly phosphorylated tau and synaptic loss. Rifampicin inactivated microglia, and restored cognition almost completely. In addition, rifampicin reduces p62 in the brain, does not affect LC3 conversion, and may restore lysosomal function.[84] Rifampicin was also reported to increase ATP6V0A1, which promotes lysosomal function.[85]

2.14 P300 inhibitors (tau acetylation inhibitors)

2.14.1 SMDC37892

Lysine acetyltransferase p300/CBP acetylates tau and regulates its degradation and toxicity. A new p300 inhibitor, SMDC37892, increased autophagic flux and reduced tau secretion and propagation in a mouse model of tauopathy and in human and rodent neocortex culture. In contrast, highly activated p300/CBP blocked the autophagic flux and promoted tau secretion in neurons.[86]

2.14.2 Cilostazole

Cilostazole is a clinically used to prevent the recurrence of cerebral infarction and as a treatment for arteriosclerosis obliterans. In a study using N2a cells, cilostazole upregulated Beclin1, Atg5, and LC3-II through the upregulation of SIRT1. This led to a decrease in Aβ and APP-C-terminal fragments.[87] The authors also found that cilostazol inhibited P300 and the major tau kinase GSK3β. The decreased expression of SIRT1 was reversed by cilostazol.[88]

2.15 Multiple mechanisms

2.15.1 Vitamin D

Vitamin D is an extensively examined endogenous compound confirmed to induce autophagy in several cell models via multiple pathways, some acting at the protein level and others at the gene transcription level.[89] Mammals such as humans can endogenously produce cholecalciferol, which is vitamin D3, from 7-dehydrocholesterol by ultraviolet irradiation (295–300 nm) in the skin. Vitamin D3 can also be absorbed from food. Vitamin D3 is hydrolyzed in the liver to 25-hydroxyvitamin D (25(OH)D), which is further converted to 1,25-dihydroxyvitamin D 3 (1,25(OH)2D3) in the kidney. The molecular mechanism of autophagy induction by 1,25(OH)2D3 is VDR (vitamin D receptor) activation. 1,25(OH)2D3 increases the Ca^{2+} concentration in the cytoplasm through nongenomic action via VDR. In addition, genomic action induces autophagy by inducing the expression of Atg proteins such as cathelicidin, mammalian autophagy-related 16 like-1 (Atg16L1), and LC3II.

In AD, vitamin D deficit is associated with AD risk, and in vitro and in vivo studies revealed that vitamin D therapy not only ameliorates neuroinflammation but also reduces Aβ load by increasing its breakdown and inhibiting its production.[90] Phase II clinical trials and previous trials that examined vitamin D supplements for the treatment or prevention of AD had variety results. This inconsistency may probably be attributable to differences in the types of vitamin D, variable follow-up periods, different endpoint measurements, and fluctuating target populations.[90]

2.16 ABL1 inhibitor

2.16.1 Nilotinib and bosutinib

Tyrosine kinase inhibitors nilotinib and bosutinib inhibit ABL1 (ABL proto-oncogene 1, nonreceptor tyrosine kinase), and degraded phosphorylated tau via autophagy and restored neurotransmitter imbalance in P301L mice.[91]

2.17 Others

2.17.1 GTM-1

GTM-1, a derivative of quinolone, was demonstrated to improve memory disturbances and Aβ pathology in mouse models of AD by promoting autophagy in an mTORC-independent manner.[65,92] In particular, spatial memory of 3×Tg AD mice was improved by GTM-1.[65]

2.17.2 Bexarotene

The FDA-approved drug for cutaneous T cell lymphoma bexarotene alters autophagy markers and reduces autophagy flux in neurons expressing P301L-tau. Bexarotene also improved mitochondrial dysfunction in P301L-Tau neurons.[93]

2.17.3 Lonafarnib

Lonafarnib, a farnesyl transferase inhibitor, reduced tau inclusion, behavioral abnormalities, brain atrophy, and microgliosis in rTg4510 tauopathy mice.[94] This phenomenon accompanied the reduction of the small G protein Rhes. Lysosomal activation was also noted.

2.17.4 ROCK inhibitor

Our previous study suggested that rho-ROCK (Rho-associated coiled-coil protein kinase) inhibitors, H1152, Y-27632, and fasudil significantly reduce oligomer tau and activate autophagy based on tauopathy cell culture models (M1C cells) and tauopathy mouse models (rTg4510), although the autophagy activation pathways remain unknown.[95]

2.18 Gene therapies

In drosophila models, calpain inhibition by siRNA-activated autophagy and improved tau- induced toxicity.[98] The microRNA miR-132/212 reduced tau in mice. Depletion of microRNA (miR)-132/212 increased Aβ and promoted tau deposition by suppressing autophagy.[96,97] Conversely, the miR-132 mimics reduced tau in mice.[96]

3. Caspase activation and autophagy

Autophagy is upregulated in the central nervous system[117] to maintain homeostasis by removing abnormal proteins and organelles, inhibiting the accumulation of aggregated proteins, supporting energy demand, and maintaining neuroplasticity. There is evidence that autophagy has neuroprotective effects and is particularly important in neurons, which are post mitotic cells.

Autophagy has a strong relationship with programmed cell death (apoptosis), initiated primarily by mitochondrial membrane permeabilization (MMP). If only a few mitochondria exhibited MMP, the de-polarized mitochondria are selectively removed by autophagy, leading to mitophagy. However, if the mitophagy threshold is exceeded, MMP can be fatal. When the apoptosis regulatory protein Bcl-2 is released from the activated autophagy protein complex, these molecules are thought to block the apoptosis pathway. Conversely, induction of apoptosis is thought to be associated with autophagy inactivation. For example, caspase-3 inhibits autophagy activity by cleaving Beclin1, and the C-terminal fragment of Beclin1 leads to the amplification of apoptosis through mitochondria.

Autophagy is widely believed to a survival-promoting mechanism, but autophagy and apoptosis are interdependent. Of note, apoptosis and autophagy share common regulators, such as Beclin1, Bcl-2, p53, and Atg5 which can interact and promote neuronal death. Inhibition of autophagy promotes neuronal apoptosis.[118] Autophagy has been linked to cell death promotion as a consequence of excessive activation and also during cell removal and excitotoxicity of neuron. Therefore, several studies reported that autophagy inhibition leads to increased neuronal survival in the case of hypoxic/ischemic brain injury in mice. The function of autophagy in cell death and its detailed mechanism remains unknown, and it is not clear whether autophagy-induced cell death is associated with apoptosis or a separate process.[119]

4. Conclusion

As mentioned earlier, autophagy dysfunction that perturbs the effective clearance of misfolded proteins and cytosolic oligomers has been found in AD and many other neurodegenerative disorders, including Parkinson's disease, amyotrophic lateral sclerosis, and Huntington's disease. Therefore, the regulation of autophagy may be effective for the prevention and therapy of AD by preventing the aggregation of Aβ and tau. However, crosstalk between autophagy and caspase has been reported.[118] Caspases are capable of increasing autophagy under certain conditions.[120] Autophagy is reported to be a "double-edged sword". It protects cells from apoptosis and necrosis by breaking down toxic substances, but abnormal activation of autophagy can lead to autophagic stress, a relatively persistent imbalance in which the rate of autophagosome vacuole formation exceeds the rate of autophagosome vacuole degradation.[121] Previous studies demonstrated that autophagy has a dual function in neurodegenerative diseases. In the early stages of AD, autophagy is active, promoting the clearance of abnormally folded proteins and inhibiting further progression of AD.[121] On the other hand, in the progressed stages of AD, autophagy is in a state of autophagic stress, where the clearance of autophagosomes cannot keep up with the formation of autophagosomes. Therefore, proper activation of autophagy is essential for the treatment of AD.[119]

References

1. Irvine GB, El-Agnaf OM, Shankar GM, Walsh DM. Protein aggregation in the brain: the molecular basis for Alzheimer's and Parkinson's diseases. *Mol Med.* 2008;14:451–464.
2. Bloom GS. Amyloid-beta and tau: the trigger and bullet in Alzheimer disease pathogenesis. *JAMA Neurol.* 2014;71:505–508.
3. Morishima-Kawashima M, Hasegawa M, Takio K, et al. Proline-directed and non-proline-directed phosphorylation of PHF-tau. *J Biol Chem.* 1995;270(2):823–829.
4. Hamano T, Enomoto S, Shirafuji N, et al. Autophagy and tau protein. *Int J Mol Sci.* 2021;22(14):7475.
5. Nowacek A, Kosloski LM, Gendelman HE. Neurodegenerative disorders and nano-formulated drug development. *Nanomedicine (Lond).* 2009;4:541–555.
6. Metcalf DJ, Garcia-Arencibia M, Hochfeld WE, Rubinsztein DC. Autophagy and misfolded proteins in neurodegeneration. *Exp Neurol.* 2012;238:22–28.
7. Hamano T, Gendron TF, Causevic E, et al. Autophagic-lysosomal perturbation enhances tau aggregation in transfectants with induced wild-type tau expression. *Eur J Neurosci.* 2008;27(5):1119–1130.
8. Spilman P, Podlutskaya N, Hart MJ, et al. Inhibition of mTOR by rapamycin abolishes cognitive deficits and reduces amyloid-beta levels in a mouse model of Alzheimer's disease. *PLoS ONE.* 2010;5, e9979.
9. Caccamo A, Majumder S, Richardson A, et al. Molecular interplay between mammalian target of rapamycin (mTOR),amyloid-beta, and Tau: effects on cognitive impairments. *J Biol Chem.* 2010;285:13107–13120.
10. Ozcelik S, Fraser G, Castets P, et al. Rapamycin attenuates the progression of tau pathology in P301S tau transgenic mice. *PLoS ONE.* 2013;8, e62459.
11. Lopez A, Lee SE, Wojta K, et al. A152T tau allele causes neurodegeneration that can be ameliorated in a zebrafish model by autophagy induction. *Brain.* 2017;140(4):1128–1146.
12. Xue Z, Zhang S, Huang L, et al. Upexpression of Beclin-1-dependent autophagy protects against beta-amyloid-induced cell injury in PC 12 cells. *J Mol Neurosci.* 2013;51:180–186.
13. Lee J, Yanckello LM, Ma D, et al. Neuroimaging biomarkers of mTOR inhibition on vascular and metabolic functions in aging brain and Alzheimer's disease. *Front Aging Neurosci.* 2018;10:225.
14. Singh M, Jensen MD, Lerman A, et al. Effect of low-dose rapamycin on senescence markers and physical functioning in older adults with coronary artery disease: results of a pilot study. *J Frailty Aging.* 2016;5:204–207.

15. Jiang T, Yu JT, Zhu XC, et al. Temsirolimus promotes autophagic clearance of amyloid-β and provides protective effects in cellular and animal models of Alzheimer's disease. *Pharmacol Res.* 2014;81:54–63.
16. Jiang T, Yu JT, Zhu XC, et al. Temsirolimus attenuates tauopathy in vitro and in vivo by targeting tau hyperphosphorylation and autophagic clearance. *Neuropharmacology.* 2014;85:121–130.
17. Frederick C, Ando K, Leroy K, et al. Rapamycin ester analog CCI-779/Temsirolimus alleviates tau pathology and improves motor deficit in mutant tau transgenic mice. *J Alzheimers Dis.* 2015;44:1145–1156.
18. Cassano T, Magini A, Giovagnoli S, et al. Early intrathecal infusion of everolimus restores cognitive function and mood in a murine model of Alzheimer's disease. *Exp Neurol.* 2019;311:88–105.
19. Silva MC, Nandi GA, Tentarelli S, et al. Prolonged tau clearance and stress vulnerability rescue by pharmacological activation of autophagy in tauopathy neurons. *Nat Commun.* 2020;11(1):3258.
20. Salama M, Elhussiny M, Magdy A, et al. Dual mTORC1/mTORC2 blocker as a possible therapy for tauopathy in cellular model. *Metab Brain Dis.* 2018;33(2):583–587.
21. Congdon EE, Wu JW, Myeku N, et al. Methylthioninium chloride (methylene blue) induces autophagy and attenuates tauopathy in vitro and in vivo. *Autophagy.* 2012;8(4):609–622.
22. Gauthier S, Feldman HH, Schneider LS, et al. Efficacy and safety of tau-aggregation inhibitor therapy in patients with mild or moderate Alzheimer's disease: a randomised, controlled, double-blind, parallel-arm, phase 3 trial. *Lancet.* 2016;388(10062):2873–2884.
23. Wilcock GK, Gauthier S, Frisoni GB, et al. Potential of low dose leuco-methylthioninium bis (hydromethanesulphonate) (LMTM) monotherapy for treatment of mild Alzheimer's disease: cohort analysis as modified primary outcome in a phase III clinical trial. *J Alzheimers Dis.* 2018;61(1):435–457.
24. Ma QL, Zuo X, Yang F, et al. Curcumin suppresses soluble tau dimers and corrects molecular chaperone, synaptic, and behavioral deficit in aged human tau transgenic mice. *J Biol Chem.* 2013;288:40564065.
25. Wang C, Zhang X, Teng Z, et al. Downregulation of PI3K/Akt/mTOR signaling pathway in curcumin-induced autophagy in APP/PS1 double transgenic mice. *Eur J Pharmacol.* 2014;740:312–320.
26. Steele JW, Gandy S. Latrepirdine (Dimebon(R)), a potential Alzheimer therapeutic, regulates autophagy and neuropathology in an Alzheimer mouse model. *Autophagy.* 2013;9:617–618.
27. Chau S, Herrmann N, Ruthirakuhan MT, et al. Latrepirdine for Alzheimer's disease. *Cochrane Database Syst Rev.* 2015;4:CD009524.
28. Liu SJ, Yang C, Zhang Y, et al. Neuroprotective effect of β-asarone against Alzheimer's disease: regulation of synaptic plasticity by increased expression of SYP and GluR1. *Drug Des Devel Ther.* 2016;10:1461–1469.
29. Deng M, Huang L, Ning B, et al. β-asarone improves learning and memory and reduces acetyl cholinesterase and beta-amyloid 42 levels in APP/PS1 transgenic mice by regulating Beclin-1-dependent autophagy. *Brain Res.* 2016;1652:188–194.
30. Grossi C, Rigacci S, Ambrosini S, et al. The polyphenol oleuropein aglycone protects TgCRND8 mice against Aß plaque pathology. *PLoS ONE.* 2013;8(8), e71702.
31. Luccarini I, Grossi C, Rigacci S, et al. Oleuropein aglycone protects against pyroglutamylated-3 amyloid-ss toxicity: biochemical, epigenetic and functional correlates. *Neurobiol Aging.* 2015;36:648–663.
32. Martorell M, Forman K, Castro N, et al. Potential therapeutic effects of oleuropein aglycone in Alzheimer's disease. *Curr Pharm Biotechnol.* 2016;17:994–1001.

33. Calon F. Omega-3 polyunsaturated fatty acids in Alzheimer's disease: key questions and partial answers. *Curr Alzheimer Res.* 2011;8(5):470–478.
34. Lim GP, Calon F, Morihara T, et al. A diet enriched with the omega-3 fatty acid docosahexaenoic acid reduces amyloid burden in an aged Alzheimer mouse model. *J Neurosci.* 2005;25(12):3032–3040.
35. Milanovic D, Petrovic S, Brkic M, et al. Short-term fish oil treatment changes the composition of phospholipids while not affecting the expression of Mfsd2a omega-3 transporter in the brain and liver of the 5xFAD mouse model of Alzheimer's disease. *Nutrients.* 2018;10(9):1250.
36. Chen JL, Luo C, Pu D, et al. Metformin attenuates diabetes-induced tau hyperphosphorylation in vitro and in vivo by enhancing autophagic clearance. *Exp Neurol.* 2019;311:44–56.
37. Kickstein E, Krauss S, Thornhill P, et al. Biguanide metformin acts on tau phosphorylation via mTOR/protein phosphatase 2A (PP2A) signaling. *Proc Natl Acad Sci U S A.* 2010;107(50):21830–21835.
38. Li J, Deng J, Sheng W, Zuo Z. Metformin attenuates Alzheimer's disease-like neuropathology in obese, leptin-resistant mice. *Pharmacol Biochem Behav.* 2012;101:564–574.
39. Pierzynowska K, Gaffke L, Cyske Z, et al. Autophagy stimulation as a promising approach in treatment of neurodegenerative diseases. *Metab Brain Dis.* 2018;33:989–1008.
40. Kruger U, Wang Y, Kumar S, Mandelkow EM. Autophagic degradation of tau in primary neurons and its enhancement by trehalose. *Neurobiol Aging.* 2012;33:2291–2305.
41. Schaeffer V, Lavenir I, Ozcelik S, et al. Stimulation of autophagy reduces neurodegeneration in a mouse model of human tauopathy. *Brain.* 2012;135:2169–2177.
42. Liu R, Barkhordarian H, Emadi S, et al. Trehalose differentially inhibits aggregation and neurotoxicity of beta-amyloid 40 and 42. *Neurobiol Dis.* 2005;20:74–81.
43. Du J, Liang Y, Xu F, et al. Trehalose rescues Alzheimer's disease phenotypes in APP/PS1 transgenic mice. *J Pharm Pharmacol.* 2013;65:1753–1756.
44. Vingtdeux V, Giliberto L, Zhao H, et al. AMP-activated protein kinase signaling activation by resveratrol modulates amyloid-beta peptide metabolism. *J Biol Chem.* 2010;285:9100–9113.
45. Gomes BAQ, Silva JPB, Romeiro CFR, et al. Neuroprotective mechanisms of resveratrol in Alzheimer's disease: role of SIRT1. *Oxid Med Cell Longev.* 2018;2018:8152373.
46. Morselli E, Galluzzi L, Kepp O, et al. Autophagy mediates pharmacological lifespan extension by spermidine and resveratrol. *Aging (Albany NY).* 2009;1(12):961–970.
47. Turner RS, Thomas RG, Craft S, et al. A randomized, double-blind, placebo-controlled trial of resveratrol for Alzheimer disease. *Neurology.* 2015;85:1383–1391.
48. Porquet D, Grinan-Ferre C, Ferrer I, et al. Neuroprotective role of trans-resveratrol in a murine model of familial Alzheimer's disease. *J Alzheimers Dis.* 2014;42:1209–1220.
49. Moussa C, Hebron M, Huang X, et al. Resveratrol regulates neuro-inflammation and induces adaptive immunity in Alzheimer's disease. *J Neuroinflammation.* 2017;14(1):1.
50. Ayasolla KR, Singh AK, Singh I. 5-aminoimidazole-4-carboxamide-1-beta-4-ribofuranoside (AICAR) attenuates the expression of LPS- and Abeta peptide-induced inflammatory mediators in astroglia. *J Neuroinflammation.* 2005;2:21.
51. Du LL, Chai DM, Zhao LN, et al. AMPK activation ameliorates Alzheimer's disease-like pathology and spatial memory impairment in a streptozotocin-induced Alzheimer's disease model in rats. *J Alzheimers Dis.* 2015;43(3):775–784.
52. Zhu Z, Yan J, Jiang W, et al. Arctigenin effectively ameliorates memory impairment in Alzheimer's disease model mice targeting both beta-amyloid production and clearance. *J Neurosci.* 2013;33:13138–13149.
53. Zhang J, Yang Z, Xie L, et al. Statins, autophagy and cancer metastasis. *Int J Biochem Cell Biol.* 2013;45(3):745–752.

54. Li HH, Lin CL, Huang CN. Neuroprotective effects of statins against amyloid beta-induced neurotoxicity. *Neural Regen Res.* 2018;13(2):198–206.
55. Ostrowski SM, Wilkinson BL, Golde TE, Landreth G. Statins reduce amyloid-beta production through inhibition of protein isoprenylation. *J Biol Chem.* 2007;282(37):26832–26844.
56. Hamano T, Yen SH, Gendron T, et al. Pitavastatin decreases tau levels via the inactivation of Rho/ROCK. *Neurobiol Aging.* 2012;33(10):2306–2320.
57. Kajinami K, Tsukamoto K, Koba S, et al. Statin Intolerance Clinical Guide 2018. *J Atheroscler Thromb.* 2020;27(4):375–396.
58. Shimada K, Motoi Y, Ishiguro K, et al. Long-term oral lithium treatment attenuates motor disturbance in tauopathy model mice: implications of autophagy promotion. *Neurobiol Dis.* 2012;46:101–108.
59. Sarkar S, Floto RA, Berger Z, et al. Lithium induces autophagy by inhibiting inositol monophosphatase. *J Cell Biol.* 2005;170:1101–1111.
60. Forlenza OV, de Paula VJ, Machado-Vieira R, et al. Does lithium prevent Alzheimer's disease? *Drugs Aging.* 2012;29(5):335–342.
61. Motoi Y, Shimada K, Ishiguro K, Hattori N. Lithium and autophagy. *ACS Chem Nerosci.* 2014;5(6):434–442.
62. Forlenza OV, De-Paula VJ, Diniz BS. Neuroprotective effects of lithium: implications for the treatment of Alzheimer's disease and related neurodegenerative disorders. *ACS Chem Nerosci.* 2014;5(6):443–450.
63. Matsunaga S, Kishi T, Annas P, et al. Lithium as a treatment for Alzheimer's disease: a systematic review and meta-analysis. *J Alzheimers Dis.* 2015;48:403–410.
64. Li L, Zhang S, Zhang X, et al. Autophagy enhancer carbamazepine alleviates memory deficits and cerebral amyloid-β pathology in a mouse model of Alzheimer's disease. *Curr Alzheimer Res.* 2013;10(4):433–441.
65. Zhang L, Wang L, Wang R, et al. Evaluating the effectiveness of GTM-1, rapamycin, and carbamazepine on autophagy and Alzheimer disease. *Med Sci Monit.* 2017;23:801–808.
66. Chauhan S, Ahmed Z, Bradfute SB, et al. Pharmaceutical screen identifies novel target processes for activation of autophagy with a broad translational potential. *Nat Commun.* 2015;6:8620.
67. Currais A, Prior M, Dargusch R, et al. Modulation of p25 and inflammatory pathways by fisetin maintains cognitive function in Alzheimer's disease transgenic mice. *Aging Cell.* 2014;13:379–390.
68. Kim S, Choi KJ, Cho SJ, et al. Fisetin stimulates autophagic degradation of phosphorylated tau via the activation of TFEB and Nrf2 transcription factors. *Sci Rep.* 2016;6:24933.
69. Song JX, Malampati S, Zeng Y, et al. A small molecule transcription factor EB activator ameliorates beta-amyloid precursor protein and tau pathology in Alzheimer's disease models. *Aging Cell.* 2020;19(2), e13069.
70. Liu D, Pitta M, Jiang H, et al. Nicotinamide forestalls pathology and cognitive decline in Alzheimer mice: evidence for improved neuronal bioenergetics and autophagy procession. *Neurobiol Aging.* 2013;34:1564–1580.
71. Gong B, Pan Y, Vempati P, et al. Nicotinamide riboside restores cognition through an upregulation of proliferator-activated receptor-γ coactivator 1α regulated β-secretase 1 degradation and mitochondrial gene expression in Alzheimer's mouse models. *Neurobiol Aging.* 2013;34(6):1581–1588.
72. Phelan MJ, Mulnard RA, Gillen DL, Schreiber SS. Phase II clinical trial of nicotinamide for the treatment of mild to moderate Alzheimer's disease. *J Geriatr Med Gerontol.* 2017;3:021.

73. Grill J. *Nicotinamide as an Early Alzheimer's Disease Treatment (NEAT)*; 2017. https://clinicaltrials.gov/ct2/show/NCT03061474.
74. Anekonda TS, Quinn JF, Harris C, et al. L-type voltage-gated calcium channel blockade with isradipine as a therapeutic strategy for Alzheimer's disease. *Neurobiol Dis.* 2011;41:62–70.
75. Matsunaga S, Kishi T, Iwata N. Memantine monotherapy for Alzheimer's disease: a systematic review and meta-analysis. *PLoS ONE.* 2015;10, e0123289.
76. Song G, Li Y, Lin L, Cao Y. Anti-autophagic and anti-apoptotic effects of memantine in a SH-SY5Y cell model of Alzheimer's disease via mammalian target of rapamycin-dependent and -independent pathways. *Mol Med Rep.* 2015;12:7615–7622.
77. Hirano K, Fujimaki M, Sasazawa Y, et al. Neuroprotective effects of memantine via enhancement of autophagy. *Biochem Biophys Res Commun.* 2019;518(1):161–170.
78. Tian Y, Bustos V, Flajolet M, Greengard P. A small-molecule enhancer of autophagy decreases levels of Abeta and APP-CTF via Atg5-dependent autophagy pathway. *FASEB J.* 2011;25:1934–1942.
79. Madeo F, Bauer MA, Carmona-Gutierrez D, Kroemer G. Spermidine: a physiological autophagy inducer acting as an anti-aging vitamin in humans? *Autophagy.* 2019;15(1):165–168.
80. Morselli E, Mariño G, Bennetzen MV, et al. Spermidine and resveratrol induce autophagy by distinct pathways converging on the acetylproteome. *J Cell Biol.* 2011;192(4):615–629.
81. Yang Y, Chen S, Zhang Y, et al. Induction of autophagy by spermidine is neuroprotective via inhibition of caspase 3-mediated Beclin 1 cleavage. *Cell Death Dis.* 2017;4, e2738.
82. Inestrosa NC, Tapia-Rojas C, Griffith TN, et al. Tetrahydrohyperforin prevents cognitive deficit, Aβ deposition, tau phosphorylation and synaptotoxicity in the APPswe/PSEN1ΔE9 model of Alzheimer's disease: a possible effect on APP processing. *Transl Psychiatry.* 2011;1(7), e20.
83. Cavieres VA, González A, Muñoz VC, et al. Tetrahydrohyperforin inhibits the proteolytic processing of amyloid precursor protein and enhances its degradation by Atg5-dependent autophagy. *PLoS One.* 2015;10(8), e0136313.
84. Umeda T, Ono K, Sakai A, et al. Rifampicin is a candidate preventive medicine against amyloid-and tau oligomers. *Brain.* 2016;139:1568–1586.
85. Liang Y, Zheng D, Peng S, et al. Rifampicin attenuates rotenone-treated microglia inflammation via improving lysosomal function. *Toxicol In Vitro.* 2020;63, 104690.
86. Chen X, Li Y, Wang C, et al. Promoting tau secretion and propagation by hyperactive p300/CBP via autophagy-lysosomal pathway in tauopathy. *Mol Neurodegener.* 2020;15:2.
87. Lee HR, Shin HK, Park SY, et al. Cilostazol upregulates autophagy via SIRT1 activation: reducing amyloid-β peptide and APP-CTFβ levels in neuronal cells. *PLoS One.* 2015;10(8), e0134486.
88. Lee HR, Shin HK, Park SY, et al. Attenuation of β-amyloid-induced tauopathy via activation of CK2α/SIRT1: targeting for cilostazol. *J Neurosci Res.* 2014;92(2):206–217.
89. Levine B, Packer M, Codogno P. Development of autophagy inducers in clinical medicine. *J Clin Invest.* 2015;125(1):14–24.
90. Banerjee A, Khemka VK, Ganguly A, et al. Vitamin D and Alzheimer's disease: neurocognition to therapeutics. *Int J Alzheimers Dis.* 2015;2015, 192747.
91. Hebron ML, Javidnia M, Moussa CE. Tau clearance improves astrocytic function and brain glutamate-glutamine cycle. *J Neurol Sci.* 2018;391:90–99.
92. Chu C, Zhang X, Ma W, et al. Induction of autophagy by a novel small molecule improves aβ pathology and ameliorates cognitive deficits. *PLoS ONE.* 2013;8(6), e65367.

93. Huuskonen MT, Loppi S, Dhungana H, et al. Bexarotene targets autophagy and is protective against thromboembolic stroke in aged mice with tauopathy. *Sci Rep.* 2016;6:33176.
94. Hernandez I, Luna G, Rauch JN, et al. A farnesyltransferase inhibitor activates lysosomes and reduces tau pathology in mice with tauopathy. *Sci Transl Med.* 2019;11:eaat3005.
95. Hamano T, Shirafuji N, Yen SH, et al. Rho-kinase ROCK inhibitors reduce oligomeric tau protein. *Neurobiol Aging.* 2020;89:41–54.
96. Smith PY, Hernandez-Rapp J, Jolivette F, et al. miR-132/212 deficiency impairs tau metabolism and promotes pathological aggregation in vivo. *Hum Mol Genet.* 2015;24:6721–6735.
97. Hernandez-Rapp J, Rainone S, Goupil C, et al. microRNA-132/212 deficiency enhances A production and senile plaque deposition in Alzheimer's disease triple transgenic mice. *Sci Rep.* 2016;6:30953.
98. Menzies F, Garcia-Arencibia M, Imarisio S, et al. Calpain inhibition mediates autophagy-dependent protection against polyglutamine toxicity. *Cell Death Differ.* 2015;22:433–444.
99. Binder JL, Chander P, Deretic, et al. Optical induction of autophagy via transcription factor EB (TFEB) reduces pathological tau in neurons. *PLoS ONE.* 2020;15, e0230026.
100. Polito VA, Li H, Martini-Stoica H, et al. Selective clearance of aberrant tau proteins and rescue of neurotoxicity by transcription factor EB. *EMBO Mol Med.* 2014;6:1142–1160.
101. Wang H, Wang R, Carrera I, et al. TFEB overexpression in the P301S model of tauopathy mitigates increased PHF1 levels and lipofuscin puncta and rescues memory deficits. *eNeuro.* 2016;3:9340–9351.
102. Uddin MS, Stachowiak A, Mamun AA, et al. Autophagy and Alzheimer's disease: from molecular mechanisms to therapeutic implications. *Front Aging Neurosci.* 2018;10:04. https://doi.org/10.3389/fnagi.2018.00004.
103. Jiang S, Bhaskar K. Degradation and transmission of tau by autophagic-endolysosomal networks and potential therapeutic targets for tauopathy. *Front Mol Neurosci.* 2020;13, 586731.
104. Schmukler E, Pinkas-Kramarski R. Autophagy induction in the treatment of Alzheimer's disease. *Drug Dev Res.* 2020;81(2):184–193.
105. Jung CH, Ro SH, Cao J, et al. mTOR regulation of autophagy. *FEBS Lett.* 2010;584:1287–1295.
106. Harrison DE, Strong R, Sharp ZD, et al. Rapamycin fed late in life extends lifespan in genetically heterogeneous mice. *Nature.* 2009;460:392–395.
107. Guerreiro R, Bras J. The age factor in Alzheimer's disease. *Genome Med.* 2015;7:106.
108. Sun YX, Ji X, Mao X, et al. Differential activation of mTOR complex 1 signaling in human brain with mild to severe Alzheimer's disease. *J Alzheimers Dis.* 2014;38:437–444.
109. Iyer AM, van Scheppingen J, Milenkovic I, et al. mTOR Hyperactivation in down syndrome hippocampus appears early during development. *J Neuropathol Exp Neurol.* 2014;73:671–683.
110. Tang Z, Ioja E, Bereczki E, et al. mTor mediates tau localization and secretion: implication for Alzheimer's disease. *Biochim Biophys Acta.* 2015;1853(7):1646–1657.
111. Domise M, Didier S, Marinangeli C, et al. AMP-activated protein kinase modulates tau phosphorylation and tau pathology in vivo. *Sci Rep.* 2016;6:26758.
112. Okuma T, Kishimoto A. A history of investigation on the mood stabilizing effect of carbamazepine in Japan. *Psychiatry Clin Neurosci.* 1998;52(1):3–12.
113. Settembre C, Di Malta C, Polito VA, et al. TFEB links autophagy to lysosomal biogenesis. *Science.* 2011;332:1429–1433.
114. Martini-Stoica H, Xu Y, Ballabio A, et al. The autophagy-lysosomal pathway in neurodegeneration: a TFEB perspective. *Trends Neurosci.* 2016;39(4):221–234.

115. Cifuentes D, Poittevin M, Bonnin P, et al. Inactivation of nitric oxide synthesis exacerbates the development of Alzheimer disease pathology in APPPS1 mice (amyloid precursor protein/presenilin-1). *Hypertension*. 2017;70(3):613–623.
116. Cerpa W, Hancke JL, Morazzoni P, et al. The hyperforin derivative IDN5706 occludes spatial memory impairments and neuropathological changes in a double transgenic Alzheimer's mouse model. *Curr Alzheimer Res.* 2010;7(2):126–133.
117. Boland B, Kumar A, Lee S, et al. Autophagy induction and autophagosome clearance in neurons: relationship to autophagic pathology in Alzheimer's disease. *J Neurosci*. 2008;28(27):6926–6937.
118. Tsapras P, Nezis IP. Caspase involvement in autophagy. *Cell Death Differ.* 2017;24(8):1369–1379.
119. Krishnan S, Shrestha Y, Jayatunga DPW, et al. Activate or inhibit? Implications of autophagy modulation as a therapeutic strategy for Alzheimer's disease. *Int J Mol Sci*. 2020;21(18):6739.
120. Han J, Hou W, Goldstein LA, et al. A complex between Atg7 and Caspase-9: a novel mechanism of cross-regulation between autophagy and apoptosis. *J Biol Chem*. 2014;289(10):6485–6497.
121. Tung YT, Wang BJ, Hu MK, et al. Autophagy: a double-edged sword in Alzheimer's disease. *J Biosci*. 2012;37(1):157–165.

CHAPTER 15

Drug discovery in Alzheimer's disease using metal chelators: Warning toward their usage

Tatsuro Mutoh
Department of Neurology and Neuroscience, Fujita Health University Hospital, Toyoake-City, Aichi, Japan
Department of Internal Medicine Fujita Health University Chubu International Airport Clinic, Tokoname-City, Aichi, Japan

1. Introduction

Alzheimer's disease (AD) is a neurodegenerative disorder affecting aged populations leading to cognitive impairments and dementia. Although AD is postulated as a multifactorial disorder such as genetic, environmental, and life styles, its real cause remains to be fully identified.[1] Neuropathologically, senile plaque and neurofibrillary tangles (NFTs) have been known as hallmarks for this disorder. The production of senile plaque is now known to be derived from abnormal amyloid peptides, amyloid β (Aβ), which is produced by the action of β-secretase and γ-secretase, whereas α-secretase is responsible for the formation of normal proteolytic fragment of amyloid precursor protein (APP) in normal subjects.[2] On the other hand, NFTs are produced by the aggregation of hyperphosphorylated tau protein due to abnormal regulation of protein kinase activities such as glycogen synthase kinase-3β (GSK-3β), cyclin-dependent kinase 5 (CDK5), and extracellular signal-regulated kinase 1/2 (ERK1/2) in AD brains.[3,4] The hyperphosphorylation of the tau protein can elicit destabilization of microtubule and its breakdown, therefore perturb axonal transport, and make tau far more prone to the aggregation into NFTs.[5–7]

Neuroinflammation has been recently highlighted as another strong risk factor for AD. The genetic (e.g., *trem2*, *cd33*, *apoe*), and systemic factors (e.g., diabetes mellitus, obesity, and traumatic brain injuries) can cause neuroinflammation in the brain of AD patients.[8] Accumulated Aβ triggers a proinflammatory response from the brain's resident immune cells such as microglia and astrocytes, leading to the phagocytosis of plaques as well

Autophagy Dysfunction in Alzheimer's Disease and Dementia
https://doi.org/10.1016/B978-0-323-89906-2.00005-8

as their proteolytic degradation. It also stimulates proinflammatory cytokine productions from these immune cells in the brain. Receptors on microglia, such as triggering receptor expressed on myeloid cells 2 (TREM2) and toll-like receptors (TLRs), can recognize Aβ and therefore trigger a phagocytic pathway activation, leading to the degradation of extracellular Aβ with cooperation of the actions of neprilysin and insulin degrading enzyme, although these Aβ degrading enzyme activities are downregulated in AD brains.[9–12]

Another important aspect of the pathogenesis of AD is an abnormal heavy metal homeostasis in patients with neurodegenerative disorders including AD and Parkinson's disease (PD).

2. Abnormal metals in AD patients

Previous studies have highlighted the significance of abnormal metal homeostasis in the brain of AD patients,[13] although its pathogenesis remains to be fully elucidated. A dysfunction of metal trafficking in the brain, where metals are redistributed into inappropriate compartments.[14] This hypothesis, originating from early clioquinol-related findings, is based upon a triad of transition metals: Fe, Cu, and Zn.[15–17] Moreover, not only clioquinol (5-chloro-7-iodo-8-quinolinol) but also its derivative PBT2 (5,7-dichloro-2-((dimethylamino)methyl) 8-quinolinol) has been known to exhibit potential as a therapeutic medicine for AD due to their metal protein-attenuating functions (Fig. 1). Actually, another derivative, PBT434, an orally bioavailable 8-hydroxyquinazolin-4(3*H*)-one, binds iron sufficiently to abolish pathological reaction with α-synuclein, but with an affinity that is designed not to disrupt physiological iron homeostasis is expected to become an orphan drug for the treatment of PD.[18] Researchers also found a significant decrease in serum manganese level in AD subjects vs. healthy individuals but concluded that neither lead nor manganese levels

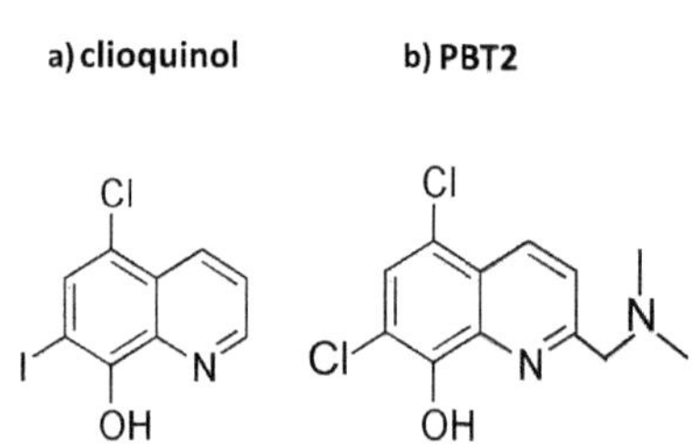

Fig. 1 Chemical structure of clioquinol and PBT2. (A) Clioquinol. (B) PBT2.

represented biomarkers in the studied cohort,[19] although manganese has been linked with the important regulators of brain immunity, and it can have profound effects on microglial and astrocytic functions, regulating the proinflammatory responses yielding neurotoxic effects on neurons.[20,21] In addition, Zn is an essential trace mineral and second in abundance in mammalian tissues. It plays a critical role for immunity, growth, and development. It is also a cofactor for more than 300 enzymes and is essential for the correct functioning of over 2000 transcription factors.[22] In pathological aging and neurodegenerative disorders such as AD wherein the abnormal deposition of misfolded Aβ peptide into plaques, which bind Zn, it results in a significant decrease in intracellular Zn.[22] Moreover, Zn dyshomeostasis has been linked to a range of neurological abnormalities including (but not limited to) depression and schizophrenia, amyotrophic lateral sclerosis (ALS), Down's syndrome, and multiple sclerosis.[22] Concerning Cu, it has been also suggested that individuals suffering from AD have higher serum levels of copper ions than healthy subjects, with higher levels of labile copper in the serum correlating with poor cognitive performance and increased rates of conversion from mild cognitive impairment to AD.[23–25] Although total copper content is lower than control brains, AD brains exhibit a higher proportion of redox-active exchangeable copper ions correlating with increased oxidative stress and AD pathology.[26] Moreover, exposure of copper-Aβ complex inhibits phagocytic activities of microglia and increases the release of proinflammatory cytokines such as TNF-α and IL-1β. These sequences caused the impairment of phagocytosis of accumulated Aβ and exacerbated neuroinflammation.[27]

In addition to these, iron also can induce severe brain damage through the production of reactive oxygen species (ROS). Change in iron levels in AD patients is known to be detected as early as the mild cognitive impairment stage of the disease.[28] Intriguingly, APP is able to act as a modulator of iron functions. It can oxidize Fe^{2+} and load Fe^{3+} into transferrin and interacts with the iron transporter ferroportin.[29] In normal conditions, iron and its regulatory proteins such as ferritin and transferrin are mainly present in oligodendrocytes. On the contrary, substantial amounts of iron and ferritin are detected within senile plaques, NFTs, and blood vessels in AD patients,[30,31] intriguingly in AD patients many microglial cells contain iron and ferritin.[32] Furthermore, it should be noted that the affinity for Aβ to iron is about eight orders of magnitude stronger than that of transferrin. Therefore, its accumulation in Aβ may provoke significant impairment of iron homeostasis in AD patients.[33]

Physiologically, range of these metal ions between deficiency and toxicity is relatively small and therefore needs to be tightly controlled. It has been suggested that dysregulation in the homeostasis of essential metals and exposure to nonessential metals have significant impact on the pathological process of AD.

3. Clioquinol and SMON (subacute myelo-optico-neuropathy)

3.1 SMON (subacute myelo-optico-neuropathy)

Clioquinol was originally developed in 1899 as an antiseptic and subsequently used to treat diarrhea as well as certain fungal and protozoal infections of gastrointestinal tract (GI tract). Intriguingly, clioquinol was getting recognized in some patients showing myelopathy, optic neuritis, and neuropathy between 1958 and 1970 in Japan. Therefore, Japanese physicians called these special conditions as subacute myelo-optico-neuropathy (SMON). At first, multiple hypotheses on the etiology of SMON were considered: intoxication by industrial waste or pesticides; a metabolic disorder or vitamin deficiency such as vitamin B12; and involvement of viral or bacterial infections,[34] none of which were confirmed by subsequent studies. Subsequently, many researchers and physicians noticed that good correlation of the incidence of the disorder and the amount of prescribed clioquinol medication because many patients with SMON exhibited green-colored tongue and green-pigmented urine and feces indicating iron (Fe(III)) chelate of clioquinol. Over 10,000 patients in Japan were affected by SMON before therapeutic use of clioquinol discontinued in clinical practices in Japan. After the ban of the sale of clioquinol in September 1970, there was a drastic disappearance of new cases of SMON though nearly 3000 patients still suffer from its sequelae in 2002.[35] In 1975, Sobue et al. reported the clinical details of near one thousand SMON cases.[36] According to them, approximately 80% of the patients exhibited some sort of improvements 7–12 months after the onset of the disease. The abdominal symptoms observed at the time of the onset of the disease decreased markedly in the course of the disease. Among neurological impairments, the prognosis of motor symptoms is more favorable than sensory disturbances. Approximately, 60% of the cases showed more or less favorable during the course of the disorder. Forty percent of the cases with visual disturbances completely recovered or improved, whereas 9% cases got worse. Furthermore, the patients who had been administered clioquinol over long period showed a higher rate of severe or moderated motor, sensory, and visual disturbances, compared with the group with its short-term administration. It was also observed that a high rate of relapse in the group of long-term administration of clioquinol. About 10% of the total cases were unable to walk or need assistance in walking, whereas overall 65% of them returned to their jobs in 1 year or more after the onset.

On the other hand, neuropathologic studies on SMON patients and experimental reproduction of the disease in animals produced by the administration of clioquinol have been offered the profound implication of the etiology of this disease. Common pathologic features seen both in SMON patients and in dogs and cats chronically intoxicated with clioquinol were distal dominant axonopathy with secondary demyelination, mainly in the long spinal tracts (posterior and lateral column of the spinal cord) and optic tracts. Moreover, typical abdominal symptoms present in patients after clioquinol ingestion could also be reproduced experimentally in dogs.[37] Other researchers have reported the neuropathological findings of 11 previously reported cases of SMON. All cases showed sensory impairment of the lower extremities, although exaggerated deep tendon reflex was observed in 64% of the cases (7/11). Furthermore, 82% of the patients exhibited dorsal funiculi degeneration of the spinal cord, whereas lateral funiculi alteration was only observed 2 cases out of 11 patients.[38]

3.2 In vivo concentration of clioquinol is extremely important

SMON patients were typically administered clioquinol at approximately 1.5 g/day.[39] In human subjects, the single oral intake of 1.5 g of clioquinol resulted in peak plasma concentrations of approximately 20 μg/mL (65 μM) after 4 h, whereas multiple doses (3 × 0.5 g/day, 3 days) led to a plasma concentration of about 30 μg/mL (98 μM).[40] Less than 1% of clioquinol is detectable in urine and about 95% is present glucuronate as well as 3% as sulfate conjugates. An experimental study has revealed that bile salts increased bioavailability of clioquinol by acting as supporting agents and increasing intestinal reabsorption.[41] We should keep in mind these high plasma concentrations of clioquinol in human who are taking this medicine, although it currently remains unclear whether affected neurons were directly exposed to this concentration of clioquinol. At this point, it is extremely important and potentially essential to note that clioquinol has become utilized for the potential treatment of neurodegenerative disorders including AD and Hunthington's disease (HD)[42] as mentioned earlier. Intriguingly, the pilot phase 2 trial employing clioquinol for AD revealed blood steady-state level of clioquinol was from 4.03 ± 2.10 (13 μM), 6.74 ± 3.7 (22 μM), and 7.6 ± 2.12 μg/mL (25 μM) against total dosage of clioquinol of 250, 500, and 750 mg, respectively.[43] These values were quite lower than those found in patients with SMON. Lower concentration of clioquinol (less than 10 μM) has been reported to induce autophagy in a zinc-dependent manner in a culture system and, therefore, contributes to the clearance of aggregated proteins without inducing cell death.[44] Therefore, you can appreciate that patients with SMON were exposed to such high concentration of clioquinol in periphery and possibly in the brain, because it is lipophilic compound and therefore it is postulated to cross the BBB. Unfortunately, so far, no report of in vivo concentration of clioquinol in the brains of patients with SMON has been available.

Here it should be pointed out that our recent study has revealed that 20 μM clioquinol-induced astrocytic cell death through the impairment of autophagic degradation, resulting in the decreased intracellular ATP levels, increased ROS levels, and finally cell death (Fig. 2). This cell death seemed to be independent of apoptosis because there was no change in Annexin V staining, TUNEL staining, or active caspase-3 levels in the cells.[45] Intriguingly, 20 μM clioquinol but not less than 20 μM significantly caused incomplete autophagy execution by impairing the lysosomal functions resulting in the

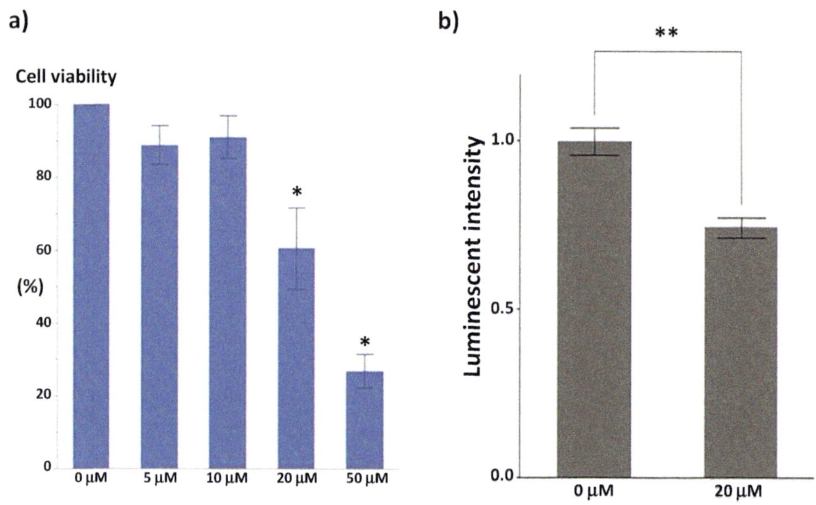

Fig. 2 Clioquinol-induced KT-5 astrocytic cell death. (A) Various concentrations of clioquinol-induced cell death of KT-5 cells were monitored with MTT assay. (B) Clioquinol caused a significant downregulation of intracellular ATP levels in KT-5 cells. Intracellular ATP levels were monitored with ATP assay kit (a Luminescent ATP Detection Assay Kit (Abcam, Cambridge, MA, USA)). *$P<0.05$, **$P<0.01$.

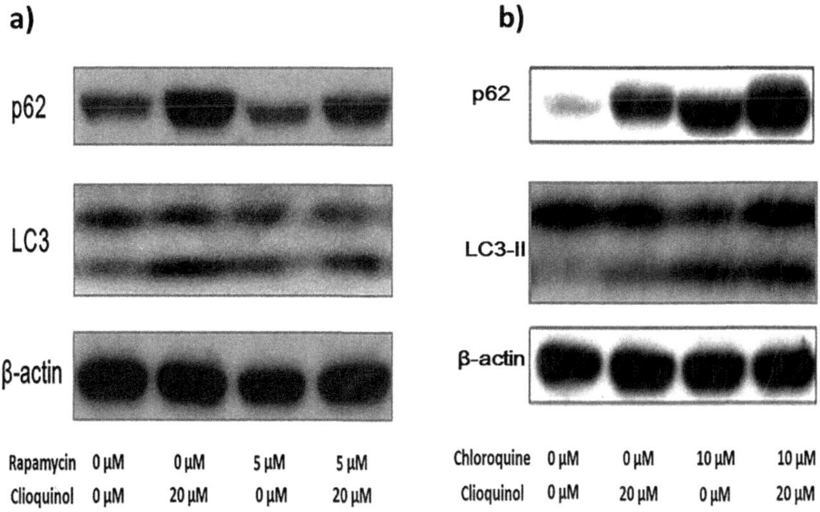

Fig. 3 The effects of rapamycin and chloroquine on clioquinol-induced changes in the expression levels of autophagy-related proteins. (A) An activator of autophagy, rapamycin, inhibited clioquinol-induced increase in LC3-II and p62 expression, and protected clioquinol-induced cell damage. This finding strongly suggests that autophagic process in KT-5 cells is a cell-protective response. (B) An inhibitor of autophagy, chloroquine, augmented clioquinol-induced increase in LC3-II and p62 expression, and augmented clioquinol-induced cell death. This finding also indicates that autophagy is a cell-protective response.

significant inhibition of lysosomal acid hydrolase activities (β-galactosidase, β-hexosaminidase, α-galactosidase, β-glucosidase, and cathepsin D). In fact, autophagosomal accumulation was observed in 20 μM clinoquinol-treated KT-5 cells.[45] Therefore, we postulated that clioquinol-induced cell death in astrocytes appears to be closely related to a failure of full execution of autophagy. Actually, LC3-II, p62, and Beclin 1 expressions were upregulated in 20 μM clinoquinol-treated KT-5 cells without a marked reduction of p62 protein even in the later stage of autophagy. Moreover, rapamycin, a stimulator for autophagy, prevented clinoquinol-induced astrocytic cell death, whereas chloroquine, an inhibitor of autophagy, accelerated clioquinol toxicity in the cells[45] (Fig. 3). Recent investigations supported the above findings because cell death closely related to autophagy might be present.[46]

3.3 Molecular action of clioquinol

3.3.1 Metal chelating activity

One of the first identified mechanism of action is related to interact with metals. Under this notion, several researchers reported the results to identify efficient metal chelators among existing antibiotics and antiinflammatory drugs.[47,48] Clioquinol was found to be a strong candidate and to dissolve Aβ deposits in postmortem human tissues.[49] Although clioquinol is a copper/zinc ionophore, it increased the intracellular level of zinc, but not copper.[50] Nonetheless, the question what is the molecular basis of clioquinol's action that is originally thought to be related to Cu/Zn or other metal chelating activity is still debated and considered important questions to be addressed fully. In addition to clioquinol, carnosine is another example of metal chelator, especially Zn-chelating activity, which is endogenous dipeptide abundantly present in the skeletal muscles and in the brain, and has numerous beneficial effects such as antioxidant, anticrosslinking, and antiglycation activities. Therefore, carnosine can be used as an endogenous metal chelator for the treatment of neurodegenerative disorders such as AD, PD, and HD.[51]

3.3.2 Its effects on autophagy

Autophagy is an important cellular response to clear aggregated proteins and accumulated lipids in the cells. Deprivation of growth promoting stimuli such as growth factors, amino acids, and cellular ATP activates autophagy via mTORC1 and AMPK signaling pathways, and these stimuli regulate biogenesis of autophagosomes, although there is another pathway independent of mTORC1. Conjugation systems promote formation of lipidated membrane-bound LC3-II on nascent autophagosomal membranes, which engulf autophagic cargoes, leading to the formation of autophagosomes. Fusion events between autophagosomes and endosomes result in the formation of amphisomes, which lead mature into autolysosomes following fusion with lysosomes. Autolysosomes that can also form by direct fusion of autophagosomes with lysosomes are digestive organelles where acid hydrolases degrade autophagosomal substrates followed by the efflux of degraded material into cytoplasm.[52] Because the accumulation of aggregated proteins in neurodegenerative disorders such as AD could be attributed to defective clearance of misfolded, aggregated proteins; therefore, some researchers postulated that clioquinol might have effects on autophagy responses. In fact, it was found that clioquinol induces autophagy or that it can reverse the arrested autophagy.[44,45] Clioquinol has been reported to upregulate phosphatidylethanolamine conjugation of protein LC3-I (called as LC3-II) in the cells.[53] This effect seems to be derived from the ionophore properties of clioquinol rather than its zinc chelating actions because another zinc chelator was shown to reduce LC3-II accumulation. Intriguingly, the addition of zinc increased LC3-II accumulation, suggesting the role of zinc in the induction of autophagy.[44]

3.3.3 Its effect on Trk-neurotrophin signaling

Neurons are nondividing cells in vivo, and nerve growth factor (NGF) is one of the cardinal growth factors to support neuronal survival and differentiation.[54] Trk is a high-affinity NGF receptor that exhibits tyrosine kinase activity, and its activation represents the initial step in the intracellular signal transduction pathway of neuronal differentiation and survival.[55] Following ligand binding, the Trk receptor is activated by homodimerization. The degree of tyrosine autophosphorylation correlates well with the biological effects in responsive cells. Our previous studies of the possible neurotoxicity by clioquinol on pheochromocytoma cells; PC12 cells have revealed that clioquinol inhibited NGF-induced Trk autophosphorylation in a dose-dependent manner. In fact, NGF treatment of PC12 cells causes a neuronal differentiation into cholinergic neurons. The above inhibitory activity was further confirmed by the data of the inhibition of NGF-induced mitogen-activated protein kinase (MAPK) phosphorylation, which is located in the downstream of NGF-Trk intracellular signaling pathway. Clioquinol also caused neurite retraction induced by NGF and cell death. Importantly, differentiated cells with NGF were more vulnerable than naïve cells to clioquinol. These results strongly suggest that clioquinol may cause the perturbation of the intracellular survival pathway by inhibiting Trk-initiated signaling pathway. Intriguingly, presence of ZnCl in the culture medium exerted more potent neurotoxic effects on these cells.[56] Our subsequent study has revealed that clioquinol reduced the level of histone acetylation, and the histone deacetylase (HDAC) inhibitor, Trichostatin A, upregulated acetylated histones and prevented the neutrotoxicity caused by clioquinol in PC12 cells.[57] In addition, treatment with HDAC inhibitor decreased neurite retraction and restored the inhibition of NGF-induced Trk autophosphorylation by clioquinol. Thus, it seemed that clioquinol induces neuronal cell death via deacetylation of histones following the impairment of Trk-dependent neuroprotective pathway. It is important to mention that these observed effects can be seen under no supplemented metal ions in the culture system,[56] indicating new pathway of clioquinol-induced cytotoxicity.

4. Conclusion

We have been discussing a possible utilization of metal chelators for the treatment of neurodegenerative disorders such as AD. This rationale is derived from abnormal affinities of aggregated proteins such as Aβ in AD with metal ions. Chelating these metals can minimize the accumulation of aggregated proteins in patients' brains. On the other hand, we also should pay more attention on the concentration of these chelators in vivo because historically we now know that one of the chelators, clioquinol, caused severe neurological disorder, SMON. They might produce unexpected neurotoxic effects on brain cells.

Acknowledgments

The work from the authors' lab was partly supported by the grant-in-aid from the ministry of Education, Science, and Technology of Japan and from the ministry of Health, Labor, and Welfare of Japan and by a grant from Smoking Research Foundation Japan to T.M.

Conflict of interest

The author declares no conflict of interest in relation to the present work.

References

1. Blennow K, de Leon MJ, Zetterberg H. Alzheimer's disease. *Lancet*. 2006;368:387–403.
2. Bateman RJ, Xiong C, Benzinger TL, et al. Clinical and biomarker changes in dominantly inherited Alzherimer's disease. *N Engl J Med*. 2012;367:795–804.
3. Lee MS, Tsai LH. Cdk5: one of the links between senile plaques and neurofibrillary tangles? *J Alzheimers Dis*. 2003;5(2):127–137.
4. Zheng WH, Bastianetto S, Mennicken F, Ma W, Kar S. Amyloid beta peptide induces tau phosphorylation and loss of cholinergic neurons in rat primary septal cultures. *Neuroscience*. 2002;115:201–211.
5. Cowan CM, Bossing T, Page A, Shepherd D, Mudher A. Soluble hyper-phosphorylated tau causes microtubule breakdown and functionally compromises normal tau in vivo. *Acta Neuropathol*. 2010;120:593–604.
6. Jeganathan S, Hascher A, Chinnathambi S, Biernat J, Mandelkow EM, Mandelkow E. Proline-directed pseudo-phosphorylation at at8 and phf1 epitopes induces a compaction of the paperclip folding of tau and generates a pathological (mc-1) conformation. *J Biol Chem*. 2008;283:32066–32076.
7. Zhang B, Maiti A, Shively S, et al. Microtubule-binding drugs offset tau sequestration by stabilizing microtubules and reversing fast axonal transport deficits in a tauopathy model. *Proc Natl Acad Sci U S A*. 2005;102:227–231.
8. Newcombe EA, Camats-Perna J, Silva ML, Valmas N, Huat TJ, Medeiros R. Inflammation: the link between comorbidities, genetics, and Alzheimer's disease. *J Neuroinflammation*. 2018;15:276.
9. Liu S, Liu Y, Hao W, et al. Tlr2 is a primary receptor for Alzheimer's amyloid beta peptide to trigger neuroinflammatory activation. *J Immunol*. 2012;188:1098–1107.
10. Zhao Y, Wu X, Li X, et al. Trem2 is a receptor for beta-amyloid that mediates microglial function. *Neuron*. 2018;97:1023–1031.
11. Hellstrom-Lindahl E, Ravid R, Nordberg A. Age-dependent decline of neprilysin in Alzheimer's disease and normal brain: inverse correlation with a beta levels. *Neurobiol Aging*. 2008;29:210–221.
12. Zhao Z, Xiang Z, Haroutunian V, Buxbaum JD, Stetka B, Pasinetti GM. Insulin degrading enzyme activity selectively decreases in the hippocampal formation of cases at high risk to develop Alzheimer's disease. *Neurobiol Aging*. 2007;28:824–830.

13. Sensi SL, Granzotto A, Siotto M, Squitti R. Copper and zinc dysregulation in Alzheimer's disease. *Trends Pharmacol Sci*. 2018;39:1049–1063.
14. Ayton S, Lei P, Bush AI. Metallostasis in Alzheimer's disease. *Free Radic Biol Med*. 2013;62:76–89.
15. Cuajungco MP, Faget KY. Zinc takes the center stage: its paradoxical role in Alzheimer's disease. *Brain Res Brain Res Rev*. 2003;41(1):44–56.
16. Grasso G, Pietropaolo A, Spoto G, et al. Copper(I) and copper(II) inhibit Aβ peptides proteolysis by insulin-degrading enzyme differently: implications for metallostasis alteration in Alzheimer's disease. *Chemistry*. 2011;17(9):2752–2762.
17. Squitti R. Metals in Alzheimer's disease: a systemic perspective. *Front Biosci (Landmark Ed)*. 2012;17:451–472.
18. Finkelstein DI, Billings JL, Adlard PA, et al. The novel compound PBT434 prevents iron mediated neurodegeneration and alpha-synuclein toxicity in multiple models of Parkinson's disease. *Acta Neuropathol Commun*. 2017;5(1):53.
19. Hare DJ, Faux NG, Roberts BR, Volitakis I, Martins RN, Bush AI. Lead and manganese levels in serum and erythrocytes in Alzheimer's disease and mild cognitive impairment: results from the Australian imaging, biomarkers and lifestyle flagship study of ageing. *Metallomics*. 2016;8(6):628–632.
20. Kirkley KS, Popichak KA, Afzali MF, Legare ME, Tjalkens RB. Microglia amplify inflammatory activation of astrocytes in manganese neurotoxicity. *J Neuroinflammation*. 2017;14:99.
21. Wang D, Zhang J, Jiang W, et al. The role of nlrp3-casp1 in inflammasome-mediated neuroinflammation and autophagy dysfunction in manganese-induced, hippocampal-dependent impairment of learning and memory ability. *Autophagy*. 2017;13:914–927.
22. Hancock SM, Finkelstein DI, Adlard PA. Glia and zinc in ageing and Alzheimer's disease: a mechanism for cognitive decline? *Front Aging Neurosci*. 2014. https://doi.org/10.3389/fnagi.2014.00137.
23. Squitti R, Ghidoni R, Siotto M, et al. Value of serum nonceruloplasmin copper for prediction of mild cognitive impairment conversion to Alzheimer disease. *Ann Neurol*. 2014;75:574–580.
24. Ventriglia M, Bucossi S, Panetta V, Squitti R. Copper in Alzherimer's disease: a meta-analysis of serum, plasma, and cerebrospinal fluid studies. *J Alzheimers Dis*. 2012;30:981–984.
25. Wang ZX, Tan L, Wang HF, et al. Serum iron, zinc, and copper levels in patients with alzheimer's disease: a replication study and meta-analysis. *J Alzheimers Dis*. 2015;47(3):565–581.
26. James SA, Volitakis I, Adlard PA, Duce JA, Masters CL, Cherny RA. Elevated labile cu is associated with oxidative pathology in Alzherimer disease. *Free Radic Biol Med*. 2012;52:298–302.
27. Kitazawa M, Hsu HW, Mdeiros R. Copper exposure perturbs brain inflammatory responses and impairs clearance of amyloid-beta. *Toxicol Sci*. 2016;152:194–204.
28. Smith MA, Zhu X, Tabaton M, et al. Increased iron and free radical generation in preclinical Alzheimer disease and mild cognitive impairment. *J Alzheimers Dis*. 2010;19:363–372.
29. Rogers JT, Randall JD, Cahill CM, Eder PS, Huang X, Gunshin H. An iron-responsive element type ii in the 5′-untranlated region of the Alzheimer's amyloid precursor protein transcript. *J Biol Chem*. 2002;277:45518–45528.
30. Smith MA, Harris PL, Sayre LM, Perry G. Iron accumulation in Alzheimer disease is a source of redox-generated free radicals. *Proc Natl Acad Sci U S A*. 1997;94:9866–9868.
31. Good PF, Perl DP, Bierer LM, Schmeidler J. Selective accumulation of aluminum and iron in the neurofibrillary tangles of Alzheimer's disease: a laser microprobe (lamma) study. *Ann Neurol*. 1992;31:286–292.

32. Connor JR, Snyder BS, Beard JL, Fine RE, Mufson EJ. A histochemical study of iron, transferrin, and ferritin in Alzheimer's diseased brains. *J Neurosci Res.* 1992;31:75–83.
33. Jiang D, Li X, Williams R, et al. Ternary complexes of iron, amyloid-beta, and nitrilotriacetic acid: binding affinities, redox properties, and relevance to iron-induced oxidative stress in alzheimer's disease. *Biochemistry.* 2009;48:7939–7947.
34. Kono R. Subacute myelo-optico-neuropathy, a new neurological disease prevailing in Japan. *Jpn J Med Sci Biol.* 1971;24:195–216.
35. Konagaya M, Matsumoto A, Takase S, et al. Clinical analysis of longstanding subacute myelo-optico-neuropathy: sequelae of clioquinol at 32 years after its ban. *J Neurol Sci.* 2004;218:85–90.
36. Sobue I, Aoki K, Ohtani M. Prognosis of SMON patients. *Jpn J Med Sci Biol.* 1975;28(Suppl):203–217.
37. Tateishi J. Subacute myelo-optico-neuropathy: clioquinol intoxication in humans and animals. *Neuropathology.* 2000;20(Suppl):S20–S24.
38. Konno H, Takase S, Fukui T. Neuropathology of longstanding subacute myelo-optico-neuropathy (SMON): an autopsy case of SMON with duration of 28 years. *No To Shinkei.* 2001;53(9):875–880 [in Japanese].
39. Egashira Y, Matsuyama H. Subacute myelo-optico-neuropathy (SMON) in Japan. With special reference to the autopsy cases. *Acta Pathol Jpn.* 1982;32(Suppl 1):101–116.
40. Jack DB, Riess W. Pharmacokinetics of iodochlorhydroxyquin in man. *J Pharm Sci.* 1973;62(12):1929–1932.
41. Hayashi M, Fuwa T, Awazu S, Hanano M. Various factors affecting intestinal absorption of idodochlorhydroxyquin in rat and man. *Chem Pharm Bull.* 1976;24:2603–2609.
42. Nguyen T, Hamby A, Massa SM. Clioquinol down-regulates mutant huntingtin expression in vitro and mitigates pathology in a Huntington's disease mouse model. *Proc Natl Acad Sci U S A.* 2005;102:11840–11845.
43. Ritchie CW, Bush AI, Mackinnon A, et al. Metal-protein attenuation with iodochlorhydroxyquin (clioquinol) targeting Ab amyloid deposition and toxicity in Alzheimer's disease. *Arch Neurol.* 2003;60:1685–1691.
44. Park MH, Lee SJ, Byun HR, et al. Clioquinol induces autophagy in cultured astrocytes and neurons by acting as a zinc ionophore. *Neurobiol Dis.* 2011;42:242–251.
45. Mizutani Y, Maeda T, Murate K, Watanabe H, Mutoh T. Clioquinol kills astrocyte-derived KT-5 cells by the impairment of the autophagy-lysosome pathway. *Arch Toxicol.* 2021;95(2):631–640. https://doi.org/10.1007/s00204-020-02943-8.
46. Hotchkiss RS, Strasser A, McDunn JE, Swanson PE. Cell death. *N Engl J Med.* 2009;361:1570–1583.
47. Helmuth L. Neuroscience. An antibiotic to treat Alzheimer's? *Science.* 2000;290(5495):1273–1274.
48. Bush AI. Metal complexing agents as therapies for Alzheimer's disease. *Neurobiol Aging.* 2002;23(6):1031–1038.
49. Cherny RA, Atwood CS, Xilinas ME, et al. Treatment with a copper-zinc chelator markedly and rapidly inhibits beta-amyloid accumulation in Alzheimer's disease transgenic mice. *Neuron.* 2001;30(3):665–676.
50. Katsuyama M, Kimura E, Ibi M, et al. Clioquinol inhibits dopamine-β-hydroxylase secretion and noradrenaline synthesis by afecting the redox status of ATOX1 and copper transport in human neuroblastoma SH-SY5Y cells. *Arch Toxicol.* 2021;95:135–148.
51. Kawahara M, Tanaka K-I, Kato-Negishi M. Zinc, carnosine, and neurodegenerative diseases. *Nutrients.* 2018;10:147. https://doi.org/10.3390/nu10020147.
52. Ward C, Martinez-Lopez N, Otten EG, et al. Autophagy, lipophagy and lysosomal lipid storage disorders. *Biochim Biophys Acta.* 2016;1861:269–284.
53. Tanida I, Ueno T, Kominami E. LC3 and autophagy. *Methods Mol Biol.* 2008;445:77–88.

54. Thoenen H. Neurotrophins and neuronal plasticity. *Science.* 1995;270(5236):593–598. https://doi.org/10.1126/science.270.5236.593.
55. Kaplan DR, Miller FD. Neurotrophin signal transduction in the nervous system. *Curr Opin Neurobiol.* 2000;10:381–391.
56. Asakura K, Ueda A, Kawamura N, Ueda M, Mihara T, Mutoh T. Clinoquinol inhibits NGF-induced Trk autophosphorylation and neurite outgrowth in PC12 cells. *Brain Res.* 2009;1301:110–115.
57. Fukui T, Asakura K, Hikichi C, et al. Histone deacetylase inhibitor attenuates neurotoxicity of clioquinol in PC12 cells. *Toxicology.* 2015;331:112–118.

CHAPTER 16

Development of autophagy enhancers for Parkinson's disease therapy

Yukiko Sasazawa[a], Nobutaka Hattori[b], and Shinji Saiki[b]
[a]Research Institute for Diseases of Old Age, Juntendo University Graduate School of Medicine, Bunkyo-ku, Tokyo, Japan
[b]Department of Neurology, Juntendo University Graduate School of Medicine, Bunkyo-ku, Tokyo, Japan

1. Features of Parkinson's disease

Parkinson's disease (PD), the second most common progressive neurodegenerative disorder after Alzheimer's disease, depletes dopamine in the central nervous system. PD affects a large number of motor and nonmotor features and occurs in approximately 0.5%–1% of individuals between 65 and 69 years of age, rising to 1%–3% among persons 80 years of age and older.[1–3] Motor symptoms of PD are characterized by a clinical syndrome universally known as parkinsonism, which includes three cardinal features: rigidity, slowness, and tremor. In addition to motor symptoms, most PD patients develop nonmotor symptoms including mood disorders, apathy, anhedonia, depression, cognitive dysfunction, and hallucinosis.[4]

2. Current PD therapy

The most common therapies for PD involve supplementing dopamine to enhance dopamine levels, mimicking the role of dopamine, or inhibiting dopamine degradation. The amino acid tyrosine is converted into dopamine by the actions of two enzymes. First, tyrosine hydroxylase (TH) converts tyrosine to L-dihydroxyphenylalanine (L-DOPA/levodpa), which is subsequently converted into dopamine by the enzyme DOPA decarboxylase, also called aromatic amino acid decarboxylase (AADC). Subsequently, dopamine is metabolized to 3-metoxytyramine by the enzyme catechol-O-methyltransferase (COMT), which is in turn converted to 3-methoxy-4-hydroxyacetaldehyde by monoamine oxidase-B (MAO-B)

Autophagy Dysfunction in Alzheimer's Disease and Dementia
https://doi.org/10.1016/B978-0-323-89906-2.00003-4

and then converted to biologically inactive metabolite homovanillic acid (HVA) by aldehyde dehydrogenase (ALDH). Alternatively, dopamine is metabolized to 3,4-dihydroxyphenylacetaldehyde (DOPAL) by MAO-B, and then to 3,4-dihydroxyphenylethanol (DOPET) by alcohol dehydrogenase (ADH) or to 3,4-dihydroxyphenylacetic acid (DOPAC) by ALDH. DOPAC is then degraded to HVA by COMT. COMT also mediates L-DOPA degradation into 3-*O*-methyldopa.

Metabolic pathway of dopamine and the targets of therapeutic drugs clinically used for PD patients are described in Fig. 1. Because dopamine itself fails to pass through the blood-brain barrier, administration of the dopamine precursor levodopa is the gold-standard treatment. Because exogenously administered levodopa is extensively converted into dopamine by AADC in the periphery, levodopa is always prescribed as a mixture with a peripheral decarboxylase inhibitor to reduce its peripheral conversion into dopamine. COMT inhibitors and MAO-B inhibitors contribute to prolonging the benefits of levodopa, while dopamine agonists are useful adjuncts to levodopa. Levodopa exerts a revolutionary effect; however, chronic use can result in adverse effects such as levodopa-induced dyskinesia (involuntary movement). Moreover, as current therapies cannot suppress the loss of dopaminergic neurons, novel therapeutics are needed to overcome PD.

3. α-Synuclein accumulation in PD

The primary pathologic feature of PD is the presence of Lewy bodies, a neuronal inclusion consisting of mainly α-synuclein protein. Increasing evidence suggests that misfolded α-synuclein protein spreads from cell-to-cell in a prion-like fashion[5]; however, it remains to be fully elucidated from where the initial α-synuclein inclusions break out. Recently, it has been hypothesized that α-synuclein inclusions initially appear in nerve terminals of the enteric nervous system before spreading through autonomic connections to the dorsal motor nucleus of the vagus and intermediolateral cell columns of the sympathetic system.[6,7] However, autopsies have revealed that a minority of cases with Lewy pathology do not have inclusions in the dorsal motor nucleus of the vagus,[8] suggesting that PD probably does not start in the enteric nervous system in all cases. The identification of familial cases with α-synuclein gene (*SNCA*) multiplications revealed a strong correlation between α-synuclein levels and disease severity[9,10]; thus, impaired clearance of α-synuclein is suggested to result in its accumulation in both sporadic and familial cases of PD.

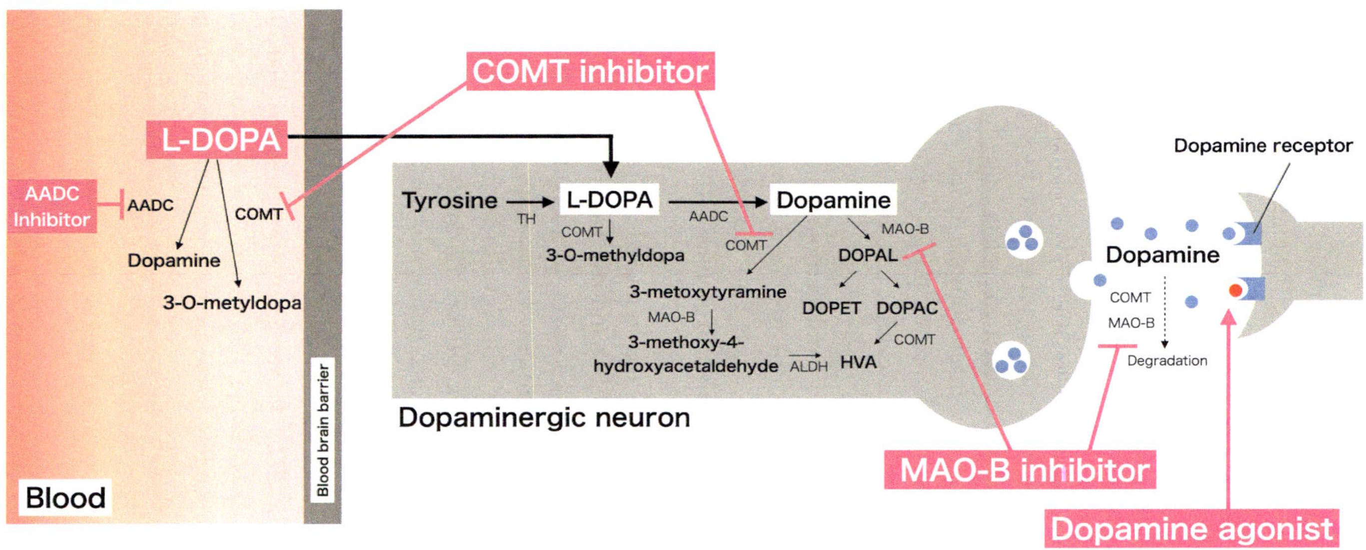

Fig. 1 Dopamine metabolism and the target of current PD therapy. Dopamine is synthesized from tyrosine through sequential reactions catalyzed by TH and AADC, and it is degraded into the main inactive metabolites DOPAC and HVA by the enzymes MAO, COMT, and ALDH. The current therapeutic drugs for PD are for enhancement dopamine level by supplementation of dopamine precursor or inhibition of the enzymes involved in dopamine degradation.

4. Autophagy impairment in PD

α-Synuclein levels are maintained by multiple degradative pathways, including ubiquitin-proteasome system (UPS) and autophagy. In this review, we focus on autophagy degradation pathway. Autophagy is an evolutionarily conserved system for lysosomal degradation of cytoplasmic components including organelles. The significance of autophagy for maintenance of the nervous system was demonstrated by studies showing that mice specifically lacking *Atg5* or *Atg7* (two genes essential for autophagy) in the central nervous system showed behavioral defects similar to those observed in neurodegenerative diseases, such as progressive motor impairment and cytoplasmic inclusion bodies.[11,12] In mammalian cells, there are three types of autophagy: microautophagy, macroautophagy, and chaperone-mediated autophagy (CMA). In general, macroautophagy is referred to as just "autophagy". Various genetic or pharmacological approaches in vivo and in vitro have shown that wild-type α-synuclein are degraded by macroautophagy and CMA.[13,14] On the other hand, pathogenic mutant of α-synuclein (A53T and A30P) is defective in the degradation by CMA.[14] In addition, oligomeric form of α-synuclein was also resistant to CMA-dependent degradation in isolated lysosomes, in contrast to wild type or dimeric α-synuclein.[15]

Recent studies have identified genes related to the autophagy-lysosomal pathway as causative for familial PD and risk factors for sporadic PD. PD-associated genes can be classified roughly into two types: one involved in protein/vesicle trafficking or protein degradation, which contributes to protein quality management, the other involved in the maintenance of mitochondrial function (Table 1). Of these, the genes related to autophagy-lysosomal pathway are detailed later. Autosomal mutations in the *LRRK2/PARK8* gene, which encodes a multidomain protein kinase with multiple functions, are the most common genetic determinants underlying familial forms of PD.[16] In addition, *LRRK2/PARK8* has been identified as a risk factor for sporadic PD.[17] Growing evidence suggests that LRRK2 is involved in endosomal-lysosome trafficking.[18] In addition, mutations in *PARK17*, the gene encoding vacuolar protein sorting protein-associated protein 35 (involved in endosome-to-Golgi trafficking system, and associated with autophagy by controlling Atg9a vesicle trafficking[19]), were identified to cause a rare form of autosomal dominant PD.[20,21] Mutations in the gene *ATP13A2/PARK9*, which encodes a transmembrane lysosomal P-type ATPase, cause familial Kufor-Rakeb syndrome (a parkinsonism with

Table 1 PD-causative genes associated to protein quality control and mitochondrial function.

	PARK	Gene	Inheritance	Possible function
	PARK1/ PARK4	*SNCA*	AD	Synapse vescle trafficking
Associated to protein quality control	*PARK5*	*UCHL1*	AD	Ubiquitin-protein hydrolase
	PARK8	*LRRK2*	AD	Protein kinase
	PARK9	*ATP13A2*	AR	Lysosomal ATPase
	PARK17	*VPS35*	AD	Endosomal transport
	PARK19	*DNAJC6*	AR	Endocytosis
Associated to mitochondrial function	*PARK2*	*PRKN*	AR	Mitophagy
	PARK6	*PINK1*	AR	Mitophagy
	PARK7	*DJ-1*	AR	Antioxidant protein
	PARK13	*HTRA2*	AR	Mitochondrial serine protease
	PARK22	*CHCHD2*	AD	Mitochondrial maintenance
	PARK23	*VPS13C*	AR	Mitochondrial maintenance

AD, autosomal dominant; *AR*, autosomal recessive.

dementia) and early-onset PD.[22] Mutations in *ATP13A2* impair lysosomal function, which causes autophagy dysfunction and impaired degradation of lysosomal substrates.[23] A recent study showed that ATP13A2 functions as lysosomal polyamine exporter whose dysfunction results in lysosome-dependent cell death, which may be implicated in neurodegeneration.[24] Ubiquitin-specific peptidase 24 (USP24), which is located in the *PARK10* gene, is associated with PD predisposition and age of onset[25]; however, mechanisms underlying its involvement or a specific gene mutation responsible for its association with PD have not been identified. However, a recent study showed that USP24 acts as a novel negative regulator of autophagy by affecting ubiquitination and stability of the unc51-like kinase 1 (ULK1) protein, a serine-threonine kinase that initiates autophagy.[26] Findings from this study also indicated that USP24 mRNA and protein levels are elevated in the substantia nigra pars compacta (SNpc) of patients with idiopathic PD compared with age-matched unaffected controls, suggesting that suppression of autophagy by USP24 also occurs in at least some cases of nonfamilial PD.[26] A recent meta-analysis of genome-wide association studies identified PD risk loci that may play roles in autophagy and lysosomal function, including *GBA* (glucocerebrosidase), *TMEM175* (transmembrane protein 175), *CTSB* (cathepsin B), *ATP6V0A1* (ATPase H^+ transporting V0 subunit a1), and *GALC* (galactosylceramidase).[27]

Because defects in autophagy/lysosomal pathways are closely related to PD pathology, chemical agents that modulate autophagy and facilitate α-synuclein clearance are expected to yield novel types of disease-modifying therapeutic drugs for PD.

5. Targeting macroautophagy for potential PD therapy

Various small compounds have demonstrated therapeutic effects in PD models by modulating autophagy pathways. Herein, we review representative therapeutic targets in autophagic pathways for PD, as well as small compounds capable of regulating these pathways (Fig. 2).

5.1 Targeting mTOR

Mammalian target of rapamycin (mTOR), an evolutionarily-conserved serine/threonine kinase, plays a central role in autophagy. Activation of mTORC1 by nutrients and growth factors results in inhibition of autophagy through phosphorylation of multiple autophagy-related proteins, as well as downregulation of ULK1 and human vacuolar protein sorting 34 (VPS 34) complexes, which are both important for the initiation and formation of autophagosomes. mTORC1 also phosphorylates and inhibits nuclear localization of transcription factor EB (TFEB/TFE3), a master regulator of genes related to lysosomal biogenesis and autophagy. In postmortem brains of patients with PD and Lewy body disease, both upregulation of mTOR protein levels and α-synuclein accumulation were observed.[28] Additionally, maneb and paraquat, agrichemicals that selectively target dopaminergic neurons, significantly increased mTOR levels and induced synucleinopathy in mice.[29]

Rapamycin, the most widely applied small-molecule mTORC1 inhibitor, binds to immunophilin and FK-binding protein (*FKBP12*) to form a complex that interacts with mTOR to prevent its binding of RAPTOR, which blocks the signal transduction of mTORC1.[3031] Rapamycin increased the formation of autophagosomes and promoted α-synuclein clearance in a 1-methyl-4-phenylpyridinium ion (MPP^+)-treated PD cellular model, MPTP-intoxicated PD model mice,[32] and α-synuclein transgenic PD model mice.[28] Moreover, rapamycin improved the motor function of mice transgenic for A53T α-synuclein, a common mutation in PD.[33] In addition, some

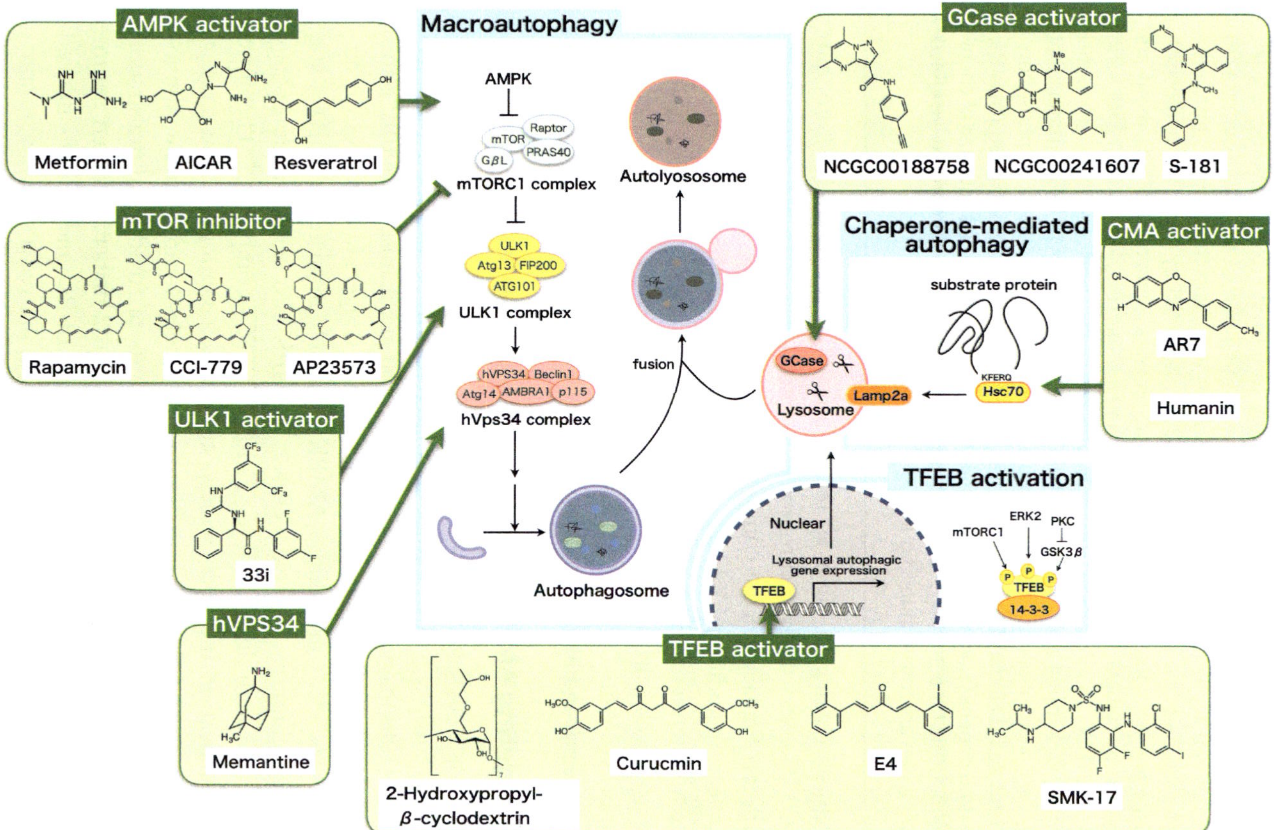

Fig. 2 See figure legend on next page.

Fig. 2 Chemical structures of autophagy modulator. The first step of macroautophagy is the nucleation of specific membrane structures known as the phagophores that subsequently elongate and close on itself to form double-membrane vesicles, called autophagosomes. The next stage involves the fusion of autophagosomes with lysosomes and subsequent formation of autolysosomes, and the contents inside the autophagosomes and their inner membrane are degraded by lysosomal hydrolases. In chaperone-mediated autophagy, substrate proteins containing KFERQ motif are recognized by cytoplasmic chaperone hsc70 and are then uptaken by the LAMP2 and degraded. TFEB is a master regulator of lysosomal function and autophagy. In normal condition, TFEB is phosphorylated by mTORC1 or ERK2 and binds to 14-3-3 protein that retain the transcription factor in the cytoplasm. The inhibition of the TFEB phosphorylation induces TFEB translocation from the cytoplasm to the nucleus and regulates the expression of genes involved in sequential steps of autophagy. The structures of chemical modulators for autophagy-lysosomal pathway that exhibit neuroprotective effects for PD models are shown in this figure.

---

analogs of rapamycin, such as temsirolimus (CCI-779) and deforolimus (AP23573), exert similar neuroprotective effects. Although there are beneficial effects of rapamycin, this compound may cause a wide range of unwanted side effects including peripheral edema, hypertriglyceridemia, hypertension, hypercholesterolemia, increased creatinine, constipation, abdominal pain, diarrhea, headache, urinary tract infection, anemia, nausea, arthralgia, pain, and thrombocytopenia.[34] Thus, rapamycin and its analogs appear to be unsuitable for long-term and high-dose therapy due to their side effects.

5.2 Targeting AMP-activated protein kinase (AMPK)

5′-AMP-activated protein kinase (AMPK) is a key regulator of metabolism and autophagy. In contrast to mTOR, AMPK activation positively regulates autophagic activity through several mechanisms. AMPK phosphorylates and directly activates ULK1, which triggers initiation of the autophagic cascade.[35] In addition, AMPK negatively regulates mTORC1 and blocks its inhibitory effect on ULK1.[35] AMPK also stimulates autophagy initiation through differential regulation of VPS34 complexes.[36] Therefore, AMPK activation has been widely studied as a neuroprotective strategy for PD treatment. Metformin, 5-aminoimidazole-4-carboxamide ribonucleotide (AICAR), and resveratrol are agonists of AMPK. Administration of metformin, a drug widely used to treat type 2 diabetes, elicited neuroprotective effects in both a drosophila model of PD and MPTP-induced PD model mice.[37,38] AICAR, an AMP analog, showed a similar neuroprotective effect.[39] Resveratrol activates AMPK by interacting with its direct target SIRT1 to exert neuroprotective effects in both in vitro and in vivo models.[40,41] Treatment of PC12 cells overexpressing wild-type or mutant α-synuclein, as well as rotenone-exposed SH-SY5Y cells, with resveratrol resulted in phosphorylation of AMPK at Thr172, which activated SIRT1 deacetylation enzymatic activity and enhanced α-synuclein degradation by activating autophagy through the AMPK/SIRT1 signaling pathway.[42] Furthermore, metformin decreased Ser129-phosphorylated α-synuclein levels in α-synuclein-overexpressing SH-SY5Y cells.[43]

5.3 Targeting ULK1 and VPS34

ULK1, the yeast Atg1 ortholog, is the initiating enzyme of autophagic processes. Specifically, the ULK complex formed by ULK1, mAtg13, FIP200, and Atg101 is required to initiate autophagy. This complex subsequently phosphorylates Beclin-1, which triggers the lipid kinase activity of VPS34, leading to the upregulation of phosphatidylinositol 3-phosphate and recruitment of autophagy proteins for autophagosome formation.[44,45] The small molecule 33i (BL-918), which was identified as a potent activator of ULK1 by structure-based drug design, could activate ULK1 and the rest of the ULK complex.[46] Specifically, 33i induced autophagy by enhancing ULK complex formation in SH-SY5Y cells and displayed a cytoprotective effect in an MPP^+-treated PD cellular model. Importantly, 33i protected against MPTP-induced motor dysfunction and loss of dopaminergic neurons by targeting ULK1-modulated autophagy in mouse models of PD.[46]

Memantine, which is approved for moderate-to-severe Alzheimer's disease, antagonizes *N*-methyl-D-aspartic acid receptor 1 (NMDAR1) to inhibit excessive calcium influx induced by chronic overstimulation of the NMDA receptor. The novel function of memantine as an autophagy enhancer was identified by a screen for autophagy modulators among 796 clinically approved drugs. Memantine upregulated autophagy flux via the VPS34 complex. Importantly, this compound enhanced elimination of degraded mitochondria in iPSC-derived neurons of patients with *PARK2* or *PARK6* mutations who exhibited defective PINK1/parkin-mediated mitophagy.[47]

6. Targeting lysosome for potential PD therapy

6.1 Targeting TFEB

TFEB plays an important role in lysosomal biogenesis and autophagy. Phosphorylation by mTOR or ERK2 causes TFEB to be retained in the cytoplasm, whereas dephosphorylation permits translocation of TFEB to the nucleus to induce transcription of target genes bearing the coordinated lysosomal expression and regulation (CLEAR) motif. The importance of TFEB in α-synuclein clearance was revealed by a study that showed that genetic or pharmacological activation of TFEB-enhanced autophagy and protected nigral neurons from α-synuclein toxicity.[48] The representative TFEB activator is curcumin, a natural pigment derived from the root of the turmeric herb.[49] Recently, the curcumin derivative E4 was identified as a potent TFEB activator. Compound E4 inhibited mTORC1 activity and promoted TFEB dephosphorylation.[49] Importantly, E4 also promoted the degradation of α-synuclein and protected against MPP^+ cytotoxicity in PC12 cells.[49]

2-Hydroxypropyl-β-cyclodextrin (HPβCD) is a US Food and Drug Administration-approved excipient used to improve drug stability and bioavailability.[50] Kilpatrick et al. showed that HPβCD-activated TFEB to promote autophagic clearance of aggregated α-synuclein.[51] In addition, a recently identified autophagy inducer, SMK17, enhanced autophagy via the PRKC/PKC-TFEB pathway. SMK17 suppressed the formation of protein aggresomes observed in an MPP^+-induced PD cellular model.[52]

6.2 Targeting glucocerebrosidase (GCase)

Mutations in the gene encoding GCase (*GBA*), a lysosomal enzyme involved in sphingolipid metabolism, are the most common risk factor for developing PD. Several reports indicate modest but significant increases in intracellular α-synuclein levels evaluated by western blotting in human midbrain neurons differentiated from iPSCs of patients with PD and heterozygous *GBA* mutations.[53–55] Decreased activity of GCase causes neuronal aggregation of α-synuclein in SH-SY5Y cells.[56] Therefore, GCase is considered an attractive target for PD therapy. NCGC00188758 and NCGC00241607, two GCase activators identified by high-throughput screening,[57,58] reduced α-synuclein inclusions in dopaminergic neurons.[59,60] Recently, the small molecule S-181 was identified as a wild-type GCase activator. Treatment of PD iPSC-derived dopaminergic neurons with S-181 partially restored lysosomal function and reduced accumulation of oxidized dopamine, glucosylceramide, and α-synuclein. Moreover, treatment of mice heterozygous for the D409V *GBA1* mutation with S-181 resulted in the activation of wild-type GCase and a consequent reduction of GCase lipid substrates and α-synuclein in brain tissue.[61]

6.3 Targeting CMA

In contrast to macroautophagy, CMA involves the selective degradation of KFERQ-like motif-bearing proteins delivered to lysosomes via the chaperone heat shock cognate 71-kDa protein (HSC70) and cochaperones, including carboxyl terminus of HSC70-interacting protein, heat shock protein (HSP) 40, and HSP70-HSP90 organizing protein. After the chaperone-substrate complex binds to the lysosomal membrane receptor LAMP2A,[62] CMA substrate proteins are translocated into the lysosomal lumen and degraded by hydrolases.[63,64] Notably, because α-synuclein sequence contains a pentapeptide sequence ($_{95}$VKKDQ$_{99}$) that is consistent with a CMA-recognition motif,[65] wild-type α-synuclein was selectively translocated into lysosomes for degradation by CMA.[14] The association between PD pathogenesis and CMA dysfunction was demonstrated in several studies. Deficiency of *DJ-1/PARK7*, whose mutation causes early-onset familial PD, increased accumulation of α-synuclein by accelerating LAMP2A degradation in lysosomes.[66] Furthermore, expression levels of the CMA protein LAMP2A and HSC70 were significantly reduced in the SNpc and amygdala of patients with PD.[67,68]

Because signaling through retinoic acid receptor a (RARa) was found to inhibit CMA, synthetic derivatives of all-trans-retinoic acid (ATRA) were developed by structure-based chemical design to specifically neutralize this inhibitory effect. AR7, an ATRA derivative, enhanced CMA and suppressed paraquat-induced oligomeric species of α-synuclein.[69] Humanin, a mitochondria-associated 24-amino-acid peptide that exerts cytoprotective, cardioprotective, and neuroprotective effects in vivo and in vitro,[70] was found to be an endogenous activator of CMA.[71] Humanin directly activates CMA by interacting with HSP90 to increase substrate binding and translocation into lysosomes.[71]

Further work is needed to demonstrate efficacy of these CMA activators for PD therapy. However, in contrast to wild-type α-synuclein, pathogenic A53T and A30P α-synuclein are poorly degraded by CMA, despite a high affinity for the CMA receptor LAMP2A.[14] Thus, CMA activators may only exert neuroprotective effects in patients who do not have a mutation in α-synuclein.

6.4 Clinical trials of nilotinib for PD therapy

The tyrosine kinase Abl1 is activated in human postmortem PD brains.[72,73] Moreover, expression of α-synuclein in the mouse SN activates Abl1 via phosphorylation, and increased Abl expression further increases α-synuclein levels. These results suggest that the activity of Abl is closely related to PD pathology, making Abl an attractive target for PD therapy.

Nilotinib is a brain-penetrant tyrosine kinase inhibitor approved for treatment of Philadelphia chromosome-positive chronic myeloid leukemia. Administration of nilotinib decreased Abl activity and enhanced autophagic α-synuclein clearance in α-synuclein transgenic PD model mice.[74] These results support the use of nilotinib as a therapeutic strategy to degrade α-synuclein in PD and other α-synucleinopathies. Thus, clinical trials were performed. A phase I study of 12 patients with PD at Georgetown University in the US suggested that nilotinib may increase dopamine metabolism and potentially alleviate motor and nonmotor symptoms of PD (ClinicalTrials.gov Identifier: NCT02281474).[75] In a subsequent phase II study of 75 patients with PD at Georgetown University (ClinicalTrials.gov Identifier: NCT02954978), the 150-mg nilotinib group exhibited increased cerebrospinal fluid (CSF) levels of the dopamine metabolites homovanillic acid and 3,4-dihydroxyphenylacetic acid, while the 300-mg nilotinib group only exhibited an increase in 3,4-dihydroxyphenylacetic acid. Importantly, the 150-mg nilotinib group but not the 300-mg nilotinib group demonstrated a reduction of α-synuclein oligomers in the CSF.[76] However, another phase II study of 76 patients with PD recruited from 25 US Parkinson study group sites showed that while nilotinib demonstrated acceptable safety and tolerability (ClinicalTrials.gov Identifier: NCT03205488), it had low CSF exposure and failed to affect certain biomarkers. Unfortunately, nilotinib was concluded to be unsuitable for further testing in PD.[77]

7. Conclusions and future perspectives

PD has been recognized as not only a brain disease but also a whole-body disease because α-synuclein inclusions initially appear in nerve terminals of the enteric nervous system before spreading through autonomic connections to the dorsal motor nucleus of the vagus and intermediolateral cell columns of the sympathetic system. Therefore, enhancement of autophagy to clear α-synuclein is thought to be a promising strategy for PD therapy. In fact, use of macroautophagy-enhancing agents in preclinical models has provided important knowledge about the role of autophagy in PD, as reviewed in this chapter.

Many autophagy enhancers exert neuroprotective effects in various in vivo and in vitro models. However, the therapeutic effect of macroautophagy enhancers may be limited due to their lack of selectivity because the proteins regulating macroautophagy (e.g., mTOR and AMPK) are also involved in other signaling pathways.

Recently developed compounds selectively targeting CMA or downstream elements of the autophagic pathway, such as TFEB and GCase, may have fewer side effects and exert remarkable therapeutic effects. During the past decade, various selective forms of autophagy and their underlying molecular mechanisms were identified, including mitophagy, ERphagy, nucleophagy, lysophagy, and aggrephagy. Aggrephagy is a type of autophagy selective for ubiquitinated aggregated protein. Compounds that can enhance aggrephagy or selectively degrade α-synuclein would have a dramatic therapeutic effect.

Acknowledgment

We thank Edanz (https://jp.edanz.com/ac) for editing a draft of this manuscript.

References

1. Tanner CM, Goldman SM. Epidemiology of Parkinson's disease. *Neurol Clin.* 1996;14(2):317–335.
2. Nussbaum RL, Ellis CE. Alzheimer's disease and Parkinson's disease. *N Engl J Med.* 2003;348(14):1356–1364.
3. Hirsch L, Jette N, Frolkis A, Steeves T, Pringsheim T. The incidence of Parkinson's disease: a systematic review and meta-analysis. *Neuroepidemiology.* 2016;46(4):292–300.
4. Shalash AS, Hamid E, Elrassas HH, et al. Non-motor symptoms as predictors of quality of life in Egyptian patients with Parkinson's disease: a cross-sectional study using a culturally adapted 39-item Parkinson's disease questionnaire. *Front Neurol.* 2018;9:357.
5. Vargas JY, Grudina C, Zurzolo C. The prion-like spreading of alpha-synuclein: from in vitro to in vivo models of Parkinson's disease. *Ageing Res Rev.* 2019;50:89–101.
6. Braak H, Del Tredici K, Rub U, de Vos RA, Jansen Steur EN, Braak E. Staging of brain pathology related to sporadic Parkinson's disease. *Neurobiol Aging.* 2003;24(2):197–211.
7. Borghammer P. How does Parkinson's disease begin? Perspectives on neuroanatomical pathways, prions, and histology. *Mov Disord.* 2018;33(1):48–57.
8. Parkkinen L, Pirttila T, Alafuzoff I. Applicability of current staging/categorization of alpha-synuclein pathology and their clinical relevance. *Acta Neuropathol.* 2008;115(4):399–407.
9. Singleton AB, Farrer M, Johnson J, et al. alpha-Synuclein locus triplication causes Parkinson's disease. *Science.* 2003;302(5646):841.
10. Chartier-Harlin MC, Kachergus J, Roumier C, et al. Alpha-synuclein locus duplication as a cause of familial Parkinson's disease. *Lancet.* 2004;364(9440):1167–1169.
11. Komatsu M, Waguri S, Chiba T, et al. Loss of autophagy in the central nervous system causes neurodegeneration in mice. *Nature.* 2006;441(7095):880–884.
12. Hara T, Nakamura K, Matsui M, et al. Suppression of basal autophagy in neural cells causes neurodegenerative disease in mice. *Nature.* 2006;441(7095):885–889.
13. Spencer B, Potkar R, Trejo M, et al. Beclin 1 gene transfer activates autophagy and ameliorates the neurodegenerative pathology in alpha-synuclein models of Parkinson's and Lewy body diseases. *J Neurosci.* 2009;29(43):13578–13588.
14. Cuervo AM, Stefanis L, Fredenburg R, Lansbury PT, Sulzer D. Impaired degradation of mutant alpha-synuclein by chaperone-mediated autophagy. *Science.* 2004;305(5688):1292–1295.

15. Martinez-Vicente M, Talloczy Z, Kaushik S, et al. Dopamine-modified alpha-synuclein blocks chaperone-mediated autophagy. *J Clin Invest*. 2008;118(2):777–788.
16. Zimprich A, Muller-Myhsok B, Farrer M, et al. The PARK8 locus in autosomal dominant parkinsonism: confirmation of linkage and further delineation of the disease-containing interval. *Am J Hum Genet*. 2004;74(1):11–19.
17. Gilks WP, Abou-Sleiman PM, Gandhi S, et al. A common LRRK2 mutation in idiopathic Parkinson's disease. *Lancet*. 2005;365(9457):415–416.
18. Manzoni C, Lewis PA. LRRK2 and autophagy. *Adv Neurobiol*. 2017;14:89–105.
19. Zavodszky E, Seaman MN, Moreau K, et al. Mutation in VPS35 associated with Parkinson's disease impairs WASH complex association and inhibits autophagy. *Nat Commun*. 2014;5:3828.
20. Gan-Or Z, Dion PA, Rouleau GA. Genetic perspective on the role of the autophagy-lysosome pathway in Parkinson disease. *Autophagy*. 2015;11(9):1443–1457.
21. Trinh J, Farrer M. Advances in the genetics of Parkinson disease. *Nat Rev Neurol*. 2013;9(8):445–454.
22. Ramirez A, Heimbach A, Grundemann J, et al. Hereditary parkinsonism with dementia is caused by mutations in ATP13A2, encoding a lysosomal type 5 P-type ATPase. *Nat Genet*. 2006;38(10):1184–1191.
23. Dehay B, Ramirez A, Martinez-Vicente M, et al. Loss of P-type ATPase ATP13A2/PARK9 function induces general lysosomal deficiency and leads to Parkinson disease neurodegeneration. *Proc Natl Acad Sci U S A*. 2012;109(24):9611–9616.
24. van Veen S, Martin S, Van den Haute C, et al. ATP13A2 deficiency disrupts lysosomal polyamine export. *Nature*. 2020;578(7795):419–424.
25. Li Y, Schrodi S, Rowland C, Tacey K, Catanese J, Grupe A. Genetic evidence for ubiquitin-specific proteases USP24 and USP40 as candidate genes for late-onset Parkinson disease. *Hum Mutat*. 2006;27(10):1017–1023.
26. Thayer JA, Awad O, Hegdekar N, et al. The PARK10 gene USP24 is a negative regulator of autophagy and ULK1 protein stability. *Autophagy*. 2020;16(1):140–153.
27. Chang D, Nalls MA, Hallgrimsdottir IB, et al. A meta-analysis of genome-wide association studies identifies 17 new Parkinson's disease risk loci. *Nat Genet*. 2017;49(10):1511–1516.
28. Crews L, Spencer B, Desplats P, et al. Selective molecular alterations in the autophagy pathway in patients with Lewy body disease and in models of alpha-synucleinopathy. *PLoS ONE*. 2010;5(2), e9313.
29. Wills J, Credle J, Oaks AW, et al. Paraquat, but not maneb, induces synucleinopathy and tauopathy in striata of mice through inhibition of proteasomal and autophagic pathways. *PLoS ONE*. 2012;7(1), e30745.
30. Chen J, Zheng XF, Brown EJ, Schreiber SL. Identification of an 11-kDa FKBP12-rapamycin-binding domain within the 289-kDa FKBP12-rapamycin-associated protein and characterization of a critical serine residue. *Proc Natl Acad Sci U S A*. 1995;92(11):4947–4951.
31. Choi J, Chen J, Schreiber SL, Clardy J. Structure of the FKBP12-rapamycin complex interacting with the binding domain of human FRAP. *Science*. 1996;273(5272):239–242.
32. Dehay B, Bove J, Rodriguez-Muela N, et al. Pathogenic lysosomal depletion in Parkinson's disease. *J Neurosci*. 2010;30(37):12535–12544.
33. Bai X, Wey MC, Fernandez E, et al. Rapamycin improves motor function, reduces 4-hydroxynonenal adducted protein in brain, and attenuates synaptic injury in a mouse model of synucleinopathy. *Pathobiol Aging Age Relat Dis*. 2015;5:28743.
34. Bove J, Martinez-Vicente M, Vila M. Fighting neurodegeneration with rapamycin: mechanistic insights. *Nat Rev Neurosci*. 2011;12(8):437–452.

35. Kim J, Kundu M, Viollet B, Guan KL. AMPK and mTOR regulate autophagy through direct phosphorylation of Ulk1. *Nat Cell Biol.* 2011;13(2):132–141.
36. Kim J, Kim YC, Fang C, et al. Differential regulation of distinct Vps34 complexes by AMPK in nutrient stress and autophagy. *Cell.* 2013;152(1–2):290–303.
37. Ng CH, Guan MS, Koh C, et al. AMP kinase activation mitigates dopaminergic dysfunction and mitochondrial abnormalities in Drosophila models of Parkinson's disease. *J Neurosci.* 2012;32(41):14311–14317.
38. Patil SP, Jain PD, Ghumatkar PJ, Tambe R, Sathaye S. Neuroprotective effect of metformin in MPTP-induced Parkinson's disease in mice. *Neuroscience.* 2014;277:747–754.
39. Dulovic M, Jovanovic M, Xilouri M, et al. The protective role of AMP-activated protein kinase in alpha-synuclein neurotoxicity in vitro. *Neurobiol Dis.* 2014;63:1–11.
40. Ur Rasheed MS, Tripathi MK, Mishra AK, Shukla S, Singh MP. Resveratrol protects from toxin-induced parkinsonism: plethora of proofs hitherto petty translational value. *Mol Neurobiol.* 2016;53(5):2751–2760.
41. Lin TK, Chen SD, Chuang YC, et al. Resveratrol partially prevents rotenone-induced neurotoxicity in dopaminergic SH-SY5Y cells through induction of heme oxygenase-1 dependent autophagy. *Int J Mol Sci.* 2014;15(1):1625–1646.
42. Wu Y, Li X, Zhu JX, et al. Resveratrol-activated AMPK/SIRT1/autophagy in cellular models of Parkinson's disease. *Neurosignals.* 2011;19(3):163–174.
43. Perez-Revuelta BI, Hettich MM, Ciociaro A, et al. Metformin lowers Ser-129 phosphorylated alpha-synuclein levels via mTOR-dependent protein phosphatase 2A activation. *Cell Death Dis.* 2014;5, e1209.
44. Nazarko VY, Zhong Q. ULK1 targets Beclin-1 in autophagy. *Nat Cell Biol.* 2013;15(7):727–728.
45. Russell RC, Tian Y, Yuan H, et al. ULK1 induces autophagy by phosphorylating Beclin-1 and activating VPS34 lipid kinase. *Nat Cell Biol.* 2013;15(7):741–750.
46. Ouyang L, Zhang L, Zhang S, et al. Small-molecule activator of UNC-51-like kinase 1 (ULK1) that induces cytoprotective autophagy for Parkinson's disease treatment. *J Med Chem.* 2018;61(7):2776–2792.
47. Hirano K, Fujimaki M, Sasazawa Y, et al. Neuroprotective effects of memantine via enhancement of autophagy. *Biochem Biophys Res Commun.* 2019;518(1):161–170.
48. Decressac M, Mattsson B, Weikop P, Lundblad M, Jakobsson J, Bjorklund A. TFEB-mediated autophagy rescues midbrain dopamine neurons from alpha-synuclein toxicity. *Proc Natl Acad Sci U S A.* 2013;110(19):E1817–E1826.
49. Wang Z, Yang C, Liu J, et al. A curcumin derivative activates TFEB and protects against parkinsonian neurotoxicity in vitro. *Int J Mol Sci.* 2020;21(4), 1515.
50. Song W, Wang F, Lotfi P, Sardiello M, Segatori L. 2-Hydroxypropyl-beta-cyclodextrin promotes transcription factor EB-mediated activation of autophagy: implications for therapy. *J Biol Chem.* 2014;289(14):10211–10222.
51. Kilpatrick K, Zeng Y, Hancock T, Segatori L. Genetic and chemical activation of TFEB mediates clearance of aggregated alpha-synuclein. *PLoS ONE.* 2015;10(3), e0120819.
52. Kataura T, Tashiro E, Nishikawa S, et al. A chemical genomics-aggrephagy integrated method studying functional analysis of autophagy inducers. *Autophagy.* 2020;1–17.
53. Mazzulli JR, Xu YH, Sun Y, et al. Gaucher disease glucocerebrosidase and alpha-synuclein form a bidirectional pathogenic loop in synucleinopathies. *Cell.* 2011;146(1):37–52.
54. Schondorf DC, Aureli M, McAllister FE, et al. iPSC-derived neurons from GBA1-associated Parkinson's disease patients show autophagic defects and impaired calcium homeostasis. *Nat Commun.* 2014;5:4028.
55. Woodard CM, Campos BA, Kuo SH, et al. iPSC-derived dopamine neurons reveal differences between monozygotic twins discordant for Parkinson's disease. *Cell Rep.* 2014;9(4):1173–1182.

56. Bae EJ, Yang NY, Lee C, et al. Loss of glucocerebrosidase 1 activity causes lysosomal dysfunction and alpha-synuclein aggregation. *Exp Mol Med*. 2015;47, e188.
57. Goldin E, Zheng W, Motabar O, et al. High throughput screening for small molecule therapy for Gaucher disease using patient tissue as the source of mutant glucocerebrosidase. *PLoS ONE*. 2012;7(1), e29861.
58. Patnaik S, Zheng W, Choi JH, et al. Discovery, structure-activity relationship, and biological evaluation of noninhibitory small molecule chaperones of glucocerebrosidase. *J Med Chem*. 2012;55(12):5734–5748.
59. Aflaki E, Borger DK, Moaven N, et al. A new glucocerebrosidase chaperone reduces alpha-Synuclein and glycolipid levels in iPSC-derived dopaminergic neurons from patients with Gaucher disease and parkinsonism. *J Neurosci*. 2016;36(28):7441–7452.
60. Mazzulli JR, Zunke F, Tsunemi T, et al. Activation of beta-glucocerebrosidase reduces pathological alpha-Synuclein and restores lysosomal function in Parkinson's patient midbrain neurons. *J Neurosci*. 2016;36(29):7693–7706.
61. Burbulla LF, Jeon S, Zheng J, Song P, Silverman RB, Krainc D. A modulator of wild-type glucocerebrosidase improves pathogenic phenotypes in dopaminergic neuronal models of Parkinson's disease. *Sci Transl Med*. 2019;11(514), eaau6870.
62. Cuervo AM, Dice JF. A receptor for the selective uptake and degradation of proteins by lysosomes. *Science*. 1996;273(5274):501–503.
63. Cuervo AM, Dice JF. Lysosomes, a meeting point of proteins, chaperones, and proteases. *J Mol Med (Berl)*. 1998;76(1):6–12.
64. Cuervo AM, Dice JF, Knecht E. A population of rat liver lysosomes responsible for the selective uptake and degradation of cytosolic proteins. *J Biol Chem*. 1997;272(9):5606–5615.
65. Dice JF. Peptide sequences that target cytosolic proteins for lysosomal proteolysis. *Trends Biochem Sci*. 1990;15(8):305–309.
66. Xu CY, Kang WY, Chen YM, et al. DJ-1 inhibits alpha-Synuclein aggregation by regulating chaperone-mediated autophagy. *Front Aging Neurosci*. 2017;9:308.
67. Alvarez-Erviti L, Rodriguez-Oroz MC, Cooper JM, et al. Chaperone-mediated autophagy markers in Parkinson disease brains. *Arch Neurol*. 2010;67(12):1464–1472.
68. Murphy KE, Gysbers AM, Abbott SK, et al. Lysosomal-associated membrane protein 2 isoforms are differentially affected in early Parkinson's disease. *Mov Disord*. 2015;30(12):1639–1647.
69. Anguiano J, Garner TP, Mahalingam M, Das BC, Gavathiotis E, Cuervo AM. Chemical modulation of chaperone-mediated autophagy by retinoic acid derivatives. *Nat Chem Biol*. 2013;9(6):374–382.
70. Kariya S, Takahashi N, Ooba N, Kawahara M, Nakayama H, Ueno S. Humanin inhibits cell death of serum-deprived PC12h cells. *Neuroreport*. 2002;13(6):903–907.
71. Gong Z, Tasset I, Diaz A, et al. Humanin is an endogenous activator of chaperone-mediated autophagy. *J Cell Biol*. 2018;217(2):635–647.
72. Imam SZ, Zhou Q, Yamamoto A, et al. Novel regulation of parkin function through c-Abl-mediated tyrosine phosphorylation: implications for Parkinson's disease. *J Neurosci*. 2011;31(1):157–163.
73. Ko HS, Lee Y, Shin JH, et al. Phosphorylation by the c-Abl protein tyrosine kinase inhibits parkin's ubiquitination and protective function. *Proc Natl Acad Sci U S A*. 2010;107(38):16691–16696.
74. Hebron ML, Lonskaya I, Moussa CE. Nilotinib reverses loss of dopamine neurons and improves motor behavior via autophagic degradation of alpha-synuclein in Parkinson's disease models. *Hum Mol Genet*. 2013;22(16):3315–3328.
75. Pagan F, Hebron M, Valadez EH, et al. Nilotinib effects in Parkinson's disease and dementia with Lewy bodies. *J Parkinsons Dis*. 2016;6(3):503–517.
76. Pagan FL, Hebron ML, Wilmarth B, et al. Nilotinib effects on safety, tolerability, and potential biomarkers in Parkinson disease: a phase 2 randomized clinical trial. *JAMA Neurol*. 2020;77(3):309–317.
77. Simuni T, Fiske B, Merchant K, et al. Efficacy of nilotinib in patients with moderately advanced Parkinson disease: a randomized clinical trial. *JAMA Neurol*. 2021;78(3):312–320.

Index

Note: Page numbers followed by *f* indicate figures and *t* indicate tables.

A

Aβ40, 74–75
Aβ42, 74–75
Aβ binding alcohol dehydrogenase (ABAD), 96–98
Abnormality
  in immunological system, 214–215
  in proteolytic pathway, 214
Abnormal metals, in AD patients, 292–294
Abnormal proteins, management of, 227
Abundant intracellular tau, 173
Acetylation of tau protein, 138
Acetylcholinesterase (AChE), 209–210
Acid alpha-glucosidase (GAA), 209
Activate JNK1/Bcl-2/PI3K, 279
Adaptor protein, 247–249, 252
AGD brain homogenate injection, 178
Age-dependent reduction in autophagy, 92
Aggregated Aβ, 73, 75–77
  amyloid hypothesis, 75–76
  high-molecular-weight oligomers and off-pathways, 77, 78*f*
  inhibitor, 82–83
  oligomer hypothesis, 76
  process of, 76–77, 76*f*
  soluble low-molecular-weight oligomer, 77
Aggregated proteins, 60
Aggregated tau, 152
  autophagic impairment, 181–185, 182*f*
    CMA dysfunction in tauopathies, 183–184
    macroautophagic dysfunction in tauopathies, 181–183
    microautophagic dysfunction in tauopathies, 184–185
  intercellular spread of, 178–180
  lithium-induced autophagic promotion of, 189, 190*f*
Aggregates, 246–249, 251
Aggrephagy, 60
Aging, 94–96, 100
AICAR. *See* Aminoimidazole-4-carboxamidoribonucleoside (AICAR)
Alfy. *See* Autophagy-linked FYVE protein (Alfy)
All-trans-retinoic acid (ATRA), 315
Alzheimer's disease (AD), 111
  autophagy impaired in, 182*f*
  autophagy-lysosome system in, 145–148
  autophagy modulators, 153
  Aβ protein in aggregation, 75–77
  defective mitophagy in, 99–101
    mitochondrial dysfunction and, 101–103
    molecular mechanisms of mitophagy, 98–99
  early feature of, 95–96
  granulovacuolar degeneration in, 197–198, 198*f*
  homology of, 213–214
  pathological features of, 197
  pathological hallmarks of, 136*f*
  pathophysiology of, 91–92
  potential therapy for, 97*f*
  regulating autophagy, drugs discovery in
    activate JNK1/Bcl-2/PI3K, 279
    AMPK activator, 275–276
    Atg-dependent pathway, 279–280
    ATP60V01, 280
    calcium channel antagonist, 279
    carbamazepine, 277
    direct mTORC inhibitors, 263–273
    drugs targeting TFEB, 277–278
    gene therapies, 282
    imidazoline receptor agonists, 277
    increased lysosome, 278–279
    IP3 inhibitors, 276
    mTORC-dependent autophagy inducers, 273–274
    multiple mechanisms, 281
    negative regulator ABL1, 281
    P300 inhibitors (tau acetylation inhibitors), 280–281
    VPS34 activator, 279

Alzheimer's disease (AD) *(Continued)*
seed-dependent tau aggregation, 189
using metal chelators, drug discovery in, 291–292
abnormal metals in AD patients, 292–294
molecular action of clioquinol, 297–299
subacute myelo-optico-neuropathy (SMON), 294–295
in vivo concentration of clioquinol, 295–297, 296–297*f*
Alzheimer's disease (AD), Aβ protein in
aggregation
amyloid hypothesis, 75–76
high-molecular-weight oligomers and off-pathways, 77, 78*f*
oligomer hypothesis, 76
process of, 76–77, 76*f*
soluble low-molecular-weight oligomer, 77
decomposition of, 78–79
vs. disease
FAD and genetic mutation, 80–81
sporadic AD, 79–80
down syndrome, 82
excretion of, 79
intervention strategy
Aβ aggregation inhibitor, 82–83
immunotherapy, 83–84
secretase activity regulator, 82
summary of, 84–85
normal function of, 79
production and formation, 74–75
APP, 74
Aβ40 and Aβ42, 74–75
α-Secretase, 74
β-Secretase, 74
γ-Secretase, 74
regulation of, 78
Amino acid tyrosine, 305–306
Aminoimidazole-4-carboxamidoribonucleoside (AICAR), 276
AMP-activated protein kinase (AMPK), 51–52, 275–276, 312–313
Amyloid, 111–112, 115, 117–118, 120, 125
Amyloid β (Aβ), 91–92, 210–211, 263
aggregation inhibitor, 82–83
tau and, 147–148
Amyloid-beta precursor protein (APP), 96
Amyloid β (Aβ), in Alzheimer's disease
aggregation, 75–77
amyloid hypothesis, 75–76
high-molecular-weight oligomers and off-pathways, 77, 78*f*
oligomer hypothesis, 76
process of, 76–77, 76*f*
soluble low-molecular-weight oligomer, 77
decomposition of, 78–79
vs. disease
FAD and genetic mutation, 80–81
sporadic AD, 79–80
down syndrome, 82
excretion of, 79
intervention strategy
Aβ aggregation inhibitor, 82–83
immunotherapy, 83–84
secretase activity regulator, 82
summary of, 84–85
normal function of, 79
production and formation, 74–75
Aβ40 and Aβ42, 74–75
APP, 74
α-secretase, 74
β-secretase, 74
γ-secretase, 74
regulation of, 78
Amyloid fibrils, 73
Amyloid hypothesis, 75–76
Amyloid plaques, 73
Amyloid precursor protein (APP), 74, 91–92, 111–112
Anatomical absorption pathway, 79
Anti-AD therapeutic agents, 82
Antiinflammatory effects, 275
APELS. *See* Autophagy-endolysosomal system (APELS)
AP-lysosomal pathway (ALP), 185–189
APP. *See* Amyloid precursor protein (APP)
Arctic mutation, 81
Arctigenin, 276
Aromatic amino acid decarboxylase (AADC), 305–306

β-Asarone, 274
Asn-368 residue, 175–176
ATG. *See* Autophagic genes (ATG)
Atg-dependent pathway, 279–280
ATG proteins. *See* Autophagy-related (ATG) proteins
ATGs. *See* Autophagy genes (ATGs)
ATPase domain of HSP70, 161
ATP production, 94
ATRA. *See* All-trans-retinoic acid (ATRA)
Australian mutation, 81
Autoimmune abnormalities, 211
Autolysosome acidification, 278–279
Autolysosomes, 159, 298
Autophagic genes (ATG), 181–183
Autophagic modulators, tau propagation, 185–189
Autophagic pathways, to lysosome, 15–17
Autophagic receptors, 58–61
  aggrephagy, 60
  ER-phagy, 61
  mitophagy, 59–60
Autophagic vacuolar myopathy, 208–210, 208*t*
  AVSF myopathy, 209–210
  Pompe disease, 209
  rimmed vacuolar myopathy, 209
Autophagic vacuoles (AVs), 145–146
  autophagy and, 207
  formation, 149*f*
Autophagic vacuoles with sarcolemmal features (AVSF) myopathy, 209–210, 218–219
Autophagosomes, 15, 98–99, 143, 144–145*f*, 159–160, 181–183, 200
  "backed up" accumulation of, 185–189
  biogenesis, 160
  formation, 17
  transport in neurons, 17–18
Autophagy (AP), 61–62, 92, 151, 159–160, 173–174, 199–200, 227–229, 228*f*
  and autophagic vacuoles, 207
  caspase activation and, 282–283
  concept of, 141, 142*f*
  disruption of, 151–153
  dysfunction of, 147–148, 159
  effects on, 298
  elongation and closure, 143
  flux, 264*f*
  fusion of lysosomes, 145
  and Huntington's disease, 245–246
    macroautophagy, 246–249
  impairment in Parkinson's disease (PD), 308–310
  impairment promotes tau aggregation, 181–185
  inducers, 279
  initiation of, 141–143, 144–145*f*
  in Lewy body diseases *vs.* MSA, 235*t*
  machinery, 47–49, 48*f*
  modulators, 153
  negative regulator ABL1, 281
  nucleation of phagophores, 143
  physiological function of Htt, 249–251
  regulation, drug discovery in AD, 263–282
    activate JNK1/Bcl-2/PI3K, 279
    AMPK activator, 275–276
    Atg-dependent pathway, 279–280
    ATP60V01, 280
    calcium channel antagonist, 279
    carbamazepine, 277
    direct mTORC inhibitors, 263–273
    drugs targeting TFEB, 277–278
    gene therapies, 282
    imidazoline receptor agonists, 277
    increased lysosome, 278–279
    IP3 inhibitors, 276
    mTORC-dependent autophagy inducers, 273–274
    multiple mechanisms, 281
    negative regulator ABL1, 281
    P300 inhibitors (tau acetylation inhibitors), 280–281
    VPS34 activator, 279
  of synucleinopathies, 239–240
  therapeutic target, 251–252
  types of, 159
  upregulating drugs, 265*f*
Autophagy-endolysosomal system (APELS), 3–4, 14*f*
  autophagic pathways to the lysosome, 15–17
  endocytic pathways to lysosome, 10–12, 11*t*
  endosome trafficking in neurons, 12–15

Autophagy-endolysosomal system (APELS) *(Continued)*
functional roles of, 17
lysosome-mediated degradation of cellular macromolecules, 8, 9*f*
Rab GTPases, 17–18
Autophagy genes (ATGs), 121
Autophagy-linked FYVE protein (Alfy), 247–248
Autophagy-lysosomal pathway, 152
Autophagy-lysosome-endosome pathway, 200*f*
Autophagy-lysosome system, 135, 148–150
disruption of, 145–147
sporadic inclusion body myositis (s-IBM), 147
tau and Aβ, 147–148
Autophagy pathway
autophagic receptors, 58–61
aggrephagy, 60
ER-phagy, 61
mitophagy, 59–60
autophagy machinery, 47–49, 48*f*
signaling pathways of, 49–58, 50*f*
AMP-activated protein kinase (AMPK), 51–52
inositol and calcium signaling, 51
mechanistic target of rapamycin (MTOR), 49–51
microRNAs (miRNAs), 57
posttranslational and epigenetic modulators, 58
reactive oxygen species and redox signaling, 53–54
transcription factors, 56–57
unfolded protein response, 55–56
Autophagy-related (ATG) proteins, 159–160
Autosomal dominant mutations in APP, 80
Avs. *See* Autophagic vacuoles (Avs)
AVSF myopathy. *See* Autophagic vacuoles with sarcolemmal features (AVSF) myopathy

B

BACE1, 91–92
"Backed up" accumulation of autophagosomes, 185–189
BAG3-assisted selective autophagy, 163–164
BAG3-mediated tau degradation, 162–167
BAG3-mediated vacuolar-dependent tau degradation pathways
in neurons, 168*f*
BAG3 protein, 161–167, 162*f*
colocalization of, 165*f*
mediated tau degradation, 162–167
Barnes maze test, 177, 177*f*
Basal macroautophagy, 181–183
BCL2-associated athanogene 3 (BAG3), 160–161
Belgian mutation, 81
Bexarotene, 282
Biosynthetic processes, 33
Bis(ethylmaltolato) oxidovanadium (IV) (BEOV), 185–189
Boosting mitophagy, 97*f*
Bosutinib, 281
Braak staging, 180
Brain
Aβ plaques, 91–92
endocytosis in, 112–115
Brainstem-type Lewy bodies, 230, 232*f*
Bromhexine, 277–278
Bromodomaincontaining protein 4 (BRD4), 58

C

Caenorhabditis elegans, 99
Calcium channel antagonists, 279
Calcium signaling, 52–53
Canonical clathrin-mediated endocytic pathways, 119
Carbamazepine, 277
Cardiomyopathy, 217–220
Cargo to lysosomes, delivery mechanisms of, 181
Caspase activation, 282–283
Causative genes, 209
Cell membranes
direct secretion of tau, 151–152
sphingolipid composition of, 31–35
Cellular autophagic machinery, 98–99
Cellular autophagy, 151
Cellular degradation systems, integration of, 18
Cellular macromolecules, lysosome-mediated degradation of, 8, 9*f*

Central nervous system (CNS), 38
Ceramide, 32
Cerium oxide nanoparticles (nCeO2), 185–189
Chaperone-assisted selective autophagy, 160–161
Chaperone-mediated autophagy (CMA), 142*f*, 181, 182*f*, 227–228, 314–315
  dysfunction in tauopathies, 183–184
Chemical linker compounds, 61–62
CHIP. *See* C-terminus of Hsc70-interacting protein (CHIP)
Chloroquine, on clioquinol-induced changes, 297*f*
Cholesterol-lowering drugs, 84
Cilostazole, 281
Class III PI3K complex, 47–49
Clathrin-mediated endocytosis, 112–114, 117–118
CLEAR-network, 26
Clioquinol, 294–299
  chemical structure of, 292*f*
  in vivo concentration of, 295–297
  molecular action of, 297–299
Clioquinol-induced changes, 297*f*
Clioquinol-induced KT-5 astrocytic cell death, 296*f*
Clonal proliferation of inflammatory cells, 215
Clonidine, 277
CMA. *See* Chaperone-mediated autophagy (CMA)
CNS. *See* Central nervous system (CNS)
Coactivator associated arginine-methyltransferase 1 (CARM1), 58
Compromised mitophagy, 60, 92, 94–95, 100–103
COMT inhibitors, 306
Cortical Lewy bodies, 230
Creutzfeldt-Jacob disease, 174
Cross-species workflow, 101
Cryo-electron microscopy (cryo-EM) analyses, 173
C-terminus of Hsc70-interacting protein (CHIP), 138
Curcumin, 274, 278
Cyclophilin D (CypD), 96–98
Cytoplasmic aggregation-prone proteins, 60
Cytoplasmic substrates, 181–183
Cytoskeletal proteins, 199
Cytosolic cargo, 181

D

Damage-associated molecular patterns (DAMPs), 38
Danon disease, 209–210
  clinical features of, 217–218
  diagnosis and differential diagnosis, 219–220
  epidemiology of, 216–217
  muscle pathological features, 218–219, 218*f*
  outline of, 215–216
  pathogenesis and pathomechanism, 217
  therapy and prognosis, 221
  XMEA, 220–221
Decomposition of Aβ, 78–79
Decreased autophagy flux, 164
Deep tendon reflexes, 211
Defective mitochondrial accumulation, 150–151
Defective mitophagy in AD, 99–101
  mitochondrial dysfunction and, 101–103
  molecular mechanisms of mitophagy, 98–99
Degeneration, mechanism of, 213–214
Degradation, 112–114, 113*f*, 120–121, 125
Degradation mechanisms of cells
  autophagy-endolysosomal system (APELS), 14*f*
    autophagic pathways to the lysosome, 15–17
    endocytic pathways to lysosome, 10–12, 11*t*
    endosome trafficking in neurons, 12–15
    functional roles of, 17
    lysosome-mediated degradation of cellular macromolecules, 8, 9*f*
    Rab GTPases, 17–18
  integration of, 18
  neurons, with sophisticated trafficking system, 3–4
  ubiquitin-proteasome system (UPS)
    functional roles of, 7–8
    molecular mechanism for, 4–7

Del 8 mutant, 175–176
Dendritic autophagosomes, 17
Diabetes, 151
Diabetic animal models, 151
Direct secretion of tau, 151–152
Disease *vs.* Aβ
  FAD and genetic mutation, 80–81
    Aβ aggregation, 81
    Aβ production, 80
    Aβ42 production, 80–81
  sporadic AD, 79–80
DLB. *See* Lewy body diseases (DLB)
Dopamine
  metabolic pathway of, 306
  metabolism, 307*f*
Double-edged sword, 283–284
Double-membrane vesicles (autophagosomes), 181
Down syndrome, Aβ, 82
Drosophila models, 282
Drug discovery in AD, using metal chelators, 291–292
  abnormal metals in AD patients, 292–294
  in vivo concentration of clioquinol, 295–297, 296–297*f*
  molecular action of clioquinol
    autophagy effects, 298
    metal chelating activity, 297–298
    Trk-neurotrophin signaling effects, 299
  subacute myelo-optico-neuropathy (SMON), 294–295
Drugs, induce autophagy
  activate JNK1/Bcl-2/PI3K, 279
  AMPK activator, 275–276
  Atg-dependent pathway, 279–280
  ATP60V01, 280
  calcium channel antagonist, 279
  carbamazepine, 277
  direct mTORC inhibitors, 263–273
  drugs targeting TFEB, 277–278
  gene therapies, 282
  imidazoline receptor agonists, 277
  increased lysosome, 278–279
  IP3 inhibitors, 276
  mTORC-dependent autophagy inducers, 273–274
  multiple mechanisms, 281
  negative regulator ABL1, 281
  P300 inhibitors (tau acetylation inhibitors), 280–281
  VPS34 activator, 279
Dutch mutation, 81
Dysfunctional AP, 181
Dysfunction of autophagy, 159
Dysphagia, 211

E

EC. *See* Entorhinal cortex (EC)
Ectosomes, 152
Electromyogram, 211–212
Electron microscopy, 200
ELND005, 83
Elongation, 15
  of autophagy, 143, 144–145*f*
Endocytic cargoes, 113*f*
Endocytic machinery, 126
Endocytic pathways
  to lysosome, 10–12, 11*t*
  in neurons, 12–15
Endocytic systems, 11*t*
Endocytosis, 10, 124–125
  in AD brain, 120–124
  in brain, 112–115, 113*f*
  in phagocytic world, 117–120
  for therapeutic gain, 125
Endolysosomal degradation of autophagy-independent tau, 150
Endolysosome transport in neurons, 17–18
Endoplasmic reticulum (ER), 26, 228*f*
Endoplasmic reticulum stress (ERS), 147
Endosomal microautophagy, 16, 181, 182*f*
Endosomal sorting complex required for transport-III (ESCRT-III), 147
Endosomal sorting complex required for transport (ESCRT) system, 166–167, 184–185
Endosomes, 199–200
Endosome trafficking in neurons, 12–15
Entorhinal cortex (EC), 95–96, 178–179
Enzyme presequence protease (PreP), 96–98
Enzyme replacement therapy (ERT), 27–31, 209
Epigenetic modulators, 58

ER. *See* Endoplasmic reticulum (ER)
ER-phagy, 61
ERS. *See* Endoplasmic reticulum stress (ERS)
ERT. *See* Enzyme replacement therapy (ERT)
ESCRT-III. *See* Endosomal sorting complex required for transport-III (ESCRT-III)
ESCRT system. *See* Endosomal sorting complex required for transport (ESCRT) system
E3 ubiquitin ligase Parkin, 98–99
Euchromatic histone lysine methyltransferase 2 (EHMT2), 58
Everolimus, 273
Excessive phosphorylation of tau, 263
Excessive reactive oxygen species, 53–54
Excitatory neurons, 162
Excitotoxicity in neurons, 101–102
Excretion of Aβ, 79
Exosomes, 152
Exo-synaptic tau propagation, 180
Extracellular tau propagation, 180
Extracellular tau species, 180

F

FAD. *See* Familial Alzheimer's disease (FAD)
Familial ALS-related molecules, 214
Familial Alzheimer's disease (FAD), 80–81
  Aβ aggregation, 81
  Aβ production, 80
  Aβ42 production, 80–81
Fisetin, 278
Florida mutation, 81
Flubendazole, 277–278
Fragmented tau, 147
French mutation, 81
Functional roles
  of autophagy in neurons, 17
  of UPSs, 7–8
Fusion of lysosomes, 145

G

GAA. *See* Acid alpha-glucosidase (GAA)
Galactosylceramide (GalCer), 32
Galactosylsphingosine, 35
Gaucher Disease (GD), 27–31
Gene mutations, 227
  Aβ aggregation, 81
  Aβ production, 80
  highly cohesive Aβ42 production ratio, 81
Gene therapies, 282
German mutation, 81
Glial cytoplasmic inclusions, 236, 239*f*
Glial-like tau pathology, 178*f*
Global activation of autophagy, 251–252
Glucocerebrosidase (GCase), 314
Glucosylceramide (GlcCer), 32
Glycogen storage disease, 215–216
Glycogen storage disease type II, 209
Glycogen synthase kinase-3 (GSK3), 37
Glycoproteinoses, 27
Glycosylation of tau, 140
Granulovacuolar degeneration (GVD), 197–199, 198*f*
GSK3. *See* Glycogen synthase kinase-3 (GSK3)
GTM-1, 282
GV-971, 83
GVB -like structures in oligodendroglia, 200–203
GVD body (GVB), 197–199
  early stage of GCIs, 202*f*
  model of, 201*f*
  organelles, 199–200
  in α-synucleinopathy, 202*f*

H

HD. *See* Huntington's disease (HD)
Health neuronal processes, 167
"Healthy" lysosome, 33
Heat shock protein 70 (HSP70), 160–161
Hela cells, 163
High-dose immunoglobulin therapy, 210
Highly polarized cells, with sophisticated trafficking system, 3–4
High-molecular-weight oligomers and off-pathways, 77, 78*f*
Histone deacetylase 6 (HDAC6), 60, 246
Homeostasis of muscle cells, 214
Homeostasis signature, 162
Homologs, 228–229
Homology of Alzheimer's disease, 213–214
Htt, physiological function of, 249–251

Huntington's disease (HD), 245–246
  macroautophagy in, 246–249
    Alfy, 247–248
    NBR1, 249
    optineurin (OPTN), 249
    p62, 248
    toll-interacting protein (Tollip), 249
Hydrolytic enzymes, 207
Hydrophobic C-terminus of Aβ, 115
2-Hydroxypropyl-β-cyclodextrin (HPβCD), 314
Hyperphosphorylation of tau, 126
Hypertrophic cardiomyopathy, 209–210, 217–218

I

Iberian mutation, 81
IBM. *See* Inclusion body myositis (IBM)
Idiopathic inflammatory myopathy (IIM), 210
IIM. *See* Idiopathic inflammatory myopathy (IIM)
Imidazoline receptor agonists, 277
Immunohistochemical detection of pPERK, 198*f*
Immunohistochemical studies, 199–200
Immunological system, abnormality in, 214–215
Immunoreactivity of glial cytoplasmic inclusions, 239*f*
Immunotherapy, 83–84
Impaired insulin signaling, 151
Impaired mitophagy in AD, 99–100
Inclusion body myositis (IBM), 208
  clinical features, 211–212
  muscle pathological features, 212–213, 212*f*
  outline of, 210–211
  pathogenesis and pathomechanism
    abnormality in immunological system, 214–215
    abnormality in proteolytic pathway, 214
    homology of Alzheimer's disease, 213–214
Increased lysosome, 278–279
Increased salt tolerance 1 (IST1), 147
Inducible tauopathy cell model, 185–189
Induction, 15
Inflammation, 95–96, 103
Inflammatory cells, clonal proliferation of, 215
Inflammatory mediators, 119
Initiation of autophagy, 141–143, 144–145*f*
Inositol, 52–53
Inositol monophosphatase (IMPase), 276
Insoluble Aβ fibrils (fAβ), 76
Integration of cellular degradation systems, 18
Intercellular spread of tau aggregates, 178–180, 179*f*
  exo-synaptic tau propagation, 180
  extracellular tau propagation, 180
  trans-synaptic tau propagation, 179
  tunneling nanotubes (TNTs), 180
Internalized tau, 174–175
Intervention strategy of Aβ
  Aβ aggregation inhibitor, 82–83
  immunotherapy, 83–84
  secretase activity regulator, 82
  summary of, 84–85
Intracellular inositol levels, 52
Intracellular oxidative stress, 53–54
Intramural Periarterial drainage (IPAD), 79
Intraneuronal accumulation, 119
In vitro tau propagation, 174–176, 175*f*
In vivo concentration of clioquinol, 295–297
In vivo tau propagation, 178
Iowa mutation, 81
IPAD. *See* Intramural Periarterial drainage (IPAD)
IP3 inhibitors, 276
iPSC-derived human astrocytes, 124–125
IPV. *See* Isoleucine-proline-valine (IPV)
Iranian mutation, 81
Isoleucine-proline-valine (IPV), 161
IST1. *See* Increased salt tolerance 1 (IST1)
Italian mutation, 80

J

JNK signaling pathway, 162

K

Kaempferol, 101

KEAP1. *See* Kelch-like ECH associated protein 1 (KEAP1)
KEGG pathway analysis, 166
Kelch-like ECH associated protein 1 (KEAP1), 54
KFERQ-like motif-bearing proteins, 314–315
"Kiss and hop process,", 137
Krabbe disease, 35, 37

L

LAMP-2. *See* Lysosomal-associated membrane protein 2 (LAMP-2)
LAMP2A. *See* Lysosome-associated membrane protein type 2A (LAMP2A)
LANDO. *See* LC3-associated endocytosis (LANDO)
LAP. *See* LC3-associated phagocytosis (LAP)
Latrepiridine, 274
LC. *See* Locus coeruleus (LC)
LC3, 143, 144–145*f*
  lipidation in LAP, 121
LC3-associated endocytosis (LANDO), 122–126
LC3-associated phagocytosis (LAP), 121
Leuco-methylthioninium bis (LMTM), 273–274
Lewy body diseases (DLB)
  alterations of autophagy in, 235*t*
  clinical features of, 229–230
  immunoreactivity of, 232*f*
  impairment of autophagy in, 233–234
  neuroprotective role of, 232–233
  pathological features of, 230, 231*f*
  pathological staging systems for, 230–232
  primary/secondary event, 234–235
  α–Synuclein, 233
Lipid kinase PI3K-III, 143
Lithium, 276
Lithium-induced autophagic promotion of aggregated tau, 189, 190*f*
Lithium-triggered AP, 189
LMCD. *See* Lysosomal-mediated cell death (LMCD)
L-NAME, 279
Locus coeruleus (LC), 102–103
Lonafarnib, 282
London mutation, 81
Long-term fish oil (FO), 274
Long-term use of cholesterol-lowering drugs, 84
LRRK2/PARK8 gene, 308–310
LSDs. *See* Lysosomal storage disorders (LSDs)
Lysosomal acid hydrolases, 296–297
Lysosomal-associated membrane protein 2 (LAMP-2), 209–210, 215–216
Lysosomal degradation, 92
Lysosomal-dependent proteolytic system, 185–189
Lysosomal enzymes, 12
Lysosomal exocytosis, 33–34
Lysosomal hydrolytic enzymes, 146
Lysosomal impairment
  activity, 31–35
  in neuroinflammation, 38–39
Lysosomal-mediated cell death (LMCD), 25–26
Lysosomal membrane, 25–26
Lysosomal membrane deformation, 184–185
Lysosomal storage disorders (LSDs), 26–31
  classification of, 28–30*t*
  neurodegeneration in, 35–38
  neuronal damage in, 31–35
Lysosome, 25–26, 199–200, 207, 227–228
  accumulation of, 146
  autophagic pathways to, 15–17
  endocytic pathways, to, 10–12
  fusion of, 145
  for potential PD therapy, 313–316
    clinical trials of nilotinib, 315–316
    CMA, 314–315
    glucocerebrosidase (GCase), 314
    TFEB, 313–314
Lysosome-associated membrane protein type 2A (LAMP2A), 183–184
Lysosome-mediated degradation pathway, 246
Lysosome-mediated degradation system, of cellular macromolecules, 8, 9*f*
Lysosomes-neuronal degeneration
  lysosomal impairment in neuroinflammation, 38–39

lysosomal storage diseases, 26–31, 28–30*t*
neurodegeneration in, 35–38
neuronal damage in, 31–35
lysosomes, 25–26

M

Machine learning, 101
Macroautophagy, 9*f*, 15, 142*f*, 146, 159, 181–183, 182*f*
degradation of mitochondria, 59–60
dysfunction in tauopathies, 181–183
in Huntington's disease (HD), 246–249
machinery, 47–49, 48*f*
for potential PD therapy, 312*f*
AMP-activated protein kinase (AMPK), 312–313
mammalian target of rapamycin (mTOR), 310–312
ULK1 and VPS34, 313
Macrophage-like cells, 114–115
Macropinocytosis, 12–15
Mammalian cells, 246, 308
Mammalian target of rapamycin (mTOR), 26, 49–51, 185–189, 220, 228–229, 250–251, 310–312
inhibitor rapamycin, 101
MAMs. *See* Mitochondria associated membranes (MAMs)
MAO-B inhibitors, 306
MAPT. *See* Microtubule-associated protein tau (MAPT)
Mechanistic target of rapamycin complex 1 (mTORC1) pathway, 141–143, 144–145*f*, 147
Memantine, 279, 313
Meta-iodobenzylguanidine (MIBG), 234–235
Metal chelating activity, 297–298
Metal chelators, drug discovery in AD, 291–292
abnormal metals in AD patients, 292–294
in vivo concentration of clioquinol, 295–297, 296–297*f*
molecular action of clioquinol, 297–299
subacute myelo-optico-neuropathy (SMON), 294–295
Metformin, 275
Methylene blue, 273–274
MIBG. *See* Meta-iodobenzylguanidine (MIBG)
Microautophagy, 16, 142*f*, 181, 182*f*, 184–185, 227–228
dysfunction in tauopathies, 184–185
Microfluidic devices, 179
Microglia, 38, 112–115, 117–120, 118*f*, 122–126
MicroRNAs (miRNAs), 57
Microtubule-associated protein tau (MAPT), 91–92, 151
Microtubule organizing center (MTOC), 60
MiDAS. *See* Mitochondrial dysfunction-associated senescence (MiDAS)
Misfolded proteins, accumulation and deposition of, 197
Mitochondria, 35, 143
with Aβ and tau, 96–98
Mitochondria associated membranes (MAMs), 37
Mitochondrial ATP production, 35–36
Mitochondrial autophagy (termed mitophagy), 92
Mitochondrial dysfunction, 36–37
with Aβ and tau, 96–98
and defective mitophagy, 101–103
development and activities of neurons, 94–95
early feature of AD, 95–96
Mitochondrial dysfunction-associated senescence (MiDAS), 103
Mitochondrial fission and fusion mechanism, 60
Mitochondrial homeostasis, 36
Mitochondrial impairment, 35–38
Mitochondrial reactive oxygen species, 36–37
Mitogen-activated protein kinase (MAPK) phosphorylation, 299
Mitophagy, 59–60, 99
machinery, 98–99
mechanisms of, 93*f*
molecular mechanisms of, 98–99
pharmacological stimulation of, 101
and tau, 150–151
Molecular mechanisms, 33–34
of mitophagy, 98–99
for UPS, 4–7

Monomeric tau, 173–174
Monoubiquitination, 5*f*
MO(F) of Aβ, 115–117
MSA. *See* Multiple system atrophy (MSA)
MTOC. *See* Microtubule organizing center (MTOC)
mTOR. *See* Mammalian target of rapamycin (mTOR)
mTORC-dependent autophagy inducers, 273–274
mTORC1 pathway. *See* Mechanistic target of rapamycin complex 1 (mTORC1) pathway
mTORC signaling pathway, 263–265
Mucopolysaccharidoses, 27, 39
Multiple anti-Aβ42 antibodies, 84
Multiple-domain protein, 161
Multiple system atrophy (MSA)
autophagy in Lewy body diseases *vs.*, 235*t*
clinical and pathological features of, 236, 237–238*f*
impairment of autophagy in, 236
oligodendroglia in, 200–203
primary/secondary event, 238–239
Murine models of AD, 126
Muscle atrophy, 217–218
Muscle cells, 207–208
homeostasis of, 214
Muscle pathological features, 218–219, 218*f*
Muscle pathology, 210
Myopathies, 208

N

NAD$^+$. *See* Nicotinamide adenine dinucleotide (NAD$^+$)
Naked tau assemblies, 180
NBR1 autophagy receptor, 249
Neo-synthesized sphingolipids, 32
Nerve growth factor (NGF), 299
Nervous system, 94–95
Neurodegeneration process, 180
in lysosomal storage diseases (LSDs), 35–38
Neurodegenerative disorders, 173
Neurofibrillary tangles (NFTs), 135, 263, 291
Neuroinflammation, 291–292
lysosomal impairment in, 38–39
Neuronal cytoplasmic inclusions, 239*f*
Neuronal endocytosis, 126
Neuronal-specific functions, 181
Neurons, 3–4, 35–36, 160–162
APELS in (*see* Autophagy-endolysosomal system (APELS))
autophagosome and endolysosome transport in, 17–18
BAG3-mediated vacuolar-dependent tau degradation pathways in, 168*f*
of *Caenorhabditis elegans*, 99
endocytic pathways and endosome trafficking in, 12–15
mitochondria roles in, 94–95
neuropathological features of GVD in, 197–199, 198*f*
synaptic loss and excitotoxicity in, 101–102
UPS in, 7–8, 7*f*
Neuropathic phenotype, 39
Neuroprotective role of Lewy bodies, 232–233
Neurotoxic Aβ, 111–112
Neurotoxicity, 119
Newly synthesized APP, 116
NFE2L2. *See* Nuclear factor erythroid 2 like 2 (NFE2L2)
NFkB pathway activation, 119
NFTs. *See* Neurofibrillary tangles (NFTs)
NGF. *See* Nerve growth factor (NGF)
Nicotinamide, 278–279
Nicotinamide adenine dinucleotide (NAD$^+$), 100
Niemann-Pick type A disease (NPA), 33, 35
Nilotinib, 281, 315–316
Nilvadipine, 279
Nitazoxanide, 185–189
Nitration, 138
Nitrendipine, 279
NMR technology, 115
“Nonconventional” lipid rafts, 35
Noninvasive autophagic modulators, 185–189
Norepinephrine, 102–103
Normal function of Aβ, 79
Norvaline, 84
NPA. *See* Niemann-Pick type A disease (NPA)

Nuances, in phagocytosis *vs.* endocytosis, 120–121
Nuclear DNA-encoded mitochondrial proteins, 96–98
Nuclear factor erythroid 2 like 2 (NFE2L2), 54
Nucleation, 15
  of autophagy, 143, 144–145*f*

O

Off-pathways, 77, 78*f*
Oleuropein aglycone (polyphenol), 274
Oligodendroglia, in MSA, 200–203
Oligomer hypothesis, 76
Oligomerization, 83–84
Oligosaccharidosis, 27
O-linked glycosylation, 140
Olivopontocerebellar atrophy, 238*f*
Omega-3 fatty acids, 274
Omegasomes, 228–229
Optineurin (OPTN), 249
Optogenetic TFEB inducer, 278
Organelle hitch-hiking pathways, 152
Organelles, 199–200
Osaka mutation, 81
Osteoporosis-associated transmembrane protein 1 (OSTM1), 120
Oxidative phosphorylation (OXPHOS), 94

P

Paired helical filaments (PHF), 174
Parkin pathway, 98–99
Parkinson's disease (PD)
  autophagy impairment in, 308–310
  features of, 305
  α–synuclein accumulation in, 306
  targeting macroautophagy for, 310–313, 312*f*
  therapy, 305–306, 307*f*, 315–316
Pathological tau protein, 98
  pathways of, 151–152
  secretion, pathways of, 151–152
Pathophysiology of Alzheimer's disease (AD), 75, 91–92
Pathways of pathological tau protein secretion, 151–152
p62 autophagy receptor, 248
PBT2, chemical structure of, 292*f*
PD. *See* Parkinson's disease (PD)
PD-associated genes, 308–310
Phagocytosis, 12, 112–115, 117–119, 126
Phagosomes, 12
PHF. *See* Paired helical filaments (PHF)
Phosphorylated-TAR-DNA-binding protein 43 (TDP-43), 199
Phosphorylated tau, 164
Phosphorylation of tau protein, 137–138
Phospho-tau, 166–167
Physiological functions
  of APP, 74
  of tau, 137
Pick's disease, 175*f*
P300 inhibitors (tau acetylation inhibitors), 280–281
PINK1-Parkin-dependent mitophagy, 98–99
Plasma membrane (PM), 25–26, 34*f*
Plasma membrane APP, 116
Polymer oligomers, 77, 78*f*
Polyphenol, 274
PolyQ toxicity, 250
Polyubiquitination, 5*f*
Pompe disease, 208–209
Postmitotic cells, 160
Posttranslational modifications of tau protein, 58, 137–140
Potential PD therapy
  lysosome for, 313–316
    clinical trials of nilotinib, 315–316
    CMA, 314–315
    glucocerebrosidase (GCase), 314
    TFEB, 313–314
  macroautophagy for, 312*f*
    AMP-activated protein kinase (AMPK), 312–313
    mammalian target of rapamycin (mTOR), 310–312
    ULK1 and VPS34, 313
Presenilin 1 (PSEN1), 91–92
Principal pathological phenotypes, 27
Process of aggregated Aβ, 76–77, 76*f*
Professional phagocytes, 12
Proteasome complex, 6*f*, 7–8
Proteasome function, 7*f*
Proteolytic pathway, abnormality in, 214
Proteostatic stress, 162

Psychosine (galactosylsphingosine), 35
PTEN-induced kinase 1 (PINK1), 98–99
PXXP domain of BAG3, 161

R

Rab35, 166–169, 168*f*
Rab-GTPase family of proteins, 17–18, 114
Rapamycin (sirolimus), 263–273, 297*f*, 310–312
  analogs, 273
RARa. *See* Retinoic acid receptor a (RARa)
Ras homolog enriched in brain (RHEB), 51
RCAN1. *See* Regulator of calcineurin 1 (RCAN1)
Reactive oxygen species (ROS), 53–54, 293
Receptor-mediated endocytosis, 114–115
Receptors/receptor-cargo complexes, 113*f*
Redox signaling, 53–54
Reduce autophagosome accumulation, 278–279
Regulation of Aβ, 78
Regulator of calcineurin 1 (RCAN1), 183–184
Repeated muscle biopsy, 212–213
Resveratrol, 275
Reticulophagy, 61
Retinoic acid receptor a (RARa), 315
Rhapontigenin, 101
RHEB. *See* Ras homolog enriched in brain (RHEB)
Rhizoma coptidis, 185–189
Rifampicin, 280
Rilmenidine, 277
Rimmed vacuolar myopathy, 209
Rimmed vacuoles, 209
RING finger protein 5 (RNF5), 214
RNF5. *See* RING finger protein 5 (RNF5)
Robust mitophagy inducers, 101
ROCK inhibitor, 282
ROS. *See* Reactive oxygen species (ROS)
Rubicon, 121–124

S

Secretase activity regulator, 82
  α-secretase activity regulator, 74
  β-secretase activity regulator, 74, 82
  γ-secretase activity regulator, 74, 82
Secretory pathway, 152
Seed-dependent tau propagation, 174–175
  using cell-based assay, 175–176
Selective autophagy receptors, 15, 141
Self-assembly of tau, 174
Serum creatine kinase (CK) levels, 211
s-IBM. *See* Sporadic inclusion body myositis (s-IBM)
Signaling pathways of autophagy, 49–58, 50*f*
  AMP-activated protein kinase (AMPK), 51–52
  inositol and calcium signaling, 51
  mechanistic target of rapamycin (MTOR), 49–51
  microRNAs (miRNAs), 57
  posttranslational and epigenetic modulators, 58
  reactive oxygen species and redox signaling, 53–54
  transcription factors, 56–57
  unfolded protein response, 55–56
Single autophagic protein, 240
Sirtuin 2 (SIRT2), 56–57
Sirtuin 3 (SIRT3), 55–56
Small compound, 310
SMDC37892, 280
SMER28, 279
SMON. *See* Subacute myelo-optico-neuropathy (SMON)
SNAREs, 159–160
Soluble Aβ, 115–116
Soluble low-molecular-weight oligomer, 77
Sonicated recombinant tau, 175–176
Sophisticated trafficking system, highly polarized cells with, 3–4
Spatiotemporal progression of tau, 174
Spermidine, 280
Sphingolipid composition of cell membranes, 31–35
Sphingolipid-enriched membrane domains, 34–35
Sphingolipid hydrolases, 32–33
Sphingolipidoses, 27
Sphingolipids (SL), 31–33
Sporadic AD, 79–80
Sporadic inclusion body myositis (s-IBM), 147
Starvation-induced AP, 181–183

Statins, 276
Stress-induced cellular metabolism, 198–199
Subacute myelo-optico-neuropathy (SMON), 294–295
Swedish mutation, 80
Synaptic loss in neurons, 101–102
Synaptic proteins, 251
Synaptic vesicles, 251
Synaptopodin (SYNPO) functions, 163–164, 165*f*
α–Synuclein
  accumulation in, 306
  pathology of, 230–232
  substrate for autophagy, 233
Synucleinopathies, 239–240

T

Tau aggregation
  autophagic impairment, 181–185, 182*f*
    CMA dysfunction in tauopathies, 183–184
    macroautophagic dysfunction in tauopathies, 181–183
    microautophagic dysfunction in tauopathies, 184–185
  intercellular spread of, 178–180
Tauopathies, 91–92, 173–174
  CMA dysfunction in, 183–184
  macroautophagic dysfunction in, 181–183
  microautophagic dysfunction in, 184–185
  mouse model study, 179
Tau propagation, 174–180
  experimental methods for, 174–178
    in vitro tau propagation, 174–176
    in vivo tau propagation, 177–178
  pharmacological agents, 185–189, 187*t*
    autophagic modulators, 185–189
    lithium-induced autophagic promotion, 189
  by tunneling nanotubes (TNTs), 180
Tau protein, 151
  and Aβ, 147
  acetylation inhibitors, 280–281
  autophagy inhibitor chloroquine treatment, 149*f*
  degradation pathway, 148–150, 149*f*
  disruption of autophagy, 151–153
  "homeostasis signature,", 162
  hyperphosphorylation of, 126
  mitophagy and, 150–151
  pathology, 178
  physiological functions of, 137
  posttranslational modifications of
    acetylation, 138
    glycosylation, 140
    nitration, 138
    phosphorylation, 137–138
    toxicity, 140
    truncation, 140
    ubiquitination, 138
  propagation of, 151–153
  tangles, 91–92
  toxicity, 140
  truncated forms of, 98
TBC1 Domain Family Member 10B (TBC1D10B), 166–169
Termed mitophagy, 92
Tetrahydrohyperforin, 280
TFEB. *See* Transcription factor EB (TFEB)
TGN. *See* Trans-Golgi network (TGN)
Tg Tau-P301S mouse model, 185–189
TIMs. *See* Translocases located on the inner membrane (TIMs)
Toll-interacting protein (Tollip) functions, 249
Toll-like receptors, 119
TOMs. *See* Translocases located on the outer membrane (TOMs)
Tourette's disorder, 185–189
TP13A2/PARK9 gene, 308–310
TPPP. *See* Tubulin polymerization-prompting protein (TPPP)
Tramiprosate, 83
Transcription factor EB (TFEB), 49–51, 185–189, 277, 313–314
Transcription factors, 56–57
Trans-Golgi network (TGN), 116
Transient neuroinflammatory process, 38
Translocases located on the inner membrane (TIMs), 94

Translocases located on the outer membrane (TOMs), 94, 96–98
Trans-synaptic tau propagation, 179
Trehalose, 275
Triggering Receptor Expressed on Myeloid Cells 2 (TREM2), 118–119
Triple-Tg mouse model of AD, 185–189
Trk-neurotrophin signaling, 299
Truncated BAG3, 161
Truncated forms of tau, 98
Truncation of tau protein, 140
Tubulin polymerization-prompting protein (TPPP), 200–203, 202*f*
Tunneling nanotubes (TNTs), 152
  tau propagation by, 180
Typical pathological findings, 212–213
  of Alzheimer's disease (AD), 263
Tyr29-nitrated tau, 138
Tyrosine kinase Abl1, 315

U

Ubiquitination of tau protein, 138
Ubiquitin-like conjugation systems, 47–49
Ubiquitin-proteasome system (UPS), 3–4, 94–95, 135
  functional roles of, 7–8, 7*f*
  molecular mechanism for, 4–7
Ubiquitin-specific peptidase 24 (USP24), 308–310
Ubiquitin system, 5*f*
ULK activates Beclin-1, 159–160
ULK1 (unc-51-like kinase 1) complex, 141–143, 159–160, 313
Unfolded protein response (UPR), 55–56
Upregulating mitophagy, 101
UPS. *See* Ubiquitin-proteasome system (UPS)

V

Vacuolar/lysosomal membrane deformation, 184–185
Vacuolar-mediated degradation, 167–169
Vacuolar membrane ATPase activity 21 (VMA21), 209–210
Vacuole membrane protein 1 (VMP1), 143
Valosin containing protein (VCP) gene, 210–211
VDAC. *See* Voltage-dependent anion channel (VDAC)
Vitamin D, 281
VMA21. *See* Vacuolar membrane ATPase activity 21 (VMA21)
VMP1. *See* Vacuole membrane protein 1 (VMP1)
Voltage-dependent anion channel (VDAC), 52–53
VPS34 activator, 279, 313

W

WIPI2 (WD-repeat protein interacting with phosphoinositides protein 2), 159–160
Wolff-Parkinson-White syndrome, 217–218
WW domain of BAG3, 163

X

X-box binding protein 1 (XBP1), 55
Xenophagy, 16–17
X-linked myopathy with excessive autophagy (XMEA), 209–210, 220–221

Z

Zinc finger with KRAB and SCAN domains 3 (ZKSCAN3), 56

9780323899062